Remote Sensing and GIS Techniques in Hydrology

Mouhcine Batchi
University of Ibn Tofail, Morocco

Adil Moumane
University of Ibn Tofail, Morocco

IGI Global
Publishing Tomorrow's Research Today

Vice President of Editorial Melissa Wagner
Managing Editor of Acquisitions Mikaela Felty
Managing Editor of Book Development Jocelynn Hessler
Production Manager Mike Brehm
Cover Design Phillip Shickler

Published in the United States of America by
IGI Global Scientific Publishing
701 East Chocolate Avenue
Hershey, PA, 17033, USA
Tel: 717-533-8845
Fax: 717-533-8661
Website: https://www.igi-global.com E-mail: cust@igi-global.com

Library of Congress Cataloging-in-Publication Data

Names: Batchi, Mouhcine, 1981- editor. | Moumane, Adil, 1984- editor.
Title: Remote sensing and GIS techniques in hydrology / edited by Mouhcine Batchi, Adil Moumane.
Description: Hershey, PA : IGI Global Scientific Publishing, [2025] | Includes bibliographical references and index. | Summary: "The growing complexities of managing and conserving water resources in today's world have led to the adoption of innovative technologies such as Remote Sensing (RS) and Geographic Information Systems (GIS) in the field of hydrology. The need for a deeper understanding of hydrological processes, coupled with the increasing pressures of climate change, population growth, and environmental degradation, has made it imperative to employ advanced tools for data collection, analysis, and decision-making"-- Provided by publisher.
Identifiers: LCCN 2024061682 (print) | LCCN 2024061683 (ebook) | ISBN 9798369396513 (hardcover) | ISBN 9798369396520 (paperback) | ISBN 9798369396537 (ebook)
Subjects: LCSH: Hydrology--Data processing. | Hydrology--Remote sensing. | Geographic information systems. | Remote sensing.
Classification: LCC GB656.2.R44 R46 2025 (print) | LCC GB656.2.R44 (ebook) | DDC 511.480285--dc23/eng/20250206
LC record available at https://lccn.loc.gov/2024061682
LC ebook record available at https://lccn.loc.gov/2024061683

British Cataloguing in Publication Data
A Cataloguing in Publication record for this book is available from the British Library.

Table of Contents

Detailed Table of Contents

Remote sensing (RS) and geographical information system (GIS) have been extensively used for their ability to provide better scope for decision making in terms of monitoring and managing various natural resources: water, land, forest, agriculture, natural hazards, climate change, etc. In this chapter, the authors precisely discuss in detail about the various applications of remote sensing and GIS in numerous fields such as monitoring water quality, crop modelling, monitoring wildlife habitat and forest management, application in agriculture and soil science, water resource, as well as natural disaster management. This chapter will also incorporate several case studies where GIS and remote sensing were used as a powerful tool to delineate and monitor environmental changes and take the necessary steps to manage the same. This chapter will thus focus on the importance of remote sensing and GIS and its impact in sound decision-making, which in turn will secure the sustainable utilization of natural resources to reach the requirements of present as well as future generations.

The Akermoud coastal aquifer, situated in the northern region of Essaouira, Morocco, is an aquifer that has an important productive capacity, mainly used for irrigation. However, uncontrolled exploitation has increased the risk of sea water intrusion, leading to deteriorating water quality and threatening agricultural sustainability. In this research, in order to access susceptibility to sea water intrusion (SWI), six elements derived from diverse databases are used. Variables were combined in a GALDIT and GIS models, resulting in the analysis of 40 groundwater samples from wells. Results highlight the imminent threat of sea water encroachment into the coastal groundwater system. The resulting GALDIT index map indicates a notably high susceptibility index along a 3 km coastal band, between Tensift Oued and Bhaybeh Beach, enlarging southwards up to 5 km. Saline intrusion patterns are particularly observed between Zaouiet El Kourati and Ouled El Fequih villages, where the merging of saline and fresh waters amplifies salinization, affecting approximately 24% of the study area.

The tropical climate is subjected to extreme hydrological events, exacerbated by climate change (CC) and changes in land use/land cover (LULC). Evaluating CC and LULC impacts on streamflow in tropical regions is vital for understanding the risks to vulnerable ecosystems. This chapter presents a meta-analysis of 30 systematically selected case studies that examine the spatiotemporal effects of CC and LULC on streamflow. The findings highlight the complex interactions among CC, LULC, and hydrological dynamics, underscoring the critical role of hydrological models, meteorological inputs, and scenario-based analyses. Models like SWAT and HEC-HMS, coupled with climate projections from GCMs (CMIP5 and CMIP6), are essential for projecting future streamflow changes. Integrating remote sensing (RS) and geographical information systems (GIS) enhances spatial accuracy and reliability. Incorporating diverse climate and LULC scenarios, including representative concentration pathways (RCPs), provides insights into adaptive water resource management in tropical regions.

This study focuses on monitoring surface water dynamics in the Mediterranean region, a region increasingly grappling with water scarcity. The authors analyzed five dams—Al Massira and Mansour Eddahbi in Morocco, Sidi Salem in Tunisia, Sau Reservoir in Spain, and Cabril Dam in Portugal—using a custom JavaScript API developed for Google Earth Engine (GEE). This API enabled efficient processing of Landsat satellite imagery to calculate annual average water areas and extract representative water body images. The findings reveal significant reductions in water area across all studied reservoirs, underscoring the impact of prolonged drought conditions on water availability in the region. The Al Massira Dam in Morocco experienced the most pronounced decline. This research emphasizes the critical role of advanced remote sensing tools in supporting water resource management and policymaking in the Mediterranean region.

The combination of remote sensing with AI methods such as deep learning, along with tools like geographic information systems, has transformed both spatial analysis and

located on the Khellata river in terms of the storage of surface water from rainfall, by quantifying the surface area occupied by its water using satellite images obtained from remote sensing, and mapping the evolution of the dam using a GIS geographic information system over the period from 2013 to 2023. The results show that this structure has made it possible to store rainwater, with an increase of 65.22% in 2022 and 43.80% in 2023.

Mohamed El Bouazzati, Faculty of Humanities and Social Sciences, Ibn Tofail University, Kenitra, Morocco
Mouhcine Batchi, Faculty of Humanities and Social Sciences, Ibn Tofail University, Kenitra, Morocco
Faycal Fatah, Faculty of Humanities and Social Sciences, Ibn Tofail University, Kenitra, Morocco
Amina Moumane, Faculty of Humanities and Social Sciences, Ibn Tofail University, Kenitra, Morocco
Ibtissam Motib, Faculty of Humanities and Social Sciences, Ibn Tofail University, Kenitra, Morocco

This study investigates the spatio-temporal evolution of Lake Aguelmam Azigza from 1977 to 2024 using remote sensing and GIS techniques. Utilizing Landsat satellite imagery and the modified normalized difference water index (MNDWI), the study reveals a significant fluctuation in the lake's surface area over the analyzed period, with a concerning overall decline of 66.32%. The lake's surface area decreased from 49.11 hectares in 1977 to 16.54 hectares in 2024. This decline is likely attributed to climate change-induced factors such as drought and variations in precipitation, highlighting the vulnerability of Moroccan water resources. The research emphasizes the need for proactive measures, including resource monitoring and drought mitigation strategies, to ensure the sustainability of water resources in the region. Despite limitations related to data availability, the study provides valuable insights into the dynamics of Lake Aguelmam Azigza and underscores the urgency of addressing the challenges posed by climate change on water resources in Morocco and similar regions.

This work is the result of a qualitative assessment of water erosion in the Oued Lakhdar watershed, located within the larger Oum Erbia watershed. The PAP/CAR model and geomatic tools were used for this evaluation. The Oum Erbia watershed, spanning 1664 km^2, experiences severe erosion due to factors such as heavy and aggressive rainfall, steep slopes, sparse vegetation, and the predominance of soft rocks. Human activities, including deforestation, vegetation degradation, and inappropriate agricultural practices, have further exacerbated the erosion problem in the Oued Lakhdar watershed. The PAP/CAR method allowed for a qualitative assessment of erosion throughout the entire watershed and the identification of priority erosion zones. Data analysis using the PAP/CAR method revealed five main erosion states: extreme erosion (20.57%), strong erosion (18.56%), moderate erosion (23.02%), notable erosion (27.02%), and weak erosion (10.84%). In total, 62.15% of the watershed experiences moderate to extreme erosion.

Foreword

It is a great pleasure to present *Remote Sensing and GIS Techniques in Hydrology*, an insightful volume that delves into the innovative intersection of geospatial technology and hydrology. The rapid advancement of Remote Sensing and Geographic Information Systems (GIS) has revolutionized how we understand, monitor, and manage water resources. Under the editorial leadership of Mouhcine Batchi and Adil Moumane, this book emerges as a critical contribution to a field that is increasingly vital in the face of global water challenges.

In the current era of environmental uncertainty, the ability to analyze and manage water systems through advanced technologies has never been more essential. Remote Sensing and GIS offer unparalleled tools for tracking hydrological phenomena, enabling precise, real-time data to support critical decision-making in water resource management, flood risk mitigation, and adaptation to climate change. These technologies represent a turning point in how we address complex water-related issues, bridging the gap between theoretical hydrology and practical, data-driven solutions.

This volume brings together a range of cutting-edge research and real-world case studies, with each chapter offering unique insights into hydrological applications of geospatial techniques. The comprehensive coverage spans diverse regions and contexts, from the arid landscapes of Morocco to flood-prone areas across the globe. The depth of research, combined with the practical applications presented, ensures that this book will be a cornerstone for future work in this ever-evolving field.

I would like to acknowledge the Editorial Advisory Board whose expertise has shaped the quality and cohesiveness of this book. Their collective knowledge has been instrumental in curating a volume that balances academic rigor with practical utility.

This book is essential for students, researchers, and professionals alike, providing the tools and insights necessary to navigate the complex dynamics of water management in the 21st century. Beyond academia, it offers valuable perspectives for policymakers and practitioners, equipping them with the knowledge to make informed decisions in the stewardship of water resources.

I extend my congratulations to all contributors for their exceptional work and dedication. This volume stands as a testament to the power of collaboration and innovation in tackling one of the world's most pressing issues: the sustainable management of our water resources.

Jamal Al Karkouri
Faculty of Humanities and Social Sciences, University Ibn Tofail, Morocco

Preface

The growing complexities of managing and conserving water resources in today's world have led to the adoption of innovative technologies such as Remote Sensing (RS) and Geographic Information Systems (GIS) in the field of hydrology. The need for a deeper understanding of hydrological processes, coupled with the increasing pressures of climate change, population growth, and environmental degradation, has made it imperative to employ advanced tools for data collection, analysis, and decision-making. This book, *Remote Sensing and GIS Techniques in Hydrology*, presents cutting-edge research and case studies that highlight the immense potential of these technologies in tackling critical water resource challenges.

Remote Sensing (RS) involves the acquisition of information about Earth's surface and atmosphere without physical contact, typically using satellite or aerial sensors. In hydrology, RS has become a powerful tool for observing, monitoring, and modeling various components of the hydrological cycle, such as precipitation, evapotranspiration, soil moisture, groundwater levels, and surface water bodies. Through RS, hydrologists can gather large-scale, real-time data on water resources in even the most remote or inaccessible regions. This capability has led to more accurate predictions of hydrological events like floods, droughts, and water availability, while also enabling the assessment of long-term trends related to climate change and land use.

Geographic Information Systems (GIS), on the other hand, offer a platform for spatial data management, analysis, and visualization. In hydrology, GIS is instrumental in integrating and analyzing diverse datasets, including topography, land cover, and hydrological networks. By utilizing GIS, hydrologists can model the movement of water across landscapes, predict the impact of land use changes on water systems, and assess watershed dynamics. GIS also facilitates the creation of detailed maps that visualize spatial patterns and relationships, supporting decision-makers in identifying areas at risk of flooding, erosion, or water scarcity.

The combination of RS and GIS provides a comprehensive framework for understanding and managing hydrological systems. Together, these technologies allow for a multi-dimensional view of the water cycle, offering a clearer picture of the interactions between land, atmosphere, and water bodies. Furthermore, the integration of RS and GIS enhances the ability to develop hydrological models that predict water flow, sediment transport, and water quality, all of which are essential for sustainable water resource management.

In Chapter 1, "Application of Remote Sensing and GIS in Environmental Monitoring and Management," readers are introduced to the fundamental principles of RS and GIS and their application in monitoring environmental changes. The chapter emphasizes how these technologies have enabled the detection of land cover changes, water quality variations, and ecological disruptions that directly affect hydrological processes. By providing real-time data on water systems, RS and GIS have transformed environmental monitoring into a more proactive and precise science, leading to more informed decisions on resource management.

The second chapter, "Integrating GALDIT and GIS for Assessing Sea Water Intrusion Susceptibility in the Akermoud Coastal Water Table, Morocco," exemplifies how RS and GIS can be combined with hydrological models to address specific challenges such as seawater intrusion in coastal aquifers. Seawater intrusion, exacerbated by over-extraction of groundwater and rising sea levels, threatens freshwater resources in coastal areas around the world. The use of the GALDIT model, integrated with GIS, provides a scientific basis for identifying vulnerable zones, guiding the implementation of management strategies to protect coastal aquifers.

Chapter 3, "A Review on Estimation of Climate and Land Use Change Impacts on Tropical Streamflow Using Remote Sensing and GIS," explores the use of RS and GIS in understanding the effects of climate change and land use changes on streamflow in tropical regions. Remote sensing data on precipitation and vegetation, coupled with GIS-based hydrological modeling, offers insights into how streamflow patterns shift under changing environmental conditions. This is particularly important for tropical regions, where alterations in water availability can have significant socio-economic impacts, affecting agriculture, water supply, and biodiversity.

In "Detailed JavaScript API for Google Earth Engine to Enhance Monitoring of Surface Water Dynamics in the Drought-Affected Mediterranean Region," the authors demonstrate the power of cloud-based platforms for processing vast amounts of satellite data. By utilizing Google Earth Engine, researchers are able to analyze surface water dynamics in near real-time, providing critical information for managing water resources in regions affected by drought. The integration of such technologies into hydrological studies enables better monitoring and response to the effects of climate variability on water bodies.

Artificial Intelligence (AI) and machine learning are also becoming indispensable tools in the analysis of remote sensing data. In Chapter 5, "Boosting Artificial Intelligence Performance in Sentinel-2 Imagery Analysis," the study illustrates how AI-driven algorithms can enhance the detection and monitoring of water bodies from satellite imagery. By automating the process of identifying surface water features, AI allows for faster and more accurate analysis, providing valuable insights for hydrological modeling, disaster management, and water resource planning.

The sixth chapter, "Integration of Remote Sensing and GIS for Optimal Site Selection of Dams in Arid to Semi-Arid Environments," highlights the role of these technologies in infrastructure development. The combination of RS-derived data, GIS spatial analysis, and hydrological modeling allows for the identification of optimal sites for dam construction, balancing water availability with environmental and social factors. This approach is crucial in arid and semi-arid regions, where water storage infrastructure plays a key role in supporting agriculture, urban development, and ecosystem preservation.

Chapter 7, "Monitoring Agricultural Dynamics and Its Impact on Water Resources Using GIS and Remote Sensing Data," explores the intersection of agriculture and hydrology. Agricultural expansion and intensification place significant pressure on water resources, and RS and GIS are essential for monitoring these impacts. By analyzing changes in land use and water consumption, this chapter underscores the need for sustainable agricultural practices that conserve water and protect the hydrological balance of the region.

Chapter 8, "Monitoring Surface Water Variations at the Ain Kwachia Dam in Morocco's Northwest Central Plateau," employs RS and GIS to monitor water level fluctuations at a key reservoir. The insights gained from this study help inform reservoir management, ensuring the availability of water during periods of drought and flood. Understanding the dynamic nature of water bodies is critical for planning and managing water resources in regions with highly variable climates.

In Chapter 9, "Assessing the Spatio-Temporal Evolution of Lake Aguelmam Azigza Using Remote Sensing and GIS," the use of RS and GIS techniques to track long-term changes in lake systems offers a detailed picture of how water bodies evolve over time. The temporal analysis of satellite data reveals trends in lake size that may be driven by climate change, human activities, or natural variability, making it a valuable tool for managing and conserving freshwater ecosystems.

Finally, in Chapter 10, "PAP/CAR-Based Water Erosion Analysis in the Oued Lakhdar Watershed: A GIS and Remote Sensing Approach," the authors tackle the issue of soil erosion, which significantly impacts water quality and availability in the Oued Lakhdar watershed. The chapter demonstrates how RS and GIS tools, combined with the PAP/CAR model, allow for the identification of areas at high risk of erosion. This analysis is essential for implementing erosion control measures and ensuring the long-term sustainability of water resources in the watershed.

Mouhcine Batchi
University of Ibn Tofail, Morocco

Adil Moumane
University of Ibn Tofail, Morocco

Chapter 1

Application of Remote Sensing and GIS in Environmental Monitoring and Management

Jayashri Dutta
https://orcid.org/0009-0006-5303-1654
Gauhati University, India

Smitakshi Medhi
https://orcid.org/0009-0004-6106-1111
Gauhati University, India

Mayurakshi Gogoi
https://orcid.org/0009-0000-3706-1606
Gauhati University, India

Lisha Borgohain
https://orcid.org/0009-0007-5763-5583
Gauhati University, India

Nourhan Gamal Abdel Maboud
GIS Community Academy, Egypt

Hanaa Mustafa Muhameed
GIS Community Academy, Egypt

ABSTRACT

Remote sensing (RS) and geographical information system (GIS) have been extensively used for their ability to provide better scope for decision making in terms of monitoring and managing various natural resources: water, land, forest, agriculture, natural hazards, climate change, etc. In this chapter, the authors precisely discuss in detail about the various applications of remote sensing and GIS in numerous fields such as monitoring water quality, crop modelling, monitoring wildlife habitat and forest management, application in agriculture and soil science, water resource, as well as natural disaster management. This chapter will also incorporate several case

DOI: 10.4018/979-8-3693-9651-3.ch001

studies where GIS and remote sensing were used as a powerful tool to delineate and monitor environmental changes and take the necessary steps to manage the same. This chapter will thus focus on the importance of remote sensing and GIS and its impact in sound decision-making, which in turn will secure the sustainable utilization of natural resources to reach the requirements of present as well as future generations.

1. INTRODUCTION

Remote sensing and Geographical Information Systems (GIS) become a means of utmost importance in recent days for various disciplines. In the sixties, non-spatial mathematical models had developed for the decision-making process which has been changed to a displayable spatial model with the help of GIS in the last years. With the advent of Remote Sensing and GIS techniques, it has become easier to an extent to monitor natural resources with multi-spectral, multi-spatial and multi-temporal resolution (Kumar et al., 2015). Remote sensing provides a remarkable insight into the field of applied sciences with a greater extent of geospatial information that is useful in environmental change detection studies (Chamine et al., 2021). The remote sensing data of a wide range of resolutions from space-borne and air-borne platforms help government organizations, industries and public sector organizations to prepare their planning of the management of natural resources and decision-making process. However, GIS is such a powerful tool that it can collect, process, reacquire and transform a large set of data into a displayable form that can be used as a decision support system in environmental resource supervision and impact assessment in various disciplines i.e. agriculture, water, soil, mineral resources etc. (Droj, 2012).

Some satellites such as Landsat, and MODIS provide historical data also that help to carry out trend analysis in an uninterrupted manner (Parra, 2022). For the trend analysis of various meteorological parameters and change detection studies of land use land cover, satellite data products are used. With the aforementioned advantages, Remote Sensing and GIS have now become a very effective robust technique over many traditional methods of decision-making.

2. BACKGROUND

With the advancement of technology, GIS combined with remote sensing data provide a cost-effective and efficient approach for acquiring data in earth and environmental sciences. Remote sensing data can be collected using passive (e.g. spectral imaging) and active sensors (e.g. radar). The most used satellite sources for remote sensing images are the Landsat, followed by MODIS, SPOT 5 and Sentinel 1 (Parra, 2022a). The use of Unmanned Automatic Vehicles (UAVs) has also increased for their high spatial resolution and high flexibility for receiving images (Parra, 2022a). However meteorological factors (rain, wind, and clouds) might limit the use of UAVs (Parra, 2022a). The use of spatio-temporal data helps in mapping geological settings and watersheds, monitoring land use/cover, and monitoring various phases of natural calamities such as earthquake, flood, landslide, volcanic explosions, etc. (Kumar et al., 2015). From data entry to analysis and representation of data, GIS software facilitates the interpretation of geospatial data. For mapping and analyzing geographical data, ArcGIS and QGIS are the two most popular GIS software among academicians, industrialists, and government. Multi-criteria decision analysis (MCDA) coupling with GIS helps decision-makers in improving their decisions regarding different scenarios such as evaluating groundwater potential zones, landslide and flood hazard impacts, land suitability, etc. (Chaminé et al., 2021). Additional statistical methods like the analytical hierarchy process (AHP) technique with fuzzy set theory (FAHP) enhance the techniques in MCDA (Ahmed & Kilic, 2018).

3. APPLICATIONS OF REMOTE SENSING AND GIS IN DIFFERENT FIELDS OF THE ENVIRONMENT

3.1 Application in Water Resources

Water plays a key role in the existence of mankind. However, over the last few decades due to over-exploitation of water to fulfil the requirement of the increasing population, availability of potable water is declining rapidly (Kumar et.al 2015). In this regard, an effective strategy for monitoring and managing water resources is of utmost importance. Sustainable water management requires a comprehensive approach, as such GIS and remote sensing can act a very integral role in this. With the help of GIS water resource management and monitoring have reached a different level as these tools are helpful in spatial analysis and data curation, which on the other hand helps to make sound decisions (Yahya et.al 2023). GIS has been extensively used in water resource mapping, water quality mapping, flood forecasting, rainfall-runoff monitoring, drought monitoring etc. Some of these applications are discussed in detail below:

1) Application in Water Quality Management

Water quality monitoring is crucial, particularly for human consumption. Laboratory analysis can give accurate assessments for a particular location in space and time however, it cannot provide the spatial or timely assessment of the water quality in the region that is required for proper management. GIS and Remote sensing thus come into play as they can monitor various water quality parameters viz; turbidity (suspended sediments), temperature and chlorophyll (Kumar et.al 2015). Sensors (optical and thermal) are installed on aircraft, drones, boats and satellites which can give temporal and spatial information based on their reflectance and radiance and the data is helpful in monitoring the water quality changes of a region (Ritchie et.al 1994). The spatial distribution patterns of water quality parameters can be mapped using GIS for visual interpretation (Choudhury et al., 2024).

2) Application in Water Resource Mapping

Water resource maps can be made with the aid of GIS and remote sensing coupled with artificial intelligence (Hussain et.al 2022). A case study in 2019 (Ashtaker et.al 2019) tried to delineate a map of surface water flow dynamics in various regions of the upper Krishna River catchment of Maharashtra, India. Landsat data (cloud free) were used to scrutinize the surface water kinesis of the catchment for a period of 17 years (1996 to 2016) employing Normalized Difference Water Index (NDWI) in the study. To understand the kinesis of the surface water for 17 years, change detection study is implemented using the GIS tool. The results thus, demonstrated the significance of GIS and remote sensing practices for water resource mapping, particularly to identify changes over time.

3) Application in rainfall and runoff management

The rainfall, runoff and soil moisture of an area highly influence the pattern of the watershed of the region (Das 2019). GIS and Remote Sensing are the common tools that are applied in hydrogeology to delineate the watershed of an area. Moreover, it also helps to track the occurrence of floods and drought and also helps in identifying areas that are prone to severe rainfall and runoff conditions (Yahya et.al 2023). In a study in the Bojiang lake watershed, China, GIS and remote sensing coupled with the Soil Conservation Curve Number (SCN-CN) method was used to evaluate daily, monthly and yearly runoff for ungagged watershed (Matomella et.al 2019). Moreover, rainfall-runoff model using HEC-HMS software was created to determine the real discharge peak to produce a hydrograph associated to a important event which happened in an urban ungauged watershed located in NE of Tunisia (BEN Khalifa and Mosbahi 2021).

4) Application of GIS in drought monitoring

Areas impacted by drought have adversely affected humans over several decades. Due to climate change drought events seem to appear more frequently over the last few years (Rojas et.al 2021). However, with the aid of GIS and Remote Sensing drought prediction has become easier and as such necessary steps can be taken to sustainably manage the situation prior to the actual event. Elhaeesahar and Masoudi 2019 made a model of drought risk for Khuzestan, Iran. They have classified hazard zones into 5 classes viz; very severe, severe, moderate, minor and none. The outcome of the drought risk map prepared using GIS revealed that 29% of the province has a severe drought hazard zone. Thus, Remote Sensing and GIS have proved to be excellent tools in monitoring these natural hazards which in turn helps in mitigating the problem.

3.2. Application in Urban Planning

GIS plays a crucial part in developing land use and resource models. Remote sensing and GIS tools used in city planning can help and escort the concerned authorities and planners to properly develop the buildings and infrastructure facilities in the region (Xhafa et.al 2015). With the rapid expansion of the population worldwide, where approximately 53.7% of people live in urban areas, proper urban spatial planning for sustainable development is must. It was first in the USA where Landsat satellites were used to capture and record spatial and temporal data globally and with the help of GIS tool urban planning was developed (Xhafa et.al 2015). GIS is beneficial for urban planning in various ways such as (Yeh A G-O 1991):

1) Improving mapping for better access to the maps and effective thematic mapping.
2) Better regulation in the recovery of information.
3) Better quality of services.
4) Better access to geographical information for proper urban planning.

There are various functions in urban planning where GIS is extensively used. The role of urban design can be categorised basically into 4 groups viz; (Yeh A G-O and Chow M H 1996):

1) General administration
2) Development control
3) Plan making
4) Strategic planning

The first two role are routine planning carried out regularly, while rest of the two are conducted occasionally.

The scale of urban development can range from the entire city to a small block in a particular district. In urban planning, GIS is mainly used in land use landcover mapping, infrastructure and transport planning, service management, ecosystem modelling, assessment and planning of landscape etc. According to the needs and functions of urban developments various components of GIS such as modelling, data visualisation, data management and spatial analysis are used. As GIS is useful in storage, manipulation and analysis of economic, physical and social data of the city, planners can use this data to analyse the present situation and accordingly decide the urban planning (Chuvieco E 1993). The GIS technology basically used by planners in monitoring, preparing projects and implementing them, thus helping in decision-making. Thus, GIS and remote sensing plays an integral and comprehensive role in urban planning.

3.3 Application in Biodiversity and Wildlife Management

Biodiversity is the variability of species from different sources viz., marine, terrestrial and other aquatic habitats and it is the foundation of an ecosystem (Pant et al., 2020). The conventional methods for collecting data for the conservation of biodiversity and wildlife are time-consuming and costly, and remote sensing combined with GIS helps collect large spatial data in a short period. The suitable habitat and corridor selection for the wildlife using GIS tool is an advanced technique that needs only a few factors affecting the target such as topography, forest type, human intervention, water resources, land use land cover, roads etc (Goparaju et al., 2017). Monitoring of human - disturbances, climate change, habitat loss, pollution and other factors has become easier with the implementation of air-borne and space-borne remotely sensed data in GIS platforms where intervention is needed (Pant et al., 2022). There are many observation satellites and sensors (active and passive) available for evaluate biodiversity from very high elevation (Gillespie et al., 2008) (Table 1). Habitat mapping studies such as species diversity, plant diversity, species distribution, habitat heterogeneity and vegetation indices are now been carried out in a stipulated time frame with satellite data for a larger area (Kerry et al., 2022). A study led by Kumar et al., 2019 in the West Kameng and Tawang districts of Arunachal Pradesh and its adjacent areas for the estimation of aboveground biomass diversity and assessment of tree diversity used ALOS, PALSAR, Sentinal-1A space-borne satellite data. Koyama et al., 2019 prepared a comprehensive forest variability map in global tropical forests based on homogenous big data analysis using ALOS-2, PALSAR-2 dual polarization scanSAR data.

Table 1. Some of the remote sensing satellites/sensors used in biodiversity management

Types of RS	Senor / Satellite	Purpose
Active	LIDAR	Vertical and physical depiction of items such as plant height and growth stage
	SRTM	Provide elevation and topographic data used for species and diversity model
	RADARSAT-1	Used for land cover categorisation and variety models
	ALOS, PALSAR	Used for high-accuracy forest classification
	ICESat GLAS with LIDAR	Forest biomass mapping and validation
Passive	WorldView, QuickBird, IKONOS	Species mapping with visible and infrared bands
	Landsat	Biodiversity studies due to low cost and time-series
	MODIS and AVHRR	Provide global, continental and regional data for classification of land cover and diversity; meteorological data and fire merged into biodiversity studies

3.4 Application in Transport Planning

Environmental Impact Assessment (EIA) is a much-needed valuation in case of transport planning for decision-making. As transport plays a crucial role in human life, therefore its sustainability needs much attention. The baseline scenario in a transport project involves the location site and its surrounding area. A large set of data gathering and in-depth analysis is possible with aerial photography, high-resolution satellite imageries and environmental maps to assess the landscape vulnerability (Soria-Lara et al., 2020). In the investigation of air pollution caused by urban road traffic conditions, the effectiveness of emission-reducing measures, and urban transportation emission modelling; these technologies proved to be helpful. The network analysis in GIS can estimate the pollutants from road networks and their environmental consequences thereby managing the effect. The routes can be constructed as accessible for the residents to their daily working stations and healthcare centres. Thus, people could be encouraged to the public transport in daily life for sustainable urban transport to reduce pollution to some extent (Nadi and Murad., 2017). Using the remote sensing data, different areas in Limassol City were classified as low and high density and the objective was to boost the use of public transportation replacing private cars for a short trip inside the city (Giorgoudis et al., 2014). Lwin and Murayam, 2011 carried out a study for green walk space in Tsukuba City, Japan using web-based GIS applications and find out an eco-friendly walking route for residents. Moreover, RS will be helpful in energy efficiency planning and reduction of noise pollution. As a non-renewable energy source, fossil fuels become limited. Therefore, efficient energy consumption is an obligatory step in sustainability and for mankind. There are some standard tools in GIS for particular functions (Table 2).

Table 2. Some of the tools of GIS and their functions

GIS Tools	Functions
Network analysis	Traffic model for traffic route mapping
	Traffic simulation and analysis
	To define locations to determine accessibility
	To access the service of the public transport system
	To assess the pollution load from road network
Buffer analysis	To define an object based on the distance from other objects
	Accessibility of a location based on distance
Straight line allocation function	To locate the nearest healthcare centres
	To find an area not served by supermarket
Classification method	To use multiple regression model for spatial analysis

Modified from Nadi and Murad, 2017

3.5 Application in Agriculture

Agriculture is crucial for the economic stability and growth of every nation. Any large-scale farm management, farmer, and regional agricultural organization aims to produce food as economically and efficiently as possible.

3.5.1 Application of Remote Sensing in Agriculture

Agriculture and remote sensing are closely related. Using sensors on satellites, planes, drones, or other platforms, remote sensing in agriculture collects data about the earth's surface, counting information about soil, crops and ecological variables (Marshet, 2019). This technique makes use of several electromagnetic spectrum components, such as microwave, infrared, and visible light, to provide precise and comprehensive information without the need for physical touch. By giving fast and accurate data that helps decision-making throughout farming and agricultural management techniques, it acts a critical role in modern agriculture. (Remote Sensing and GIS in Agriculture, 2024). Using Remote Sensing in agriculture makes it easier to adjust farming methods to diverse obstacles since it continuously provides data on crop conditions at different sizes and throughout the developing period (Remote

Sensing in Agriculture, 2022). The following are some important applications of remote sensing in agriculture:

1) Managing and controlling crops

Remote sensing is pivotal in agriculture for its ability to continuously monitor crop health, growth stages, and overall vegetation vigour by assessing reflectance data. This technology plays a critical role in identifying the factors that can inhibit crop growth such as water shortage, nutrient deficits, pest invasions, or disease occurrences, which may not be visible to the naked eye. Optical sensing, particularly in the visible and infrared spectra, enables farmers to detect crop vigor, damage, and stress levels with precision. Recent advancements in remote sensing have empowered farmers to make timely and informed decisions about crop management based on real-time field observations. Additionally, remote sensing aids in crop identification, helping farmers identify specific crops affected by various environmental conditions such as weather patterns and pest infestations, facilitating proactive agricultural practices. (Remote Sensing and GIS in Agriculture, 2024; Remote Sensing in Agriculture, 2022)

2) Monitoring soil conditions

Monitoring soil is crucial for precision agriculture, with key parameters including soil organic matter (SOM), texture, pH level, and moisture content. Remote sensing technology acts a vital character by providing data on canopy health, growth stages, yield, biomass, and vegetative density. It's important to establish connections between crop performance and soil conditions to understand changes in crop growth patterns. Sensors are instrumental in detecting variations in soil moisture and composition, offering worthwhile insights for optimizing irrigation management and soil protection approaches. This data enables more efficient application of water and fertilizers, customized to the specific requirements of various crop zones. (Remote Sensing and GIS in Agriculture, 2024; Remote Sensing in Agriculture, 2022)

3) Observing and monitoring water conditions

With the anticipated doubling of irrigated lands by 2050 due to population growth and food demand, monitoring and assessing agricultural water resources is crucial for sustainable food security. Remote sensing technology in precision farming provides timely information on water bodies, irrigated croplands, and crop and soil water statuses, allowing for early identification of drought-affected areas and timely interventions to optimize water utilization. (Remote Sensing and GIS in Agriculture, 2024; Remote Sensing in Agriculture, 2022)

4) Observing and predicting weather and climate conditions

Precise climate and weather data are vital for making informed decisions about crop management, irrigation planning, and preparing for natural disasters. The practice of remote sensing technology in correct farming has significantly expanded spatial coverage, allowing for accurate prediction of upcoming weather conditions. This information enables more precise forecasting of crop requirements, resulting in cost savings. Furthermore, remote sensing in agriculture has advanced our comprehension of climate change by effectively quantifying changes over time in oceans, land, and the atmosphere. This capability is especially valuable for making dependable predictions about climate change's impact on farming and agriculture. (Remote Sensing in Agriculture, 2022).

5) Improving air quality

Different varieties of crops require specific air conditions to thrive. Certain types may flourish in one climate while struggling in another. It's important to understand the environmental needs of different crops to ensure their healthy growth. (Remote Sensing in Agriculture, 2022).

6) Enhancing Crop Yield Prediction and Precision Farming

In recent years, new farming methods have helped agriculture meet the increasing demand for food. Remote sensing, precision farming, and modern technologies like AI and the Internet of Things have improved crop yields and reduced environmental impact. These approaches involve detailed mapping to optimize resource usage and help forecast crop yields. Farmers and supply chain participants benefit from these advancements. (Remote Sensing and GIS in Agriculture, 2024; Remote Sensing in Agriculture, 2022)

3.5.2 Application of GIS in Agriculture

GIS, a powerful tool known as Geographic Information System, empowers users to create interactive maps with multiple layers. These maps are invaluable for visualizing complex data and performing spatial analysis. In the field of agriculture, GIS enables farmers to map field data, organize and analyze it, and remotely monitor their crops. The advancement of farm automation owes much to GPS, robotics, drones, and satellite monitoring, all of which are vital for collecting GIS data. Through visualization, GIS assists farmers in identifying trends and patterns, implementing change detection, and promptly addressing any issues. Precision agriculture heavily relies on GIS for collecting and interpreting extensive field data to make well-informed decisions. (GIS In Agriculture: Best Practices for AgriTech Leaders; 2022)

Here are some common examples of how GIS is used in agriculture:

1) Soil study: By using spatial data analysis, Geographic Information Systems (GIS) have the capability to generate maps that illustrate various soil types and moisture levels.
2) Land management and crop production: GIS is used in agriculture for mapping and monitoring crop categories, strength, and productions in various geographical areas, providing valuable insights for improving crop rotations, planting schedules, and harvest timing.
3) Tracking crop pest and diseases: GIS facilitates the detection of forms and hotspots where pest and disease outbursts occur, enabling focused interventions to mitigate their spread and minimize crop damage.
4) Farming and irrigation: GIS plays an important role in agriculture by integrating data from satellites and drones to meticulously manage fields. This enables precise management of water, fertilizers, and pesticides, boosting crop production while reducing environmental footprints. By analyzing terrain, soil characteristics, and crop water requirements, GIS assists in scheming effective irrigation systems. It identifies specific areas needing varying levels of water, thereby optimizing water usage and promoting conservation efforts. (GIS In Agriculture: Best Practices for AgriTech Leaders; 2022)

Some other examples include:

- Effective preparation of farm structure like roads, stowage services, and drainage systems.
- Land use planning and the preservation of natural resources by analyzing spatial data.
- To grasp market demands, streamline supply chains, and strategize transportation and distribution for efficient market access.

Additionally, GIS monitors environmental shifts affecting agriculture, such as climate variations, land deprivation, and water accessibility. This data is essential for implementing maintainable farming methods that can adapt to and counteract environmental impacts. (GIS In Agriculture: Best Practices for AgriTech Leaders; 2022)

3.6 Application in Soil Science

For any land use planning and implementation of various activities such as sustainable agricultural production, infrastructure building, etc., proper knowledge about the characteristics of soil, soil type, and its texture of an area is very much needed. Soil survey is important for inventory as well as the extent of distribution of different kinds of soil and prediction of their potentialities (Sreekanth & Das, 2016). The information gathered in soil surveys with hand-held GPS coordinates can be linked to remotely sensed images for further interpretation. In remote sensing, the radiations discharged or imitated from the earth's surface vary with the physical as well as chemical properties of the soil matrix, thus it helps to differentiate various types of soil (Sreekanth & Das, 2016). The collection of landscape information using RS is an effective technique to gather huge amounts of data with more coverage (Nelson, 2021). Digital Soil Mapping (DSM) can be generated by combining GPS and GIS which provides the necessary information to the researchers, analysts, and decision-makers for sustainable soil management practices. Some of the sensors used for soil surveys are given in Table 3.

Table 3. Different sensor types and purposes (Sreekanth & Das, 2016)

Sensors	Type	Purpose
Passive	Optical multispectral sensors	land use along with mineralogical studies
	Optical imaging spectroscopy sensors	soil properties like mineralogical composition, organic matter
	Optical TIR (Thermal Infrared) sensors	soil temperature approximation
	Passive microwave sensors	soil moisture approximation
Active	Synthetic aperture radar (SAR) sensors	soil moisture, texture, and salinity approximations
a.	Radar scatter meter sensors	soil moisture approximation

A case study of a salty depression situated in the Egyptian desert by Aly and others (2007) used RADARSAT-1 (Canadian RS satellite) Synthetic Aperture Radar (SAR) images with some modelling techniques to determine soil salinity based on the dielectric constants of soil. Many researchers use microwave SAR sensors to estimate soil moisture. Indian Space Research Organization (ISRO) successfully launched Radar Imaging Satellite 1 (RISAT-1A) with a microwave sensor which is specially designed to provide high-resolution images for monitoring soil moisture for agricultural applications including irrigation dynamics, crop stress assessment, flood monitoring, and soil moisture modelling etc. (irso.gov.in). Singh et al. (2019) reported that the greater the moisture content, the greater the dielectric constant which decreases the capabilities of penetration. It has been found that the L-band (microwave) shows higher penetration in soils as it has longer wavelengths.

Within recent decades, Predictive Soil Mapping (PSM) has been explored with the advancement of numerical and geostatistical models at different scales (Scull et al., 2003). Spatial interpolation generated in GIS software using geostatistical modeling methods such as kriging, semi-variance functions, and variograms helps researchers and analysts in decision-making (Hilal et al., 2024). Kaya et al. (2022) used machine learning and hybrid models (Regression Kriging) to produce digital maps (spatial resolution of 30 m) of Soil Organic Carbon (SOC) and available phosphorus content under various agricultural practices in a piedmont plain of Northeast Iran. Saline area mapping can also be generated by detecting salt scalds and halophytic vegetation with high-resolution (Sreekanth & Das, 2016).

Figure 1.(a) Sentinel-2 image used in Mediterranean soil sealing project, (b) soil sealing map generated from Sentinel-2 image

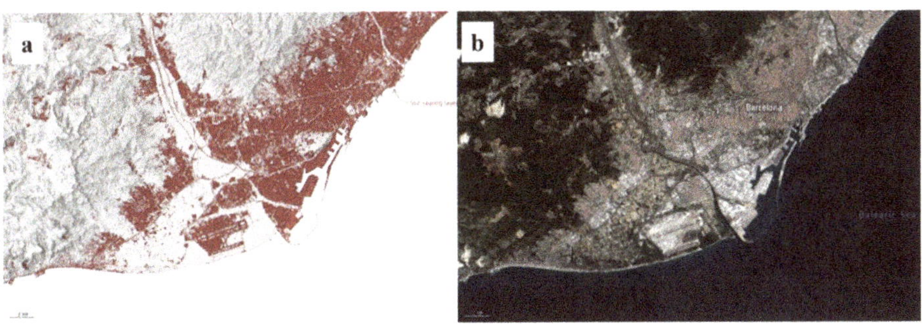

source: www.esa.int

The Mediterranean Soil Sealing project promoted by the European Space Agency (ESA) utilizes Copernicus Sentinel-2 data and generates maps depicting the extent and severity of soil sealing across the entire coastline of the Mediterranean Sea which is extremely valuable for researchers studying the changes in soil composition and permeability over time (esa. int) (Fig. 1(a) and (b)). It can be decided that the reflection factor of soil have contributed to the efficient management of the resource as it gives real-time monitoring with the help of remote sensing and GIS.

3.7 Application in Forest Management

GIS and RS techniques are significant in forest resource monitoring, forest protection, forest garnering, forest restoration, monitoring forest pests and diseases, spatio-temporal analysis of forest carbon density, degradation, forest fire monitoring etc. The higher spatial resolution satellites (Landsat and Sentinel) are most commonly used in forest management practices. Land use land cover (LULC) mapping is one of the mostly used applications of GIS and RS. Multispectral (coarse to fine resolution) and hyperspectral LiDAR (Light detection and ranging), SAR are used to map LULC, tree density, vegetation health, vegetation structure (basal area, leaf area index) etc. (Lechner et al., 2020).

Spaceborne LiDAR (Light detection and ranging) is being used to assess vegetation structure (such as true height, vertical forest structure, basal area, above-ground biomass, leaf area index, individual crowns, and gap size, etc.) increasingly as it can directly measure height by constructing three-dimensional (3D) representations (Lechner et al., 2020). The 3D maps of forest cover height and the dispersal of twigs and leaves can be prepared from the data collected from the Global Ecosystem Dynamics Investigation (GEDI) lidar instrument installed in the International Space Station which is a joint mission between National Aeronautics and Space Administration (NASA) and the University of Maryland (nasa.gov). Potapov and others (2021) created a global forest cover height map with a 30 m/pixel spatial resolution in 2019 by integrating Landsat analysis-ready time-series data (Landsat ARD) and Lidar (GEDI) data (Fig. 2). GEDI data is the most valuable tool for mapping the Canopy Height Model (CHM), Above Ground Biomass (AGB) on a large scale due to its fine spatial resolution, ability to enter the forest cover (Tamiminia et al., 2024). Passive optical remote sensing data, SAR, and LiDAR are also useful for the estimation of AGB of forests. Researchers suggested the integration of multi-source data for a continuous and highly accurate result of AGB estimation on a universal scale (Tian et al., 2023). The use of CHM provides highly accurate and efficient results in AGB mapping (Tamiminia et al., 2024).

Figure 2. Global Forest cover height map (2019) produced through the integration of GEDI data (April–October 2019) and Landsat ARD

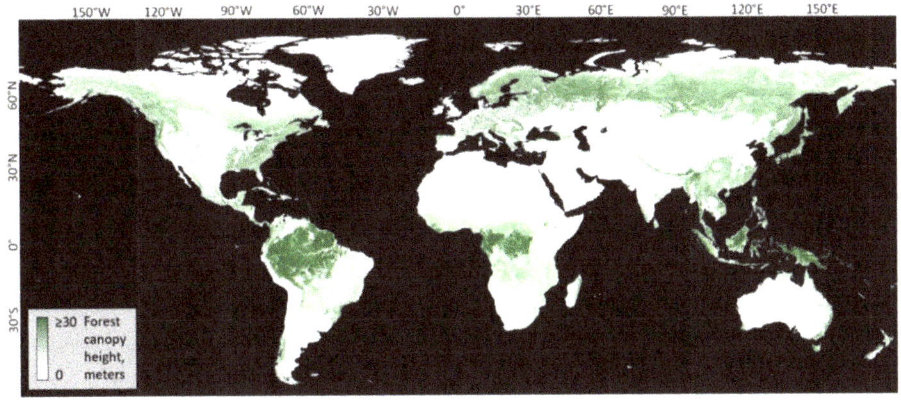

High-resolution UAVs and digital aerial photography are also good for the identification of individual species in a forest (Lechner et al., 2020). Many researchers used hyperspectral technology along with vegetation indices to assess forest pests and diseases directly related to forest ecological security (Tang et al. 2023). Wang et al. (2024) reported the potential of the UAV-based TIR (Thermal Infrared) and LiDAR data to perceive crown damage in a pine forest at the individual tree level. Liu et al. (2024) used combined Gaofen-1 and Sentinel-1 Data for visualising Coniferous Forest Distribution in a Semi-Arid Area based on Multi-Classifier Fusion and Google Earth Engine, and proposed to use these mapping methods to generate maps in a timely and correct way.

Muti-criteria decision-making (MCDM) is increasingly being used for forest road planning and AHP is one of the most commonly used techniques of MCDM. Integration of LiDAR data, MCDM, and heuristic algorithms can improve the accuracy of forest road-building mapping, (Xing et al., 2024). Satellite images, GIS and AHP are also being used in forest fire risk modelling. Some researchers use Fuzzy AHP techniques with GIS to map forest fire risk zones. Thus, the integration of GIS, satellite images, and other geographic information technology provides the required forest data quickly and efficiently to researchers, decision-makers, and forest managers.

3.8 Application in Oceanography

Satellite imagery has played a substantial role in advancing the mapping of coastal ecosystems. By capturing detailed aerial views, it enables precise estimates of land coverage and changes within these fragile environments. This technology has revolutionized our ability to monitor and understand alterations occurring in coastal ecosystems over time (Tripathy et al., 2018; Burke et al., 2001; Hochberg et al., 2003). The continuous evolution of sensor technology has led to significant improvements in remote sensing and GIS systems. These advancements enhance their applicability and appeal for managing coastal ecosystems, including estuaries, wetlands, and coral reefs. Utilizing advanced data analysis techniques further enhances the practicality and effectiveness of these systems in monitoring and preserving these vital environments (Held et al., 2003; Lyons et al., 2012; Rani et al., 2018).

Several applications of remote sensing and GIS in oceanography include:

1) Oceanographic Weather and Climate Research

Oceanographic weather and climate research involves studying the interactions between oceanic processes and the atmosphere, influencing global weather patterns and long-term climate trends. Remote sensing technologies, particularly satellite-based sensors, provide crucial data on sea surface temperature, ocean currents, sea level changes, and marine ecosystems. These sensors capture multispectral imagery that facilitates the monitoring and analysis of ocean dynamics at various scales (Wang et al., 2015; Zhang et al., 2017). GIS facilitates the mapping of low temperatures, high temperatures, temperature vulnerability, and daily thermal amplitudes through spatial analysis and integration of remote sensing data (Lagrini et al., 2019). The integration of remote sensing and GIS enhances our ability to monitor and predict changes in oceanographic weather and climate, contributing to more informed environmental management and policy decisions.

2) Spread and Migration of Oil Spills

Mapping and predicting the extent and drift of oil spills are critical tasks in environmental monitoring and disaster response. Remote sensing technologies, particularly satellite imagery, play a crucial role in these efforts. Satellite sensors can detect and monitor oil spills by capturing images across different wavelengths, allowing for the identification and mapping of oil slicks on the water surface (Leifer et al., 2012). These images are processed using Geographic Information Systems (GIS) to generate spatial maps that depict the extent of the spill. Moreover, with the aid of remote sensing and GIS data we can predict the movement and drift patterns of spilled oil. This integrated approach helps decision-makers in planning and implementing effective response strategies.

3) Exploring Ocean Colour Dynamics

Ocean colour study involves the analysis of light absorption and scattering by substances such as phytoplankton, sediments, and dissolved organic matter in the ocean. In this field remote sensing provides multispectral imagery that enables scientists to monitor and display variations in ocean colour on a global scale. Satellites equipped with sensors such as MODIS (Moderate Resolution Imaging Spectroradiometer) aboard NASA's Aqua and Terra missions capture data across visible and near-infrared wavelengths, permitting for the estimation of parameters such as chlorophyll-a concentration, a proxy for phytoplankton biomass (IOCCG, 2000; Gregg and Casey, 2007). Geographic Information Systems (GIS) is used to process and analyze these remote sensing data, to create spatial maps that illustrate changes in ocean colour over time and space. Ocean colour sensors provide essential data on oceanographic features such as currents, eddies, and mixing zones, along with the spatial and temporal distribution of phytoplankton and also help to identify potential fishing and upwelling zones, contributing significantly to studies on the global carbon cycle and other biogeochemical processes (Ahmad,2019). Synoptic maps derived from ocean colour data facilitate the analysis of coastal currents, boundary currents, and eddies. Moreover, they are instrumental in monitoring the pelagic ecosystem and detecting algal blooms (Ahmad,2019). This integrated approach enhances our understanding of marine ecosystems, ocean productivity, and the impacts of climate change on oceanic processes.

4) Monitoring Shoreline Dynamics

Shoreline change is a common issue in coastal environments, driven by ordinary processes that lead to dynamic alterations in coastal areas (Thomas et al., 2015; Burningham and French, 2017). GIS techniques integrated with remote sensing data archives enable the tracking and historical mapping of coastal shoreline dynamics (Yulianto et al, 2019). Satellite images from Landsat MSS, TM, ETM+, and ASTER acquired at different times are employed to identify shoreline positions, enabling the calculation of shoreline change rates (Maiti et al, 2009). Several approaches for coastline extraction and detection include Hand-drawn mapping, image improvement, spectral band density slicing with one or multiple bands, and both supervised and unsupervised image classification (Teodoro, 2018). GIS tools are utilized to process and analyze remote sensing data, enabling the creation of accurate shoreline maps and the quantification of shoreline change rates (Zhang et al., 2016). This integrated approach facilitates proactive decision-making in coastal zone management, disaster preparedness, and adaptation to sea-level rise.

5) Identifying Optimal Fishing Zones

Identifying optimal fishing zones is essential for sustainable fisheries management, applying innovative skills such as remote sensing and GIS for comprehensive analysis. Remote sensing tools, including satellite imagery and oceanographic sensors, provide critical data on factors like sea surface temperature, chlorophyll levels, and ocean currents. These environmental indicators are crucial for predicting fish distribution patterns and abundance (Lehodey et al., 2008). A combined method was established utilizing ocean colour monitor data for chlorophyll levels and AVHRR data for sea surface temperature (SST) to identify Potential Fishing Zones (PFZs) in Indian waters (Solanlki, 2001). This approach helps in planning and ensuring sustainable practices that preserve marine biodiversity and ecosystem health.

6) Mapping Coastal and Marine Hazards

Coastal inundation linked with severe tropical cyclones represents one of the most devastating coastal marine hazards globally, leading to significant life and property loss as well as loss of coastal infrastructure and cultivation in impacted nations (Dube et al, 2008). Satellite remote sensors and GIS have the capability to accurately visualise coastal ecosystems and their variations at the right scales and resolutions, which greatly reduces the necessity for wide field and ship measurements. This technology permits the quantitative extraction of multiple essential parameters to produce maps that evaluate various hazards and vulnerabilities (Teodoro, 2018; Mahendra et al, 2011)

3.9 Application in Natural Disaster Management

A long time-series data from Remote Sensing facilitates the temporal variation of a parameter through which a natural disaster can also be tracked. These data can be utilized after processing in a GIS environment for monitoring, early warning, awareness, preparedness, reconstruction, mitigation and relief purposes. The vulnerability and impact of a disaster can be minimized. With early detection, timely decision-making and smart responses can be made (Kumar et al., 2015). The real-time impact assessment, identification of affected areas, emergency responses and relief to the impacted population become easily accessible with the application of the GIS model. UAV aids in carrying out real-time flood risk assessment and assisting people in rescue operations. It is also helpful in identifying households impacted by disasters like landslides, storms etc. A mobile-based GIS application was employed by field staffs for electricity restoration after the cyclone hit areas in Gaja, India in 2018 (Sharma et al., 2020).

4. CASE STUDY

Derna in Response: Detecting Land Use, Land Cover, and Coastline Changes After Hurricane Daniel/Derna Flood Using Remote Sensing and GIS Techniques

On September 11, 2023, Hurricane Daniel or "Hurricane Medicane "struck northern Libya, causing catastrophic flooding in the city of Derna. The storm unleashed unprecedented rainfall, exceeding 400 milliliters in only 24-hour, which exceeded the capacity of existing dams. As a result, these structures failed, submerging Derna. The flooding is known as one of the most severe events in the city's history, occurring on September 10 and 11, 2023, and causing extensive damage to infrastructure. Historically, Derna has been prone to severe flooding, particularly during the winter months, with notable events recorded in 1969, 1995, 2018, and 2019. The recent flooding not only inundated residential areas, roads, and essential facilities, including hospitals and schools, but also led to widespread power outages and communications disruptions. The impact on basic services was significant, displacing thousands of residents and severely disrupting economic activities.

Local authorities and humanitarian organizations have mobilized immediately to evacuate affected individuals and provide assistance. However, recovery efforts from this natural disaster are ongoing, as the floods have caused significant economic losses in the short and medium term. The event requires concerted efforts to rebuild damaged infrastructure and restore commercial activities, as the long-term impacts on daily life in Derna remain profound.

The study reveals the impact of the flood and the change in the city in general, with the resulting damage to buildings and road infrastructure, in addition to the change in the city's coastline after the flood, clarifying the number of victims, the reasons for the collapse of the dams, and other studies of the city's topography and identifying the factors that contributed to the city being submerged in water.

The result confirms the danger of establishing any cities in this geographical area, the change in the coastline and the width of the submerged areas, and the port being greatly affected by this flood, which also greatly affects the economy.

The analysis highlights the prohibited area for construction in order to avoid building urban or service facilities in it.

The city of Derna is in northeastern Libya, about 300 km east of Benghazi, the capital. The town is situated on the eastern coast of the Mediterranean Sea, between the mountains and the beaches, and has around 100,000 people. Derna is considered one of the oldest cities in the region, with a rich history dating back thousands of years, witnessing the presence of the Phoenicians, Greeks, Romans, Islamic eras, and British occupation.

The city of Derna has a strategic location, characterized by a wonderful mix of mountains, hills, plains, valleys, and distinctive geomorphological features. These natural characteristics have made the city susceptible to floods for a long time, as it is surrounded by a group of mountains such as the Al-Akhdar Mountains - to the west of the city- and it is in the heart of the Derna Valley, which transports water from the mountains to the sea in the event of floods.

The city of Derna is characterized by a hot, dry summer climate and a moderate, rainy winter. The proximity of the city to the sea also affects its climate, making it a city with a moderate climate throughout the year.

Climate change had a major impact on the city, as on September 10 and 11, 2023, the city was hit by medicanes cyclone Daniel, which brought heavy rains. These heavy rains exceeded the capacity of the dams surrounding the city, causing them to collapse completely. This led to the flooding of the city with water and the destruction of large parts of the buildings and infrastructure of roads and electricity. The water also contributed to the flooding of parts of the city's coastline, resulting in a significant change in the coastline.

Hurricane Daniel: How and Where Hurricane Daniel Was Formed

Hurricane Daniel is considered one of the extreme phenomena resulting from climate change, as it is considered a “Medicane”, which is a weather phenomenon similar to tropical cyclones, which are characterized by strong winds and heavy rains. These phenomena usually form over warm waters in the Mediterranean Sea and appear as low-pressure centers that lead to severe weather conditions.

Reports show that medicanes affect several countries around the Mediterranean, such as Libya, Tunisia and Greece, and cause floods and landslides.

The hurricane formed a week before the flood over Greece and then moved to neighbor countries, causing strong winds, heavy rains, floods and loss of life in Greece, Turkey and Bulgaria. Later on, The storm reached the Libyan coast after crossing the Mediterranean Sea, causing flooding on September 10, 2023. The storm swept across eastern Libya, bringing heavy rains and strong winds of up to 80 km/h.

The storm initially affected coastal cities in the eastern region of the country, including Derna, Benghazi, and Sousse; Then it reached the cities of Marj, Shahat, and Al-Bayda

The flood was formed as a result of the sun heating the waters of the Mediterranean Sea and producing large amounts of water steam with the availability of active surface winds in the region and the disturbance of atmospheric pressure that contributed to pushing the humid air into the upper layers of the atmosphere and then condensing and forming dense thunderclouds, these clouds moved quickly to successive places from Greece to Italy, Albania, Turkey and then the Libyan coast.

Derna Valley "Wadi Derna Basin"

The total area of the Wadi Derna basin is 580 square kilometers, its perimeter is 240 km, the maximum length of the basin is 78.86 km, and the maximum width from north to south is 14.69 km (Fig. 3). The highest point is 816 meters above sea level and the lowest point is equal to the sea level.

The sources of the Wadi Derna basin begin west of the Al-Qayqab area at the water division line between the Wadi Derna basin and the Wadi Al-Kuf basin. This part of the basin is crossed by a group of valleys that generally descend from the northwest towards the southeast.

The Wadi Derna area is a flood-prone area, having experienced five major floods in October 1942, October 1959, October 1968, November 1986 and September 2011. The Derna area was classified as one of six areas in Libya at high risk of flooding.

Figure 3. Location map of the area

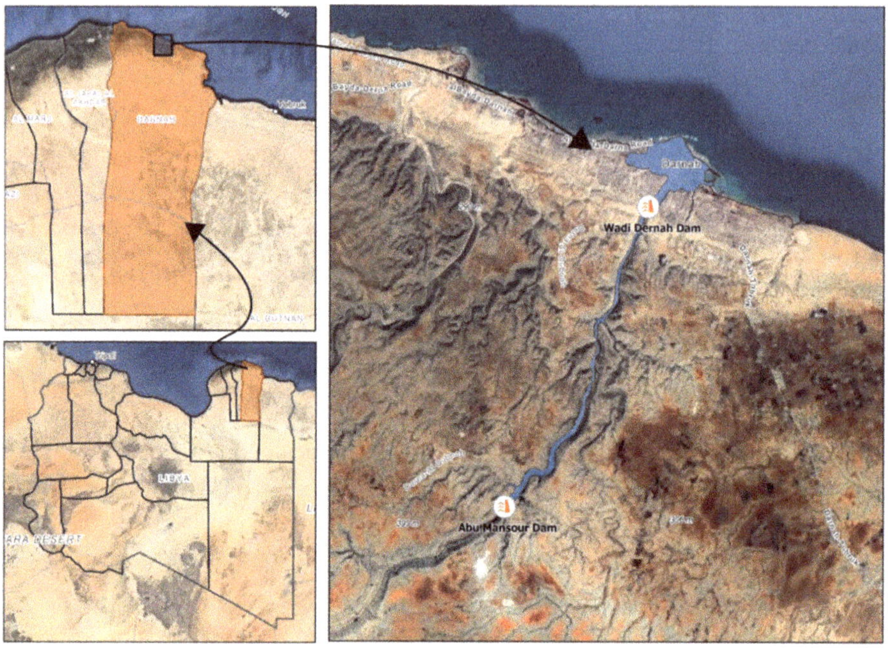

Climate Change Effect

The increasing frequency and intensity of natural disasters around the world are a direct consequence of climate change. The rise in global temperatures, shifting precipitation patterns, and more extreme weather events like hurricanes, floods, and droughts are all intricately linked to human-caused climate change (Banholzer et al., 2014). These escalating natural disasters have major impacts on local, regional, and global infrastructure, economies, and communities.

Addressing the challenge of climate change-driven disasters requires coordinated action at all levels - local, national, and international (Soergel et al., 2021). Mitigation strategies to reduce greenhouse gas emissions, such as transitioning to clean energy and sustainable agriculture, are essential. Equally important are adaptation measures to build the resilience of communities to withstand and recover from these extreme events through robust planning, preparedness, and emergency response systems.

The case of Hurricane Daniel's impact on the Libyan coastal city of Derna in 2023 provides a poignant example. The storm's power was amplified by the warming of the Mediterranean Sea, which provides the fuel for tropical systems to rapidly intensify. Additionally, sea level rise exacerbated the heavy rainfall and flooding that overwhelmed Derna's flood control infrastructure, leading to widespread inundation and damage (Fig. 4).

Addressing the increasing risks posed by climate change-driven natural disasters like this will require a multi-faceted approach. Reducing emissions to mitigate long-term climate change, combined with tailored adaptation strategies for vulnerable communities, will be crucial in safeguarding lives, livelihoods, and infrastructure into the future.

Figure 4. Hurricane Daniel's impact on the Libyan coastal city of Derna (2023)

Derna Detecting Change Using GIS and Remote Sensing

In this research we will discuss how to detect changes of landuse and coastline in Derna city – before and after the flood of Derna – using GIS software and Remote sensing technology

A) Detect Change in The City's Landuse:

After the flood, approximately 4,880 houses were damaged, and around 1,480 houses were destroyed. Additionally, about 8 km^2 of the city was submerged, which significantly impacted the city's roads and hindered the delivery of aid to the affected individuals

Figure 5. Devastating impact of the Derna flood in Libya

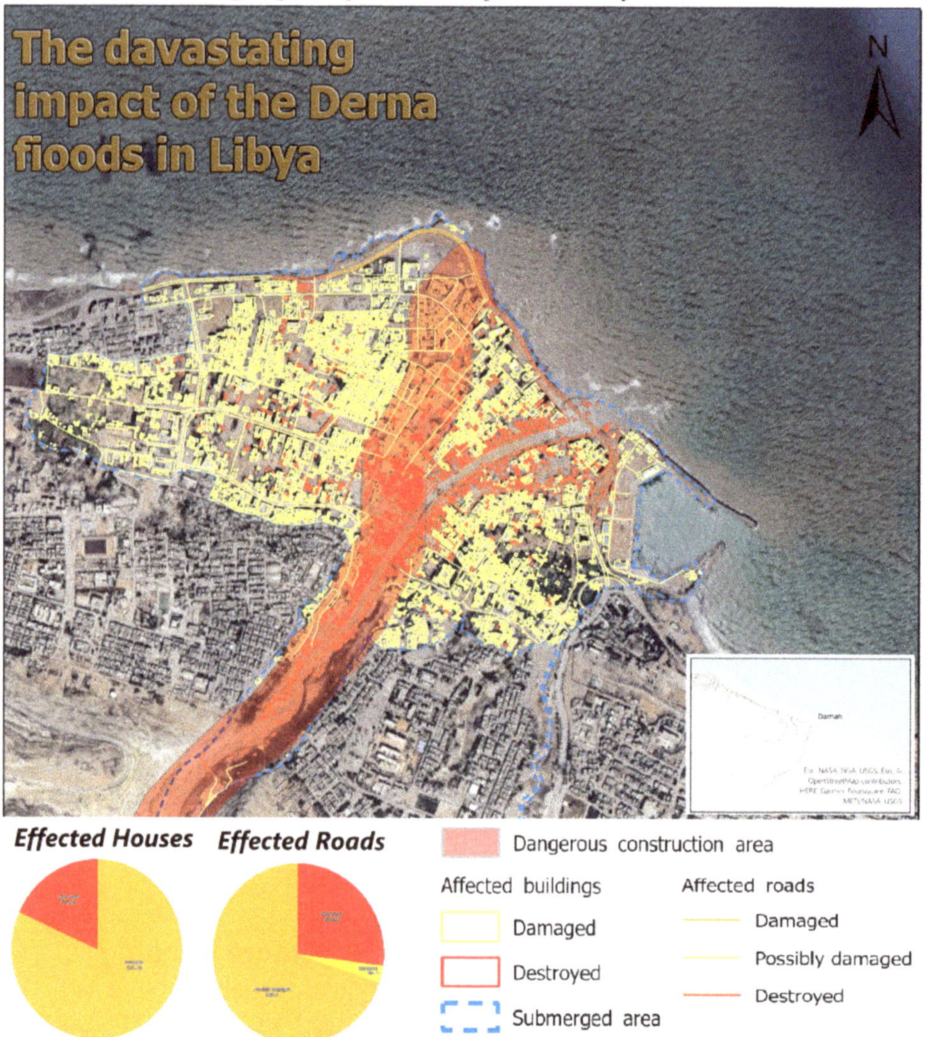

Using GIS, we created a new geodatabase that had many layers to store the changes after the flood, the geodatabase includes a road layer that details the roads affected by the flood, a building layer was developed to catalogue the buildings impacted by the flooding in the town, a dangerous construction area layer indicates locations where construction should not occur due to hazardous conditions and the sub-damaged layer reveals the areas that were specifically flooded within the city (Fig. 5).

B) Detect Change in The City's Coastline:

Using ArcGIS Desktop (ArcMap) and DSAS Tool we detected the shoreline changes, we used Landsat satellite imagery before and after the flood to detect the shoreline changes and calculate erosion and weathering of the coastline of Derna city (Fig. 6).

Figure 6. Monitoring the change of the coastline after the Derna flood (2023)

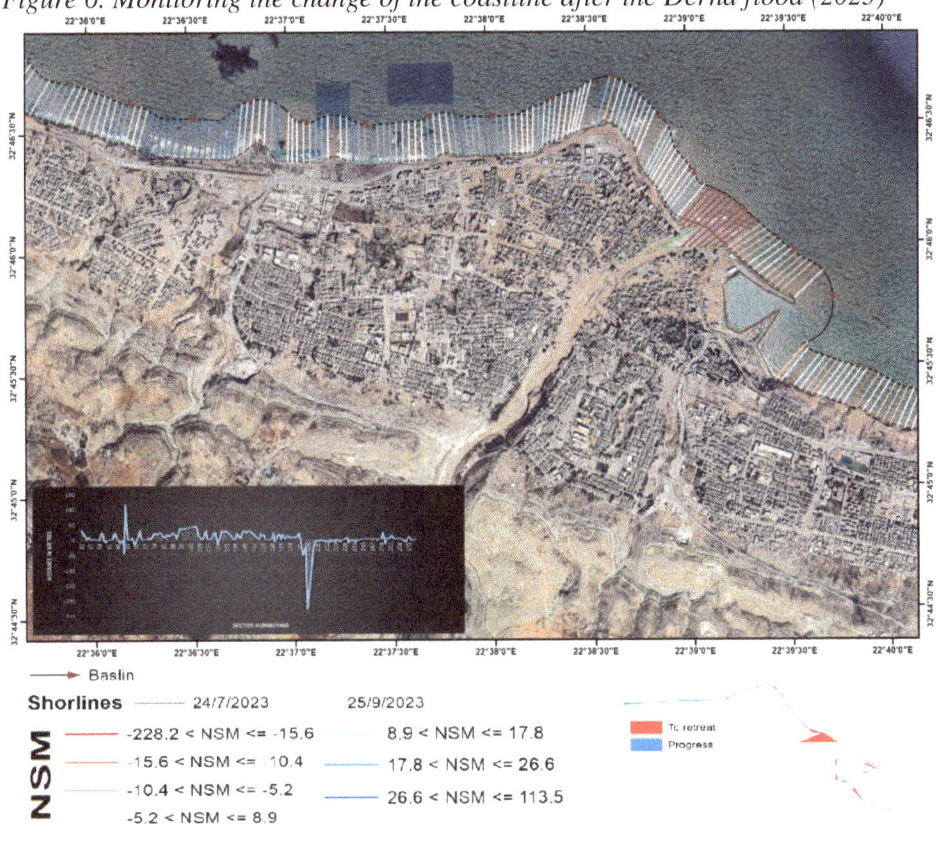

5. FUTURE ASPECTS

In an era of rapid development and environmental concerns, the remote sensing and GIS appears to be exceptionally promising in forthcoming days. As our nation continues to evolve, the demand for advanced geospatial technologies is expected to surge. These technologies will offer data for different uses by creating diverse and heterogeneous networks. The images or data that we will have in future with high resolutions and also the assembling together of different data sources will be helpful for use in new applications which are presently unemployed (Lloret, J et al., 2009). As GIS and remote sensing technologies advance and become more integrated they are poised to offer environment-friendly and cost-effective solutions in future (Sakshi, et al., 2023). Although techniques and approaches of GIS differ in the assessment stage based on the purpose of study, data availability and assessing parameters; there is a need for a standard universal model to the improvement of the assessment procedure and better comparison of the environmental impact assessment and strategic environment assessment.

6. CONCLUSION

As the world population tends to raise, the requirement for natural resources escalates. But, with the help of remote sensing and GIS technologies, we can implement strategies to manage these valuable resources effectively and efficiently. These tools enable us to monitor resource usage, identify areas of conservation concerns and optimize resource allocation contributing to sustainable resource management in the face of growing environmental challenges. Geospatial technologies act a crucial role in identifying and analyzing the various factors influencing the utilization of natural resources which helps in mitigating negative environmental impacts. Therefore, with a comprehensive understanding of these influencing factors, decisions can be made for sustainable management of natural resources, fulfilling the needs of both existing and upcoming generations.

REFERENCES

Ahmed, F., & Kilic, K. (2018). Fuzzy Analytic Hierarchy Process : A performance analysis of various algorithms CO. *Fuzzy Sets and Systems*, 1, 1–19. DOI: 10.1016/j.fss.2018.08.009

Ahmed, H. (2019). Applications of Remote Sensing in Oceanographic Research. International Journal of Oceanography & Aquaculture. DOI: 10.23880/ijoac-16000159

Aly, Z., Bonn, F. J., & Magagi, R. (2007). Analysis of the backscattering coefficient of salt-affected soils using modeling and RADARSAT-1 SAR data. *IEEE Transactions on Geoscience and Remote Sensing*, 45(2), 332–341. DOI: 10.1109/TGRS.2006.887163

Ashtekar, A. S., Mohammed-Aslam, M. A., & Moosvi, A. R. (2019). Utility of normalized difference water index and GIS for mapping surface water dynamics in sub-upper Krishna Basin. *Photonirvachak (Dehra Dun)*, 47(8), 1431–1442. DOI: 10.1007/s12524-019-01013-6

Banholzer, S., Kossin, J., & Donner, S. (2014). The Impact of Climate Change on Natural Disasters. In Singh, A., & Zommers, Z. (Eds.), *Reducing Disaster: Early Warning Systems For Climate Change*. Springer. DOI: 10.1007/978-94-017-8598-3_2

Ben Khélifa, W., & Mosbahi, M. (2022). Modeling of rainfall-runoff process using HEC-HMS model for an urban ungauged watershed in Tunisia. *Modeling Earth Systems and Environment*, 8(2), 1749–1758. DOI: 10.1007/s40808-021-01177-6

Burke, L., Kura, Y., Kassem, K., Revenga, C., Spalding, M., McAllister, D., & Caddy, J. (2001). *Coastal Ecosystems*. World Resources Institute.

Burningham, H., & French, J. (2017). Understanding coastal change using shoreline trend analysis supported by cluster-based segmentation. *Geomorphology*, 282, 131–149. DOI: 10.1016/j.geomorph.2016.12.029

Campo, A. G. D. (2012). GIS in Environmental Assessment: A Review of Current Issues and Future Needs of current issues and future needs. Journal of Environmental Assessment Policy and Management, 14(1), 1-23. .DOI: 10.1142/S146433321250007X

Chaminé, H. I., Pereira, A. J. S. C., Teodoro, A. C., & Teixeira, J. (2021). Remote sensing and GIS applications in earth and environmental systems sciences. *SN Applied Sciences*, 3(12), 2–4. DOI: 10.1007/s42452-021-04855-3

Choudhury, R., Nath, B., Rahman, M. M., Medhi, S., & Dutta, J. (2024). Hydrogeochemical characteristics of groundwater contamination in Guwahati city, Assam, India: Tracing the elemental Threads. *Journal of Environmental Management*, 359(February), 120933. DOI: 10.1016/j.jenvman.2024.120933 PMID: 38696848

Chuvieco, E. (1993). Integration of linear programming and GIS for land-use modelling. *International Journal of Geographical Information Systems*, 7(1), 71–83. DOI: 10.1080/02693799308901940

Conchedda, G., Durieux, L., & Mayaux, P. (2008). An object-based method for mapping and change analysis in mangrove ecosystems. *ISPRS Journal of Photogrammetry and Remote Sensing*, 63(5), 578–589. DOI: 10.1016/j.isprsjprs.2008.04.002

Das, S. (2019). *Extreme rainfall estimation at ungauged sites: Comparison between region-of-influence approach of regional analysis and spatial interpolation technique*. International.

Droj, G. (2012). GIS and remote sensing in environmental management. *Journal of Environmental Protection and Ecology*, 13(1), 361–367.

Elhaeesahar, M., & Masoudi, M. (2019). GIS analysis for vulnerability assessment of drought in Khuzestan province in Iran using standardized precipitation index (SPI). *Iran Agricultural Research*, 38(2), 9–16.

Fajar, Y., Suwarsono, S., Taufik, M., & Rokhis, K. (2019). The dynamics of shoreline change analysis based on the integration of remote sensing and geographic information system (GIS) techniques in Pekalongan coastal area, Central Java, Indonesia. *Journal of Degraded and Mining Lands Management*, 6(3), 1789–1782. DOI: 10.15243/jdmlm.2019.063.1789

From space to soil. (n.d.). The European Space Agency (ESA). https:// www.esa.int

Gebeyehu, M. N. (2019). Remote Sensing and GIS application in agriculture and natural resource mangement. *International Journal of Environmental Sciences and Natural Resources*, 19(2). Advance online publication. DOI: 10.19080/IJESNR.2019.19.556009

Gillespie, T. W., Foody, G. M., Rocchini, D., Giorgi, A. P., & Saatchi, S. (2008). Measuring and modeling biodiversity from space. *Progress in Physical Geography*, 32(2), 203–221. DOI: 10.1177/0309133308093606

Giorgoudis, M. D., Hadjimitsis, D. G., & Shiftan, Y. (2014). Sustainable Transport Planning using GIS and Remote sensing: an integrated approach. DOI: 10.1117/12.2077752

Giri, S., Mukhopadhyay, A., Hazra, S., Mukherjee, S., Roy, D., Ghosh, S., Ghosh, T., & Mitra, D. (2014). A study on abundance and distribution of mangrove species in Indian Sundarban using remote sensing technique. *Journal of Coastal Conservation*, 18(4), 359–367. Advance online publication. DOI: 10.1007/s11852-014-0322-3

Global Ecosystem Dynamics Investigation (GEDI). (n.d.). Earthdata, NASA. https://www.earthdata. nasa.gov/sensors/gedi

Goparaju, L., Ecology, V., & Foundation, N. H. (2021). Wildlife monitoring and management using Remote Sensing and GIS. ICT Tools for Knowledge Management & Control of Emerging Zoonoses & Animal Health Threats, 69-75.

Held, A., Ticehurst, C., Lymburner, L., & Williams, N. (2003). High resolution mapping of tropical mangrove ecosystems using hyperspectral and radar remote sensing. Int. J. Rem. Sens., 24(13), 2739-2759.

High-Resolution Operational Soil Moisture Product. (n.d.). ISRO. https://isro.gov.in

Hochberg, E. J., Andre'foue¨t, S., & Tyler, M. R. (2003). Sea surface correction of high spatial resolution Ikonos images to improve bottom mapping in near-shore environments. *IEEE Transactions on Geoscience and Remote Sensing*, 41(7), 1724–1729. DOI: 10.1109/TGRS.2003.815408

Hussain, A., Mada, R., & Kamboj, V. (2022). *Applications of remote sensing and GIS in hydrological and hydrogeological studies: integrated watershed management. In Addressing Environmental Challenges Through Spatial Planning*. IGI Global. DOI: 10.4018/978-1-7998-8331-9.ch012

IOCCG. (2000). Remote Sensing of Ocean Colour in Coastal, and Other Optically-Complex Waters. Reports of the International Ocean-Colour Coordinating Group, No. 3. Dartmouth, Canada: IOCCG. https://ioccg.org/reports/rep003/

Islam, M., Sarker, S., Morsad, G., Khan, M. F., Chowdhury, A., & Islam, S. L. U. (2023). Spatial–temporal changes of shoreline and Vegetation: Impacts on mangrove cover along the Sundarbans area, Bangladesh. *Journal of Coastal Conservation*, 28(1), 10. Advance online publication. DOI: 10.1007/s11852-023-01016-z

Kerry, G. R., Montalbo, F. J. P., Das, R., Patra, S., Mahapatra, G. P., Maurya, G. K., Nayak, V., Jena, A. B., Ukhurebor, K. E., Jena, R. C., Gouda, S., Majhi, S., & Rout, J. R. (2022). *An overview of remote monitoring methods in biodiversity conservation*. Environmental Science and Pollution Research. Springer Berlin Heidelberg. DOI: 10.1007/s11356-022-23242-y

Khaoula, L., Abdessamad, G., Abdellah, O., Khalid, E., Rhizlane, F., & Saida, E. (2019). Application of geographical information system (GIS) for the development of climatological air temperature vulnerability maps: An example from Morocco. .DOI: 10.1002/met.1871

Koyama, C. N., Watanabe, M., Hayashi, M., Ogawa, T., & Shimada, M. (2019). Mapping the spatial-temporal variability of tropical forests by ALOS-2 L-band SAR big data analysis. *Remote Sensing of Environment*, 233, 111372. DOI: 10.1016/j.rse.2019.111372

Kumar, A., Kishore, B. S. P. C., Saikia, P., Deka, J., Bharali, S., Singha, L. B., Tripathi, O. P., & Khan, M. L. (2019). Tree diversity assessment and above ground forests biomass estimation using SAR remote sensing: A case study of higher altitude vegetation of North-East Himalayas, India. *Physics and Chemistry of the Earth Parts A/B/C*, 111, 53–64. DOI: 10.1016/j.pce.2019.03.007

Kumar, N., Yamaç, S., & Velmurugan, A. (2015). Applications of Remote Sensing and GIS in Natural Resource Management. *Journal of the Andaman Science Association*, 20(1), 1–6.

Kundu, K., & Halder, P. (2019). Geomorphological Changes in Sundarbans Delta Using GIS and Remote Sensing Data. In Mandal, J., Mukhopadhyay, S., Dutta, P., & Dasgupta, K. (Eds.), *Computational Intelligence, Communications, and Business Analytics. CICBA 2018. Communications in Computer and Information Science* (Vol. 1031). Springer. DOI: 10.1007/978-981-13-8581-0_4

Kundu, K., Halder, P., & Mandal, J. (2020). Forest Cover Change Analysis in Sundarban Delta Using Remote Sensing Data and GIS. .DOI: 10.1007/978-981-13-7334-3_7

Lechner, A. M., Foody, G. M., & Boyd, D. S. (2020). Applications in Remote Sensing to Forest Ecology and Management. *One Earth*, 2(5), 405–412. DOI: 10.1016/j.oneear.2020.05.001

Lehodey, P., Bertignac, M., Hampton, J., Lewis, A., & Picaut, J. (2008). El Niño Southern Oscillation and tuna in the western Pacific. *Nature*, 389(6647), 715–718.

Leifer, I., Lehr, W. J., Simecek-Beatty, D., Bradley, E., Clark, R. N., Dennison, P. E., & Roberts, D. A. (2012). State of the art satellite and airborne marine oil spill remote sensing: Application to the BP Deepwater Horizon oil spill. *Remote Sensing of Environment*, 124, 185–209. DOI: 10.1016/j.rse.2012.03.024

Liu, L., Zhang, Q., Guo, Y., Li, Y., Wang, B., Chen, E., Li, Z., & Hao, S. (2024). Mapping coniferous forest distribution in a semi-arid area based on multi-classifier fusion and Google Earth Engine combining Gaofen-1 and Sentinel-1 Data : A Case Study in Northwestern Liaoning, China. *Forests*, 15(2), 288. DOI: 10.3390/f15020288

Lwin, K. K., & Murayama, Y. (2011). Computers, Environment and Urban Systems Modelling of urban green space walkability : Eco-friendly walk score calculator. *Computers, Environment and Urban Systems*, 35(5), 408–420. DOI: 10.1016/j.compenvurbsys.2011.05.002

Lyons, M.B., Phinn, S.R., & Roelfsema, C.M. (2012). Long term land cover and seagrass mapping using Landsat and object-based image analysis from 1972 to 2010 in the coastal environment of South East Queensland, Australia. *ISPRS J. Photogramm. Remote Sens., 71*, 34-46.

Mahendra, R. S., Mohanty, P. C., Bisoyi, H., Kumar, T. S., & Nayak, S. (2011, April). (20110. Assessment and management of coastal multi-hazard vulnerability along the Cuddalore- Villupuram, east coast of India using geospatial techniques. *Ocean and Coastal Management*, 54(4), 302–311. DOI: 10.1016/j.ocecoaman.2010.12.008

Maiti, S., & Bhattacharya, A. K. (2009). Shoreline change analysis and its application to prediction: A remote sensing and statistics based approach. *Marine Geology*, 257(1-4), 11–23. DOI: 10.1016/j.margeo.2008.10.006

Matomela, N., Tianxin, L., Morahanye, L., Bishoge, O. K., & Ikhumhen, H. O. (2019). Rainfall-runoff estimation of Bojiang lake watershed using SCS-CN model coupled with GIS for watershed management. *J Appl Adv Res*, 4(1), 16–24. DOI: 10.21839/jaar.2019.v4i1.263

Nadi, P. A., & Murad, A. K. (2017). Reviewing the use of Geographic Information System (GIS) to measure sustainable urban transport performance. Journal of Geoscience, Enineering. *Environmental Technology*, 2(2), 171–178. PMID: 28934911

Nelson, L. (2021). Predictive Digital Soil Mapping. White Paper prepared for LWS 548.

Pant, H., Verma, J., & Richhariya, S. (2022). Wildlife Conservation and Management using GIS and Remote Sensing. Geospatial Technology and its Applications in Resource Management, 279-286.

Pant, H. & Verma, J. & Surya, S. (2022). Application of Remote Sensing and GIS in Biodiversity Management. Green Technology: A Roadmap for Sustainable Development-2020, 1-18.

Parra, L. (2022a). Applied sciences Remote Sensing and GIS in Environmental Monitoring.

Parra, L. (2022b). Remote Sensing and GIS in Environmental Monitoring. *Applied Sciences (Basel, Switzerland)*, 12(16), 8045. Advance online publication. DOI: 10.3390/app12168045

Potapov, P., Li, X., Hernandez-Serna, A., Tyukavina, A., Hansen, M. C., Kommareddy, A., Pickens, A., Turubanova, S., Tang, H., Silva, C. E., Armston, J., Dubayah, R., Blair, J. B., & Hofton, M. (2021). Mapping global forest canopy height through integration of GEDI and Landsat data. *Remote Sensing of Environment*, 253(October), 112165. DOI: 10.1016/j.rse.2020.112165

Prasad, N., Plants, A., Semwal, M., Plants, A., & Roy, P. S. (2015). Recent Advances in Lichenology. *Remote Sensing and GIS for Biodiversity Conservation.*, (January). Advance online publication. DOI: 10.1007/978-81-322-2181-4

Rani, M., Rehman, S., Sajjad, H., Chaudhary, B. S., Sharma, J., Bhardwaj, S., & Kumar, P. (2018). Assessing coastal landscape vulnerability using geospatial techniques along VizianagarameSrikakulam coast of Andhra Pradesh, India. *Natural Hazards*, 94(2), 711–725. DOI: 10.1007/s11069-018-3414-9

Ritchie, J. C., Schiebe, F. R., Cooper, C. M., & Harrington, J. A.Jr. (1994). Chlorophyll measurements in the presence of suspended sediment using broad band spectral sensors aboard satellites. *Journal of Freshwater Ecology*, 9(2), 197–206. DOI: 10.1080/02705060.1994.9664887

Rojas, O. (2021). Next generation agricultural stress index system (ASIS) for agricultural drought monitoring. *Remote Sensing (Basel)*, 13(5), 959. DOI: 10.3390/rs13050959

Sardar, P., & Samadder, S. R. (2021). Understanding the dynamics of landscape of greater Sundarban area using multi-layer perceptron Markov chain and landscape statistics approach. Ecological Indicators. *Ecological Indicators*, 106914. Advance online publication. DOI: 10.1016/j.ecolind.2020.106914

Scull, P., Franklin, J., Chadwick, O. A., & McArthur, D. (2003). Predictive soil mapping: A review. *Progress in Physical Geography*, 27(2), 171–197. DOI: 10.1191/0309133303pp366ra

Sharma, S. K., Misra, S. K., & Singh, J. B. (2020). The role of GIS-enabled mobile applications in disaster management: A case analysis of cyclone Gaja in India. *International Journal of Information Management*, 51, 102030. DOI: 10.1016/j.ijinfomgt.2019.10.015

Singh, A., Meena, G. K., Kumar, S., & Gaurav, K. (2019). Evaluation of the penetration depth of L- And S-band (NISAR mission) microwave SAR signals into ground. URSI Asia-Pacific Radio Science Conference, AP-RASC. DOI: 10.23919/URSIAP-RASC.2019.8738217

Soergel, B., Kriegler, E., Weindl, I., Rauner, S., Dirnaichner, A., Ruhe, C., Hofmann, M., Bauer, N., Bertram, C., Bodirsky, B. L., Leimbach, M., Leininger, J., Levesque, A., Luderer, G., Pehl, M., Wingens, C., Baumstark, L., Beier, F., Dietrich, J. P., & Popp, A. (2021). A sustainable development pathway for climate action within the UN 2030 Agenda. *Nature Climate Change*, 11(8), 656–664. DOI: 10.1038/s41558-021-01098-3

Solanlki, H. U., Dwivedi, R. M., & Nayak, S. R. (2001). Application of Ocean Colour Monitor chlorophyll and AVHRR SST for fishery forecast: Preliminary validation results off Gujarat coast, northwest coast of India. *Indian Journal of Geo-Marine Sciences*, 30(3), 132–138.

Soria-lara, J. A., Arce-ruiz, R. M., Arranz-lópez, A., & Ariza-Álvarez, A. (2020). Environmental Impact Assessment for transport projects : A review of technical and process-related issues. *Standard Transport Appraisal Methods. Elsevier Inc.*, 6(1), 255–285. Advance online publication. DOI: 10.1016/bs.atpp.2020.07.002

Tamiminia, H., Salehi, B., Mahdianpari, M., & Goulden, T. (2024). State-wide forest canopy height and aboveground biomass map for New York with 10 m resolution, integrating GEDI, Sentinel-1, and Sentinel-2 data. Ecological Informatics, 79(April), 102404. .DOI: 10.1016/j.ecoinf.2023.102404

Tang, X., Zang, Z., Lin, H., Wang, X., & Wen, Z. (2023). Using a Vegetation Index to Monitor the Death Process of Chinese Fir Based on Hyperspectral Data. *Forests*, 14(12), 2444. DOI: 10.3390/f14122444

Teodoro, A.C. (2018) Optical Satellite Remote Sensing of the Coastal Zone Environment-An Overview. Intech open.

Thomas, T., Rangel-Buitrago, N., Phillips, M. R., Anfuso, G., & Williams, A. T. (2015). Mesoscale morphological change, beach rotation and storm climate influences along a macrotidal embayed beach. *Journal of Marine Science and Engineering*, 3(3), 1006–1026. DOI: 10.3390/jmse3031006

Tian, L., Wu, X., Tao, Y., Li, M., Qian, C., Liao, L., & Fu, W. (2023). Review of Remote Sensing-Based Methods for Forest Aboveground Biomass Estimation: Progress, Challenges, and Prospects. *Forests*, 14(6), 1–31. DOI: 10.3390/f14061086

Tripathy, B. R., Seenipandi, K., Sajjad, H., Joshi, P. K., Chaudhary, B. S., & Kumar, P. (2018). Monitoring of seasonal variability and movement of suspended sediment concentrations along the Thiruvananthapuram coast, southern India, using the Landsat OLI sensor. *Ocean Science*, 14(5), 1085–1092. DOI: 10.5194/os-14-1085-2018

Ukpere, D. R., & Ottah, C. (2017). Remote Sensing and Gis Application in Environmental Management. *Journal of Environmental Management*, 1(69), 5–24.

Wang, J., Lin, Q., Meng, S., Huang, H., & Liu, Y. (2024). Individual Tree-Level Monitoring of Pest Infestation Combining Airborne Thermal Imagery and Light Detection and Ranging.

Wang, M., Shi, W., & Son, S. (2015). Evaluation of MODIS SWIR and NIR-SWIR atmospheric correction algorithms using AERONET measurements. *Remote Sensing of Environment*, 159, 232–248. DOI: 10.1016/j.rse.2014.12.002

Xhafa, P. D. S. (2015). Geographic Information Systems (GIS). *Urban Planning*, 4138(April), 85–92.

Xing, J., Sun, S., Huang, Q., Chen, Z., & Zhou, Z. (2024). Application of Geoinformatics in Forest Planning and Management.

Yahya, B., Ahmed, K. A., & Saeed, A. M. (2023). Engineering and Technology Quarterly Reviews.

Yeh, A. G.-O. (1991). The development and applications of geographic information systems for urban and regional planning in the developing countries. *International Journal of Geographical Information Systems*, 5(1), 5–27. DOI: 10.1080/02693799108927828

Yeh, A. G.-O., & Chow, M. H. (1996). An integrated GIS and location-allocation approach to public facilities planning. *Computers, Environment and Urban Systems*, 20(4-5), 339–350. DOI: 10.1016/S0198-9715(97)00010-0

Zhang, J., Steele, M., Lindsay, R., & Schweiger, A. (2017). What drove the dramatic retreat of Arctic sea ice during summer (2007). *Geophysical Research Letters*, 34(10), L18502. Advance online publication. DOI: 10.1029/2007GL031527

Zhang, K., Douglas, B. C., & Leatherman, S. P. (2016). Global warming and coastal erosion. *Climatic Change*, 134(1-2), 249–264. DOI: 10.1007/s10584-017-2079-9

Chapter 2

Integrating GALDIT and GIS for Assessing Sea Water Intrusion Susceptibility in the Akermoud Coastal Water Table, Morocco: Implications for Sustainability

Abdellah Khouz
Cadi Ayyad University, Morocco & Universidade Aberta, Portugal

Jorge Trindade
Universidade Aberta, Portugal

Fatima El Bchari
Cadi Ayyad University, Morocco

Pedro Pinto Santos
https://orcid.org/0000-0001-9785-0180
University of Lisbon, Portugal

Eusébio Reis
University of Lisbon, Portugal

Adil Moumane
https://orcid.org/0000-0003-0296-2679
Ibn Tofail University, Morocco

Fatima Ezzahra El Ghazali
https://orcid.org/0000-0002-1980-3028
Cadi Ayyad University, Morocco

Mourad Jadoud
Chouaïb Doukkali University, Morocco

Blaid Bougadir
Cadi Ayyad University, Morocco

DOI: 10.4018/979-8-3693-9651-3.ch002

ABSTRACT

The Akermoud coastal aquifer, situated in the northern region of Essaouira, Morocco, is an aquifer that has an important productive capacity, mainly used for irrigation. However, uncontrolled exploitation has increased the risk of sea water intrusion, leading to deteriorating water quality and threatening agricultural sustainability. In this research, in order to access susceptibility to sea water intrusion (SWI), six elements derived from diverse databases are used. Variables were combined in a GALDIT and GIS models, resulting in the analysis of 40 groundwater samples from wells. Results highlight the imminent threat of sea water encroachment into the coastal groundwater system. The resulting GALDIT index map indicates a notably high susceptibility index along a 3 km coastal band, between Tensift Oued and Bhaybeh Beach, enlarging southwards up to 5 km. Saline intrusion patterns are particularly observed between Zaouiet El Kourati and Ouled El Fequih villages, where the merging of saline and fresh waters amplifies salinization, affecting approximately 24% of the study area.

I. INTRODUCTION

Groundwater is vital for ecosystem balance (Gholami et al., 2010; Humphreys, 2006) and is a primary water source in coastal areas (Amiri et al., 2016; Ma et al., 2020). However, groundwater quality in these regions is deteriorating, posing environmental challenges, especially in semi-arid and arid areas (Bahir and Ouhamdouch, 2020; Ez-zaouy et al., 2022; Moumane et al. 2021). Over-extraction of aquifers increases mineral content (Alabjah et al., 2018) and harms freshwater ecosystems (Mirzavand et al., 2020). Seawater intrusion, a major issue in coastal regions, is driven by groundwater recharge, discharge, and geological structure (Kumar, 2006; Seddique et al., 2019), disrupting the freshwater-seawater balance (Xue et al., 1999). Over-extraction lowers groundwater levels, triggering saline water intrusion (Pinder, 1973; Pulido-Leboeuf, 2004), which raises total dissolved solids (TDS) and renders water resources unusable (Sherif and Hamza, 2001). This issue is exacerbated by sea level rise and excessive coastal pumping (Werner and Simmons, 2009; Moujabber et al., 2006; Seddique et al., 2019).

Additional factors like agriculture (Zalidis et al., 2002), tourism (Ez zaouy et al., 2022), population density (Erostate et al., 2020), and climate change (Benini et al., 2016; Ez zaouy et al., 2022) further degrade groundwater quality in coastal regions. Salinization of coastal aquifers has garnered global attention, with various studies using models like GALDIT, coupled with hydro-geochemical analysis, to understand salinity causes and impacts (Hermans and Paepen, 2020; Ez zaouy et al., 2022).

In Morocco, population growth, economic activities, and climate change-induced rainfall decline strain water resources (Ez-zaouy et al., 2022). Uncontrolled groundwater extraction has led to unauthorized withdrawals (Laouina, 2006). The GALDIT model is effective for assessing seawater intrusion vulnerability, evaluating six key indicators (Chachadi and Lobo Ferreira, 2001; Hu et al., 2018; Ma et al., 2020). Additionally, research in hydro-geochemistry integrates major ion chemistry to analyze groundwater salinity and the fresh-saltwater interface (Han et al., 2011).

Localized studies are needed to address groundwater salinity in specific regions like the Akermoud coastal aquifers, where comprehensive investigations are lacking. Implementing preventive measures and understanding the processes controlling groundwater salinity and geochemical evolution are essential for effective management. This study integrates the GALDIT model and hydro-geochemical analysis to assess the degradation of the Akermoud aquifer under climate change and human activities, involving:

1. Assessing seawater intrusion and vulnerability using the GALDIT model.
2. Conducting a multi-approach study integrating hydro-geochemical parameters and in-situ measurements from 40 samples.
3. Evaluating groundwater quality and potential future impacts of over-extraction.
4. Enhancing the understanding of the Akermoud aquifer to develop effective management strategies.

II. GENERAL OVERVIEW OF SEAWATER INTRUSION

The chemical makeup of water plays a role in determining its usages such as drinking water, irrigation and industrial use. The quality of water and how it changes over time and across locations can be influenced by factors like the dissolution of formations as well as industrial waste and agricultural practices. Coastal underground water sources hold amounts of water for areas along the coast that are experiencing population growth and expanding agriculture, which increases the demand for water. However, there are challenges in using these resources due to issues like declining reserves and deteriorating water quality (Mdiker et al., 2009).

The Essaouira region is one area that is seeing a significant increase in its population as well as a flourishing agricultural sector. Most of the water used by the population for drinking and irrigation comes from underground sources. Seawater intrusion is a known phenomenon in regions where saltwater moves inland beneath freshwater causing significant changes in its chemistry. This has consequences on its suitability for several purposes (Fadili A., 2014).

II-1 Impact of Water Salinization on Seawater Intrusion

II-1-1 Definition of Seawater Intrusion

Seawater intrusion is an occurrence that happens when there is an inequity between the amount of freshwater and seawater. This results in a reduction of freshwater levels and an influx of seawater causing a decline in the quality of freshwater (Custodio, 2010).

II-1-2 Mechanisms of Seawater Intrusion

Coastal aquifer systems naturally discharge water towards the sea as part of the water cycle. The groundwater level in these freshwater aquifers is typically higher than the sea level, and the interface between the two is defined by a transition zone where hydrostatic equilibrium is maintained. Intensive exploitation of coastal aquifers can cause depressions that are filled by nearby seawater. Due to the density contrast between continental freshwater and saline water, seawater infiltrates the subsurface, leading to the phenomenon of seawater intrusion and subsequent degradation of water quality (Derradji, 2015).

It is evident that the coastal sector (northwestern part) of the studied region, where intensive pumping for water supply and irrigation occurs, is a vulnerable zone where the coastal aquifer is prone to saltwater contamination and direct contact with the Atlantic Ocean.

II-1-3 Transition Zone or Mixing Zone

The area where freshwater and seawater come together in aquifers is a feature. When seawater seeps into the aquifers it creates a zone where dense freshwater meets denser seawater. Over the years numerous studies have been conducted to determine the shape and location of this interface, between freshwater and seawater. Some of these studies date back to the century by researchers like Ghyben (1889) and Herzberg (1901). The thickness of this mixing zone can vary significantly ranging from a few meters to kilometers. It depends on factors such as the hydrodynamic properties of the aquifer including its permeability and diffusivity. In aquifers with hydraulic gradients this transition zone can stretch over several kilometers. Other external factors, like tides and excessive extraction of groundwater can also impact the thickness of this mixing zone (Custodio, 2002; Fadili, A., 2014).

The phenomenon of seawater intrusion in coastal aquifers can be explained by the relationship between the freshwater head (h) above mean sea level and the depth (hs) of the freshwater-saltwater interface below sea level. This is encapsulated by the Ghyben-Herzberg equation, which can be written as:

$$\rho.(hs + h) = \rho s.hs$$

where:
- hs: depth of the interface,
- h: hydraulic head (or piezometric level) relative to the base level,
- ρ: density of freshwater,
- ρs: density of seawater.

In general, hs is approximately 40 times h.

This hydrostatic equilibrium can be altered over time. A decrease in piezometric level can lead to the intrusion of seawater into the aquifer and a difficult-to-reverse salinization.

II-1-4 Potential Sources of Freshwater Mineralization

There are two causes of freshwater salinization that can be identified;

- **Human induced salinization;** One of the reasons, for this phenomenon is the excessive extraction of groundwater compared to its replenishment rate. This issue is further exacerbated by the growth of populations in areas.
- **Natural salinization;** There are primarily two factors that influence salinization, namely geology and climate. Various chemical elements originate from rocks in the aquifer. In semiarid regions salts in water become concentrated due to evaporation. In zones salinization can occur when seawater intrudes due to limited precipitation and subsequent low recharge of the aquifer. (Fadili, A., 2014).
- **Sources of salinity in coastal aquifers;** The main cause of salinization in aquifers is seawater intrusion. However, there are contributing factors including (i) sea spray, as a source of salinity, (ii) sea level rise, (iii) evaporation and concentration of shallow water tables, (iv) dissolution of evaporites found in aquifer formations and (v) human-induced salinization. (Bear et al., 1999; Werner and Simmons 2009; Fadili, A., 2014).

II-2 GALDIT Model

Understanding the variability of aquifers regarding seawater intrusion is an essential step for his management. Numerous International literature discusses methods for evaluating vulnerability, including the GALDIT approach (Nadjla et al. 2021). GALDIT is a numerical assessment system that takes into account factors impacting the hydrogeological system. It specifically assesses the susceptibility of aquifers to seawater intrusion. This model was created by Chachadi and Lobo Ferreira (2001) within the COASTIN project a collaboration between Europe and India (Chachadi and Labo Ferreira 2005; Nadjla et al. 2021). The GALDIT method has been implemented in areas like Goa in India and Algarve, in Portugal as an index based mapping technique (Nadjla et al. 2021).

III. STUDY AREA

The focus of this chapter is on the coastal aquifer of Akermoud, situated approximately 45 km north of the city of Essaouira, Morocco. The aquifer is composed of sandstone dunes and shell limestone of the Plio-Quaternary age. It stretches along the coastline in a 20 km wide and 40 km long band (Figure 1).

Figure 1. Location of the study area

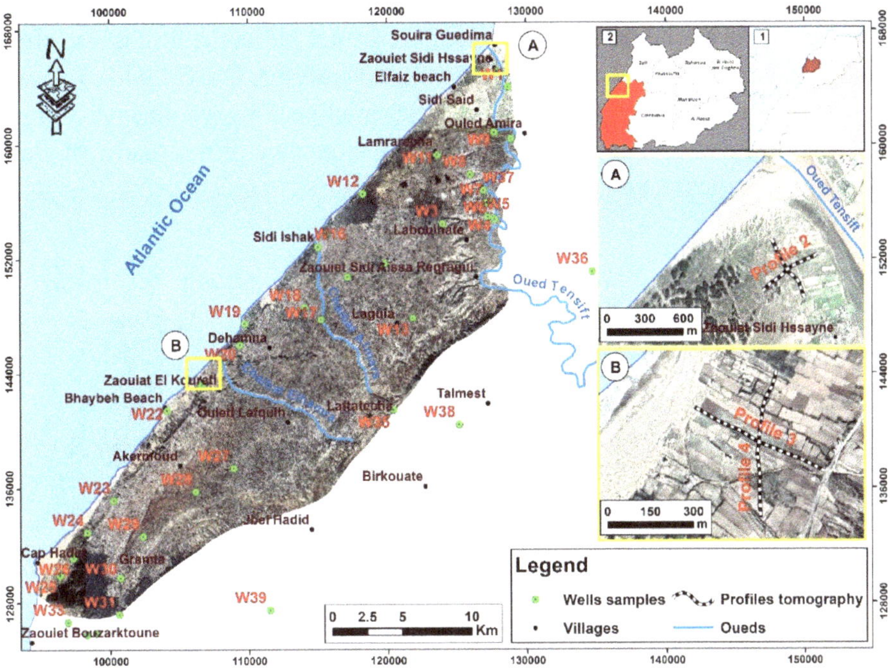

III-1 Hydrological Framework

From a hydrogeological perspective, the Essaouira basin relies heavily on a network of watercourses due to the scarcity of surface water and its limited sources. The study area is intersected by several significant rivers. The drainage basins of Oueds Chaibat Alhamra and Chaibat Almars originate from nearby mountains and the Jbel Hadid anticline, flowing into the synclinal cuvettes of the Akermoud plain (Khouz et al. 2023). These rivers tend to be short and frequently dry compared to Tensift Oued in the far north, which collects water from the High Atlas mountains.

III-2 Geological and Structural Framework

The formation of the Essaouira Basin can be attributed to the central Atlantic Rift, which initiated during the Triassic period, leading to the separation of the continents and the formation of the Atlantic Ocean. This region represents a transition from basins dominated by siliciclastic deposits to basins characterized by evaporate deposits in the Mesoproterozoic era (Salvan 1984). The Essaouira Basin, situated between the Jbel Amsittène anticline to the south and the Jbel Hadid anticline to the north (Figure 2), is a large synclinal area that is open to the ocean. While it is influenced by various anticlinal and synclinal cuvettes individually, it is also affected by multiple tectonic events, such as the Akermoud syncline (Figure 2), located between the Atlantic Ocean and the Jbel Hadid anticline. This syncline is positioned at the center of a triangular region and marks the opening of the Akermod plain's coastal syncline (Hanich L., 2001).

Figure 2. The Akermoud plain coupe (Extracted from Monition's 1953 work)

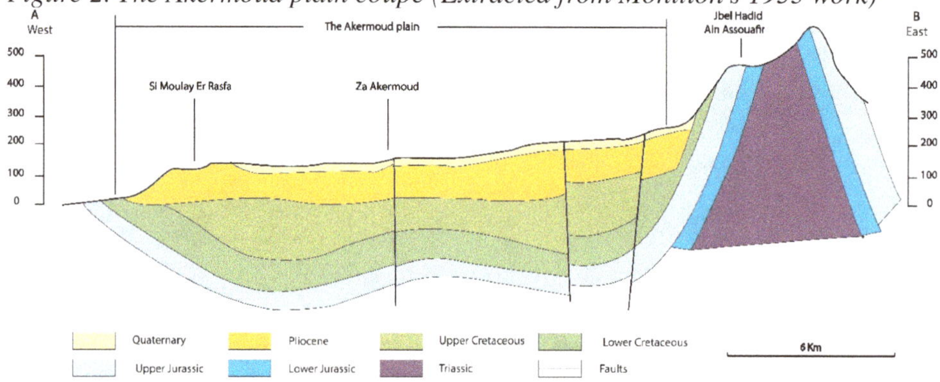

To gain a comprehensive understanding of the structure and geometry of geological formations, especially those of hydrogeological significance, a detailed study of geological data from various exploitation wells and hydrogeological reconnaissance boreholes was conducted. These data were made available to us by the ABHT and are presented in the form of descriptive sheets and lithological logs.

The well data was also used to create a geological cross-section across the entire Essaouira Basin, including the study area. The available well logs don't provide information on the ages of the encountered facies, making it impossible to establish stratigraphic correlations in terms of geological stages. Therefore, the cross-section was established through lateral facies correlation, considering three major lithological units: (i) a marl unit or predominantly marl unit, (ii) a limestone unit or predominantly limestone unit, and (iii) a suite of detrital facies encompassing silt, alluvium, dunes, etc.

The completed cross-section illustrates the deep structure of the entire Essaouira Basin to a depth of approximately 200 m. The overall shape exhibits a relatively undeformed synclinorium dominated by limestone formations with intercalated layers of marl or predominantly marl. The general picture depicted by this cross-section indicates the presence of two synclinal basins that could potentially serve as preferential zones for groundwater accumulation. The first basin, located southwest of the cross-section, is the Meskala basin, delimited by bordering normal faults. The second basin is shallower and less pronounced, extending from the northwestern half of the basin to Jebel Hadid. The coastal zone of interest (located on the left side of the Figure 2 is characterized by the presence of significant recent overburden composed of probable Plio-Quaternary deposits. The deposition of this overburden may be related to series collapse due to normal faulting.

Due to a lack of information, the cross-section could not be extended beneath Jebel Hadid, which is known as a Triassic core anticline expected to act as a barrier between the two aquifers in our study area (Figure 2).

III-3 Hydrogeological Framework

Within the coastal plain of Akermoud, the Plio-Quaternary aquifer consists of marine or terrestrial calcite with primary hydraulic conductivity resulting from its porosity, offering a significant amount of available space. The enclosing walls of this aquifer are formed by synclinal Senonian marls. This aquifer is primarily utilized in rural areas to meet domestic water needs and other essential requirements (Bahir et al., 2000).

IV. METHODOLOGY

The methodology employed in this chapter encompasses two main axes: GALDIT modeling and hydro-geochemical approaches. By utilizing the various datasets mentioned in the Figure 3, our aim is to identify the areas most susceptible to salinization in the Akermoud water table using GIS techniques. The subsequent sections will elaborate on the methodology used for the GALDIT model and the hydro-geochemical approaches.

Figure 3. Flowchart for Marine Intrusion Methodology

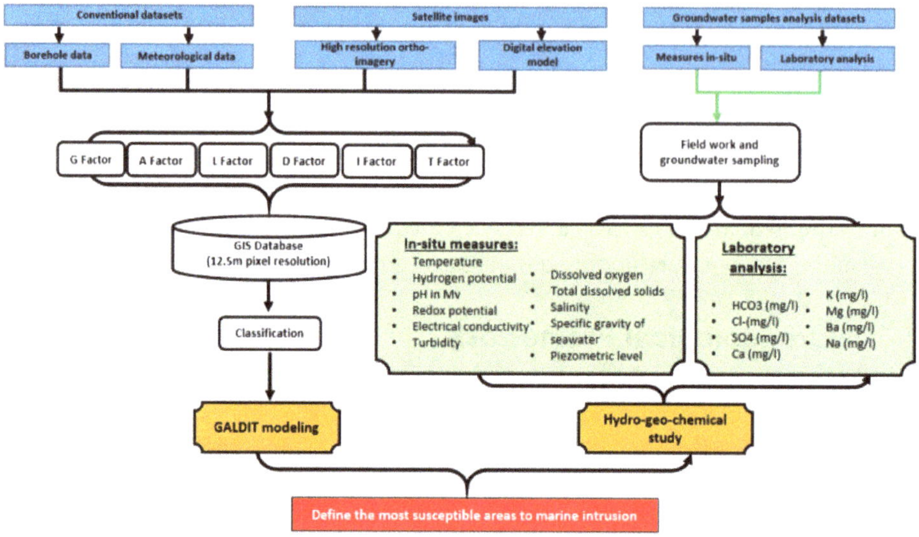

IV-1 GALDIT Modeling

The GALDIT method uses six factors related to hydrogeology, hydrology and morphology (Chachadi and Lobo Ferreira 2007) to evaluate susceptibility. These factors include the presence of groundwater (G), the aquifer's ability to transmit water (A), the groundwater level's distance from sea level (L), proximity to the shoreline (D), the impact of existing seawater intrusion (I), and the thickness of the aquifer layer (T). The significance of seawater intrusion is determined by assigning weights to each factor ranging from 1 to 4. Moreover, each factor is assigned a rating on a scale from 2.5 (indicating vulnerability) to 10 (indicating high vulnerability) as shown in Table 1.

To calculate the vulnerability index ArcGIS utilizes the raster calculator tool along, with the equation:

$$GALDIT - Index = \frac{\sum_{i=1}^{6}(W_i \cdot R_i)}{\sum_{i=1}^{6} W_i}$$

The GALDIT vulnerability index is calculated by combining the assigned ratings (R) and weights (W). The resulting index is then classified into three classes: high vulnerability (>7.5), moderate vulnerability (5 to 7.5), and low vulnerability (<5). A higher GALDIT index value indicates a greater potential for groundwater pollution and a higher vulnerability of the aquifer to seawater intrusion.

To generate the final vulnerability map, each point (borehole) is assigned a value (rating) based on Table 1. The coordinates of each borehole are used in conjunction with a GIS program to produce the map. In our study area, the hydro-geochemical parameters were obtained from sampled wells, while the other parameters were derived from the ABHT boreholes.

Table 1. Weight and ratings of the six parameters of GALDIT method

Factor	Weight	Classes	Factor variable		Rating
			Class	Range	
Groundwater occurrence/aquifer type (G)	**1**	**Class 1**	**Unconfined aquifer**		**10**
Aquifer hydraulic conductivity (Ms/l) (A)	3	Class 1	High	>9	10
		Class 2	Medium	9-6	7.5
		Class 3	Low	6-3	5
		Class 4	Very low	<3	2.5
Groundwater level above sea level (m) (L)	4	Class 1	High	<30	10
		Class 2	Medium	30–50	7.5
		Class 3	Low	50–80	5
		Class 4	Very low	>80	2.5
Distance from the shore (m) (D)	4	Class 1	High	<1000	10
		Class 2	Medium	1000–2000	7.5
		Class 3	Low	2000–6000	5
		Class 4	Very low	>6000	2.5
Impact of the existing status of seawater intrusion (I)	1	Class 1	High	>10	10
		Class 2	Medium	10–6	7.5
		Class 3	Low	6–4	5
		Class 4	Very low	<4	2.5
Thickness of the aquifer (m) (T)	2	Class 1	High	>70	10
		Class 2	Medium	70–50	7.5
		Class 3	Low	50–30	5
		Class 4	Very low	<30	2.5

The following data is required for the study:

- Inventory and coordinates of boreholes: Information and precise locations of the boreholes in the study area.

- Lithological profiles from boreholes: Detailed descriptions of the geological layers encountered in the boreholes, including the type and thickness of the aquifer.
- Depth to groundwater: Measurements of groundwater level, indicating the depth at which the water table is found.
- Hydraulic conductivity: Data obtained from pumping tests or extracted from relevant literature, providing information on the permeability of the aquifer based on its lithology.
- Topographical map: A map displaying the topography of the study area, specifically focusing on a 100-meter distance from the coastline.
- Water quality data: Data related to water quality, specifically used to calculate the Revelle coefficient (Cl/HCO_3). This information helps assess the salinity of the groundwater and its potential for seawater intrusion.

IV-2 Hydro-Geochemical Study

IV-2-1 Sampling and *In-Situ* Analysis

The work carried out in the context of this study involved sampling groundwater at 40 points distributed throughout the study area, accompanied by measurements of groundwater levels. Two field campaigns were conducted, one in 2020 (*prior* to precipitation) and another in 2021 (after direct recharge of the aquifer by precipitation). During these campaigns, the physicochemical parameters of the water were measured in situ using a HORIBA U-50 multi-parameter probe (Table 2), which provides measurements of electrical conductivity, pH, temperature, dissolved oxygen, salinity, and other parameters.

In addition to the new hydro-chemical measurements we conducted, the database provided by ABHT includes groundwater analyses conducted between 1954 and 2002. The U-50 probe used in this study simultaneously displays results for nine water characteristics parameters, namely:

- pH (Redox potential)
- DO (Dissolved oxygen)
- EC (Electrical conductivity)
- PPT (Salinity)
- TDS (Total dissolved solids)
- σT (Sea water specific gravity)
- T°C (Temperature)
- NTU (Turbidity)
- ORP(V) (Redox potential in volts)

Table 2. The specifications and typical performance characteristics of the U-50 multi-parameter probe

	Range	Resolution	Accuracy
pH	0-14	0,01pH	± 0.1pH, Automatic temperature compensation.
DO (mg/l)	0-50 mg/l	0,01 mg/l	0 à 20 mg/l ± 0.2 mg/l, et 20 à 50 mg/l ±0.5 mg/l
EC (ms/cm)EC (ms/cm)	0-9,99 S/cm	0,001 à 0,1 mS/cm	± 1% from full scale
PPT	0-70 ppt	0,1 ppt	± 3%
TDS (g/l)	0-100 g/l	0,10%	± 5 g/l.
σT	0-50 σT	0,1 σT	± 5 σ t.
T°C	-5 à 55 °C	0,01 °C	±0,15°C
NTU	0-1000 NTU	0,1 NTU	±2% of reading or 0,3 NTU
ORP (mV)	± 1999 mV	1 mV	± 15 mV.

Operational Procedure of the HORIBA U-50 Multi-Parameter Probe:

The HORIBA U50 multi-parameter probe, a water quality analyzer, consists of two components: a control unit and a probe that houses the various sensors (Figure 4).

Figure 4. Components of the HORIBA U-50 Multi-Parameter Probe

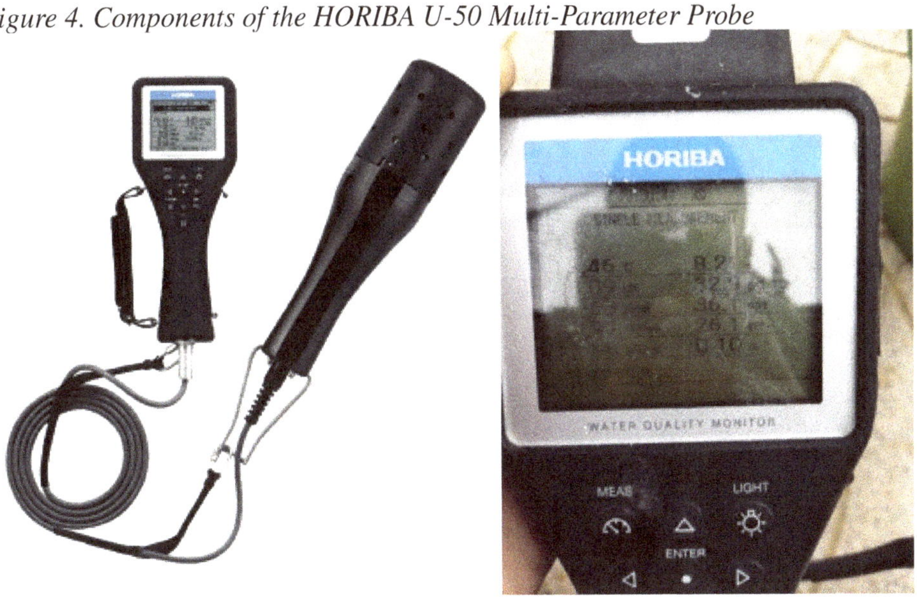

The operational procedure involves:

- Submerging the probe sensor into the water and waiting for the parameters to stabilize.
- Once the parameters have stabilized, initiating the measurement.
- The data for all the measured parameters are recorded in the device's memory and can be transferred to a computer later via a cable.

In addition to the hydro-chemical analysis, we conducted piezometric measurements using the manual piezometric probe from the Department of Geology at the Faculty of Sciences and Techniques, University Cadi Ayyad, in Marrakech. This probe is used to measure the water depth in wells, boreholes, reservoirs, etc. When the piezometric probe reaches the water surface, it emits an audible signal, and the depth reading can be observed on the graduated tape (Figure 5).

Figure 5. Examples of some wells measuring piezometric level

The collected samples were filtered first as illustrated in the Figure 6 before doing the laboratory analysis to eliminate suspended matter.

Figure 6. The samples filtration

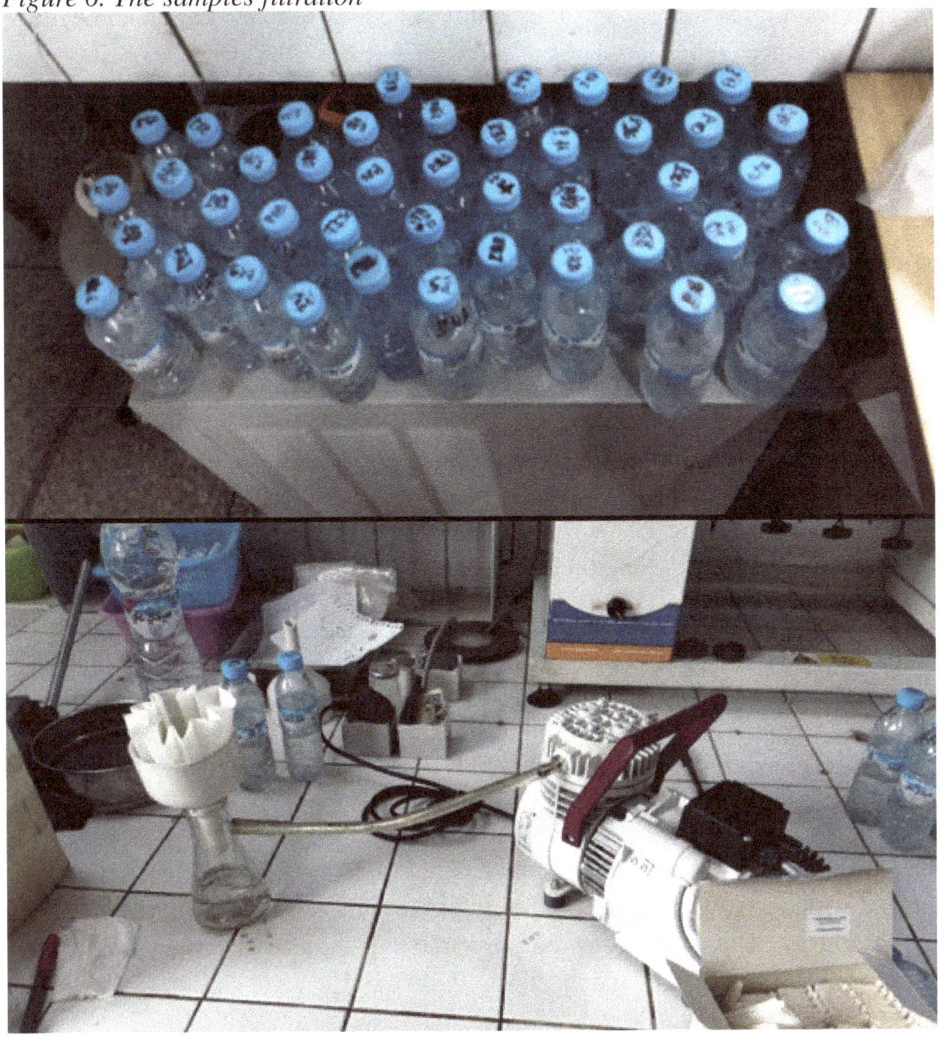

IV-3 Laboratory Analysis

IV-3-1 Alkalinity

The alkalinity of water refers to its ability to neutralize acids. In water, alkalinity is mainly connected to carbonates, bicarbonates and hydroxides. Measuring alkalinity using the titration method involves observing the decrease in pH that happens during the determination process. To determine the alkalinity in samples, from both basins, a 0.02 N sulfuric acid solution (H_2SO_4) was used for titration (Rodier 2009).

$$CO_3^{2-} + H^+ \rightarrow HCO_3^-$$

$$HCO_3^- + H^+ \rightarrow H_2CO_3$$

During titration, the pH slightly decreases. When the sample contains carbonates, an initial equivalence point can be observed around pH 8.3. This point corresponds to the conversion of carbonate and bicarbonate ions. The endpoints can be determined using the color indicators used, such as methyl orange and phenolphthalein (Rodier 2009).

The concentrations are then visualized in mg/L, based on simple calculations involving the different volumes and equilibrium concentrations.

IV-3-2 Sulfate Determination by Nephelometry

Sulfates are commonly found in natural water bodies. Their concentrations can vary depending on factors such, as the makeup of the area or potential sources of contamination. There are techniques to determine sulfate levels, all of which rely on the formation of insoluble barium sulfate ($BaSO_4$). In this study the nephelometric method was used to measure the turbidity of the solution indicating the amount of $BaSO_4$ formed. This method involves precipitating sulfates in an environment, as barium sulfate as shown by the reaction;

$$BaCl_2 \rightarrow Ba^{2+} + 2Cl^-$$

$$Ba^{2+} + SO_4^{2-} \rightarrow BaSO_4$$

The obtained precipitate is stabilized using a Tween 20 solution. The homogeneous suspensions are measured using a UV spectrophotometer.

It is necessary to prepare calibration solutions with known sulfate concentrations in order to establish a calibration curve. This curve will allow us to determine the SO_4^{2-} content in mg/L.

The calibration curve (Figure 7) showed a high correlation coefficient (R^2 = 0.99), indicating the reliability of the standards and the results.

Several samples exhibited absorbances exceeding the range threshold (0.16), requiring multiple dilutions, especially for samples affected by seawater intrusion (dilutions up to x20).

Figure 7. Calibration curve for sulfate determination

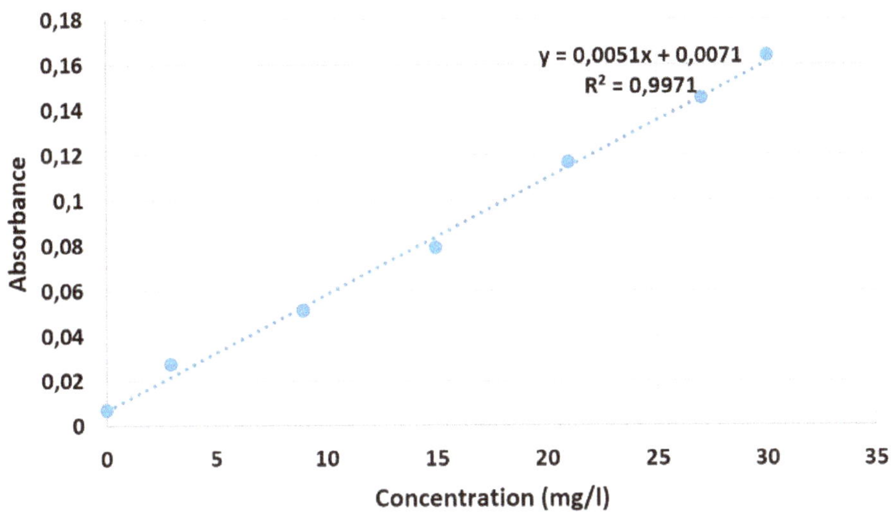

IV-3-3 Chloride Determination by Titration

The Mohr method involves precipitation titration with a solution of chloride ions of unknown concentration. After adding the potassium chromate indicator, the solution is titrated with a silver nitrate solution.

With each addition of silver ion, a precipitate of silver chloride is formed, and the solution turns red to visualize the equivalence between Ag and Cl (Rodier 2009).

$Ag^+ + Cl^- \rightarrow AgCl$ (solide)

The concentrations are then visualized in mg/L using simple calculations involving the volumes and concentrations at equilibrium.

IV-3-4 Sodium and Potassium Determination by Atomic Absorption Spectrometry

This method provides high precision and is more sensitive than flame spectrophotometry.

The solution is atomized either in a flame or in a graphite furnace at estimated temperatures of 2800°C, where it is transformed into atomic vapors. These vapors are exposed to radiation characteristic of the elements to be determined, produced by a source typically a hollow cathode lamp containing the element of interest. The radiation is absorbed by the unexcited atoms along the path of light. The concentrations are expressed in mg/L (Rodier 2009).

IV-3-4 Barium, Calcium, and Magnesium Determination by Atomic Absorption Spectrometry

Most atomic emission devices include a plasma torch (ICP torch) as this device is well suited for samples in aqueous solution. The sample, in the form of an aerosol produced by a nebulizer, is introduced at a constant flow rate of a few mg/min at the base of the torch through another small-diameter tube (1 to 2 mm). The chosen location for "light collection" from the plasma (transverse or end-on, also known as radial or axial) depends on the element and whether the analysis is based on the study of an ionic or atomic line. The temperature can range from 9,000 to 2,000 K.

IV-3-5 Ionic Balance

The ionic balance is an essential analytical tool for evaluating the quality of chemical analyses of water, and it is calculated using a simple formula:

I.B. = (Σ (cations) - Σ (anions) / Σ (cations) + Σ (anions)) x 100

The quality of the analyses is evaluated based on the following percentages:

- -5% < I.B. < 5%: Good analyses
- I.B. < ±10%: Acceptable analyses
- I.B. > ±10%: Suspicious analyses

IV-3-6 Preparation of Input Parameters

The susceptibility to Sea Water Intrusion (SWI) was evaluated using the GALDIT model, which incorporates six key parameters derived from various data sources, as illustrated in Figure 3. The Digital Elevation Model (DEM) with a 12.5 m resolution was obtained from the Alaska website.

The aquifer type (G) and thickness map were developed based on field surveys and extensive well-log profiles from the Akermoud aquifer, provided by ABHT. The hydraulic conductivity (A) of the aquifer was mapped using in-situ measurements. The groundwater depth (GW) for the Akermoud coastal aquifer was calculated by integrating static groundwater levels with topographic altitude data, collected in 2021 from 40 monitoring wells and piezometers. This data was then interpolated to provide a smooth representation of average groundwater levels.

The distance from the shore (D) was determined using Euclidean distance calculations. To evaluate the current status of SWI (I), samples from 40 groundwater boreholes and wells were analyzed. Additionally, electrical conductivity (EC) data from 16 samples were used to validate the GALDIT groundwater susceptibility model. The thickness data (T) were extracted from various well log reports.

ArcGIS Pro was employed to systematically process and prepare the thematic layers. The spatial data were georeferenced to the Lambert metric projection, zone: 1 Merchich datum. Each parameter was assigned attributes, including ratings and weights, based on their relevance to SWI. The GALDIT index was calculated using the ratings and weights recommended by Chachadi and Lobo-Ferreira (2001), considering the local hydrogeological conditions.

V. RESULTS AND DISCUSSION

V-1 GALDIT Model

V-1-1 The Akermoud GALDIT Index Parameters

Figure 8 shows the six thematic maps used to compute the GALDIT SWI susceptibility index:

- (G) - Groundwater Occurrence/Aquifer Type: The type of aquifer significantly affects the extent of SWI in freshwater. Aquifers can be classified as bounded, leaky confined, unconfined, or confined (Parizi et al., 2019; Zghibi et al., 2022). Confined aquifers with large piezometric cones due to excessive groundwater pumping are more susceptible to SWI. The Akermoud aquifer, in this case, is unconfined with an average thickness ranging from 20 to over 100 meters, consisting mainly of sand, sandstones, and conglomerates within Quaternary deposits (Figure 8).

- (A) - Aquifer Hydraulic Conductivity: The hydraulic conductivity was converted into a raster grid and weighted appropriately. In the Akermoud area, high hydraulic conductivity zones, exceeding 9 m/day, are found along the coastal strip and near the Tensift river (Figure 8). These areas, characterized by Quaternary alluvial plains with high sand or fine sandstone content, are particularly vulnerable to SWI. In contrast, southern and eastern parts of the study area have lower permeability values (0–6 m/day), indicating reduced SWI susceptibility, with formations of Cretaceous and Jurassic origin.
- (L) - Groundwater Depth: The depth of groundwater is crucial for maintaining hydrodynamic pressure along the coast, which counteracts the inland progression of SWI. Shallower depths result in lower hydrodynamic pressure, increasing SWI risk. Conversely, higher water table levels reduce SWI risk. Groundwater depth, measured from mean sea level, is categorized into four classes (Table 2): <30 m, 30–50 m, 50–80 m, and >80 m. Areas with depths less than 30 meters, especially along the coast and near the Tensift river, are highly susceptible to SWI due to significant groundwater extraction for agriculture.
- (D) - Distance from Shore: This parameter measures the influence of SWI based on proximity to the shoreline. The extent of SWI diminishes as the distance from the coast increases (Zghibi et al., 2022). The D-parameter map for Akermoud is divided into four distance classes, with ratings ranging from 2.5 (low) to 10 (high). Areas within 1,000 m of the shore are most vulnerable to SWI, with a rating of 10. Offshore distances beyond 6,000 m receive a low rating of 2.5 (Figure 8). This parameter is assigned a weight of 4 in the GALDIT groundwater index due to its significance.
- (I) - Impact of Existing SWI Status (Cl^-/HCO_3): This parameter assesses the disruption of the natural freshwater/saltwater hydraulic balance in the transition zone.

Figure 8. The six layers of GALDIT parameters of Akermoud aquifer.

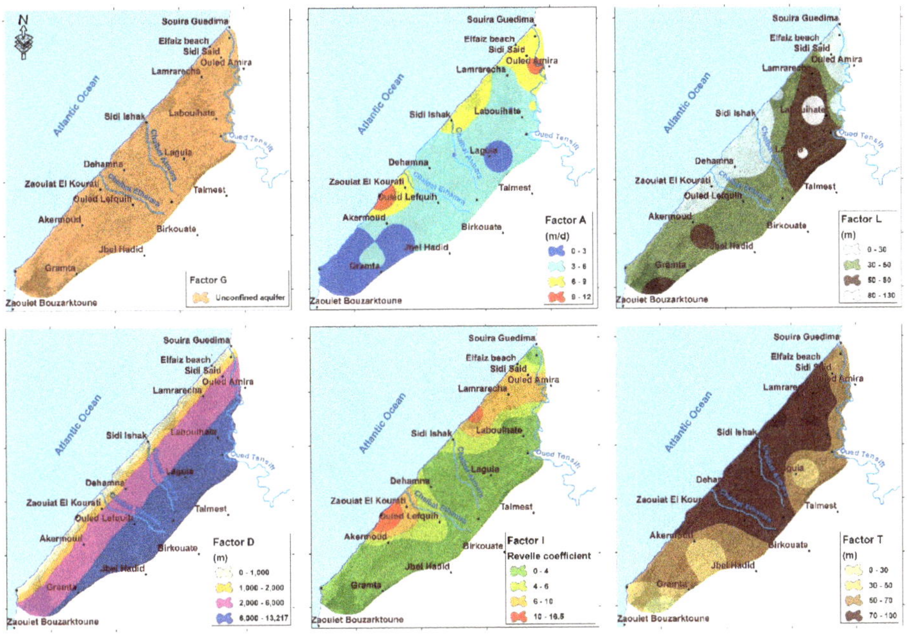

V-1-2 GALDIT GW Susceptibility Map

The GALDIT index map for the Akermoud area (Figure 9) reveals a high to very high susceptibility index, primarily observed parallel to the coastal region of the study area (discharge areas). Additionally, susceptible areas are found within a 3000-meter band inland between Tensift oued and up to 5000 meters in the southern part of Bhaybeh Beach. Further intrusion is evident between Zaouiet El Kourati and Ouled El Fequih villages, extending in an east-west direction. Notably, the Tensift Outlet experiences interference between marine and continental waters, leading to salinization of nearby wells. These vulnerable areas, characterized by sandy sandstone formations and intensive vegetable farmland, encompass approximately 24% of the study area (Table 3). This high susceptibility is attributed to multiple factors: the parameter (D) indicating a distance from the shore of less than 1 km, the parameter (I) indicating an impact of existing SWI with values above 6, the parameter (A) indicating hydraulic conductivity mostly exceeding 6 m/day, and a significant groundwater level below sea level of less than 30 meters. To ensure reliability and consistency, the results were subjected to testing and validation in subsequent analyses involving hydro-geochemical analysis.

Figure 9. Resultant SWI susceptibility maps of Akermoud aquifer

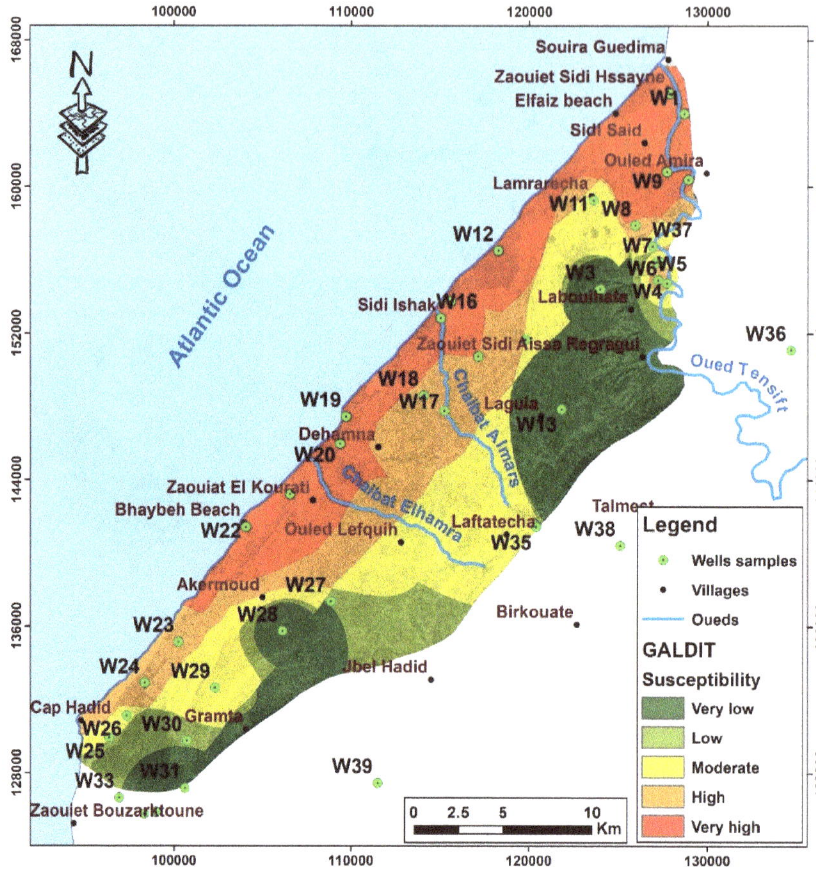

Table 3. The covered areas by the 5 classes

Susceptibility	Count pixels	Area (km²)	Percentage
Very low	709100	110.80	22.81%
Low	313466	48.98	10.09%
Medium	747336	116.77	24.04%
High	580012	90.63	18.66%
Very high	758153	118.46	24.40%

V-2 Hydro-Geochemical Contribution

The statistical analysis was conducted on 40 samples and 10 variables (EC, TDS, Ca^{2+}, Mg^{2+}, Na^{+}, K^{+}, HCO_3^{-}, Cl^{-}, SO_4^{2-}, Ba) for the Akermoud coastal aquifers situated along the Atlantic coast. To observe the spatial variation of these parameters, they were projected onto maps using GIS. Subsequently, diagrams were utilized to present the results and classify the degree of mineralization for each sample.

V-2-1 Seasonal and Spatial Variations of In-Situ Parameters

As mentioned in the methodology section, two analysis campaigns were conducted in the Akermoud aquifer: one in 2020 (before precipitation) and another in 2021 (after precipitation). There were slight differences in the sampled wells between the two campaigns, as some of them were no longer accessible. Figure 10 illustrates the spatial distribution of EC, TDS, dissolved oxygen, pH, and piezometric level for both 2020 and 2021.

For EC, which ranged from 1.36 to 9.11 in 2020 and 0.04 to 11.9 in 2021, it is worth noting that the values increased even after precipitation. This could be attributed to a greater intrusion of saline water toward the land during that year, resulting in higher EC values. Similar observations can be made for TDS, which exhibited a similar pattern of variation with different values.

Dissolved oxygen, indicating the amount of oxygen dissolved in the water, showed higher values in the northern region of Laguia and around Bhaybeh beach for both dates. The pH map, which can provide an indication of the presence of marine water or alteration of carbonic formations, revealed that pH values closer to the coastal line were more basic, with significant differences between the two dates due to variations in the sampled wells. Similarly, the piezometric level showed a change as a well located in the south of Sidi Ishak was sampled in 2020 but was no longer available in 2021.

Figure 10. Seasonal variations of in-situ measured parameters between 2020 (top row) and 2021 (bottom row)

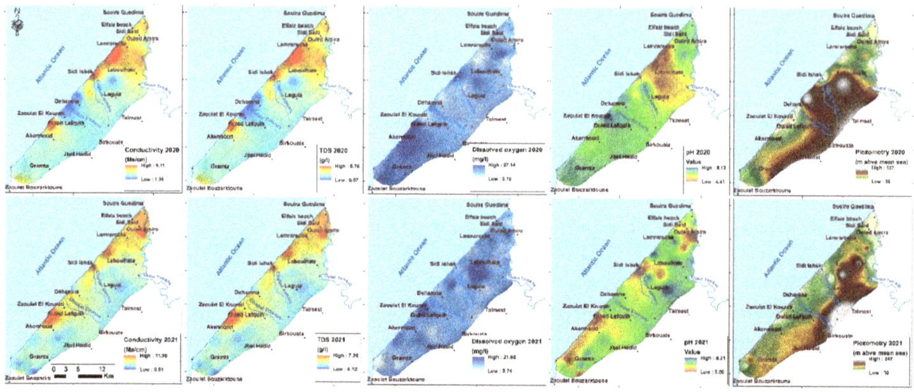

V-2-2 Spatial Variation of Laboratory Measured Parameters

After collecting water samples from the wells, various elements including Ca^{2+}, Mg^{2+}, Na^{+}, K^{+}, HCO_3^{-}, Cl^{-}, SO_4^{2-}, and Ba, were measured in different laboratories at Cadi Ayyad University. These measurements were conducted only for the second campaign, while for the first campaign, only in-situ measurements were available. Figure 11 displayed the spatial distribution of the aforementioned elements. From the Figure 11, several observations can be made:

- Bicarbonate (HCO_3), which is the predominant form of dissolved inorganic carbon in seawater and most freshwater sources, exhibited the highest values, surpassing 900 mg/l, in close proximity to Jbel Hadid. This suggests a potential lithological contamination from carbonate formations present within Jbel Hadid.
- Chloride levels, typically associated with high or elevated total dissolved solids, showed a significant increase in the Akermoud aquifer. In the northern part of the study area, after Chaibat Almars to Tensift oued, and in the south of Chaibat Elhamra around Bhaybeh beach, chloride concentrations exceeded 4500 mg/l. This indicates contamination by seawater intrusion.
- Sulfate concentrations, which should normally not exceed 250 mg/L in drinking water, exceeded 2900 mg/l near wells 11 and 17, as shown in the Figure 11 .
- Calcium, a major positive ion in natural freshwater, is widely found in minerals and constitutes 4.9% of the Earth's surface. Increased calcium concen-

trations can result from the leaching of rocks containing dissolved salts of calcium, magnesium, or iron. Groundwater in the study area receives significant calcium input from streams flowing over calcium-containing rocks and minerals found in Jbel Hadid's southern side. This explains the highest concentration of calcium between Akermoud and Zaouiat Bouzarktoune.

- Potassium (K) is one of the major components in seawater, typically present at a concentration of 380 mg/L K^+. In natural bodies of water, potassium concentrations are usually only 4% to 10% of sodium concentrations. Apart from marine contamination, the sources of potassium in the study area are likely silicate minerals, gypsum, and evaporate deposits, which release considerable amounts of potassium. This explains the higher potassium values observed in the Zaouiet El Kourati to Jbel Hadid and Lamrarecha to Tensift oued regions. Sodium (Na) is more dominant in our study area.
- Magnesium (Mg) is commonly found in groundwater that interacts with specific rocks and minerals, particularly limestone and gypsum. The dissolution of these materials releases both calcium and magnesium into the water. In our study area, magnesium mainly originates from the breakdown of dolomite, which is prevalent along Chaibat El Hamra, stemming from Jbel Hadid.
- Barium is a naturally occurring metal found in certain types of igneous and sedimentary rocks. Barium can enter groundwater and well water when rocks containing barium break down and dissolve. Barium can also combine with other chemicals to form compounds. The highest concentrations of barium are mainly located around the northern region of Jbel Hadid, suggesting potential lithological contamination.

Figure 11. Spatial variation of laboratory measured parameters (Ca^{2+}, Mg^{2+}, Na^{+}, K^{+}, HCO_3^{-}, Cl^{-}, SO_4^{2-} and Ba)

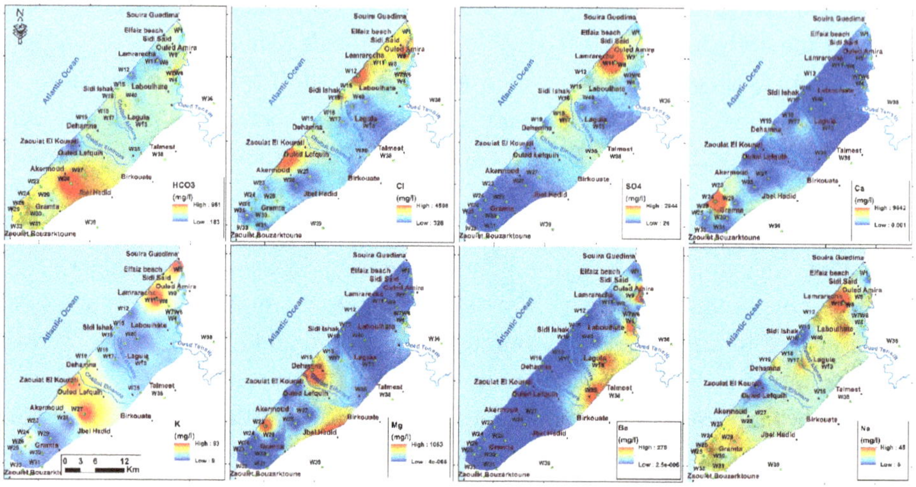

V-3 Water Classification With Conventional Diagrams

V-3-1 Piper Diagram

The results presented in Figure 12 demonstrate the dispersed distribution of samples in the piper diagrams, indicating the presence of specific hydro-chemical processes and mixing between different types of water samples. The diagrams provide insights into the groundwater characteristics of the Akermoud coastal aquifers, revealing a predominance of Ca–Cl type water followed by Na–Cl type water and mixed types.

Figure 12. Piper's trilinear diagram, illustrates the relationship between dissolved ions and water types in the Akermoud coastal aquifers

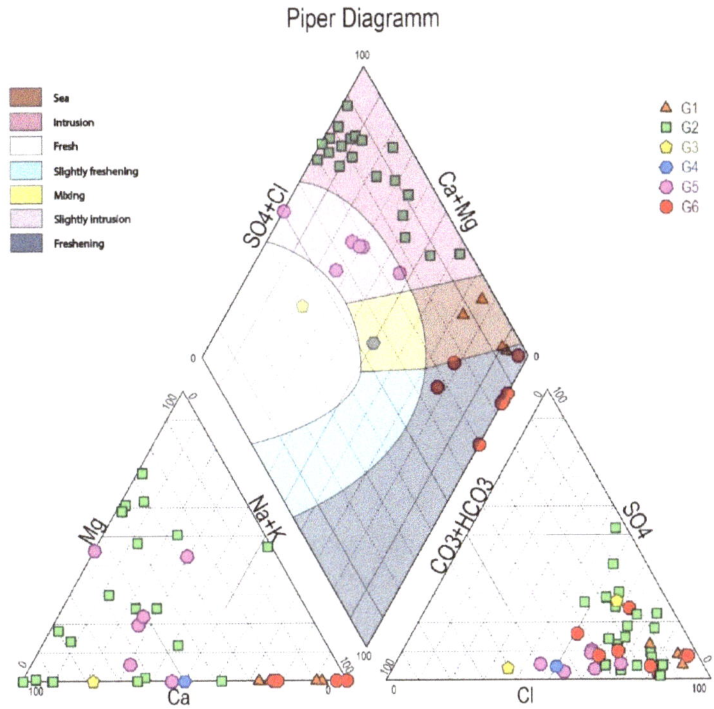

The samples were categorized into six groups:

- Sea group (G1): This group includes samples W7, W9, W12, W21, and W40.
- Intrusion group (G2): This group comprises samples W11, W33, W8, W5, W16, W18, W37, W1, W32, W6, W35, W10, W3, W19, W2, W23, W24, W4, W17, and W20.
- Fresh group (G3): Only sample W28 belongs to this group.
- Mixing group (G4): Only sample W38 falls into this group.
- Slightly intrusion group (G5): This group consists of samples W26, W25, W13, W31, W14, and W39.
- Freshening group (G6): This group includes samples W27, W22, W40, W15, W29, W30, and W34.

Based on the data presented in Figure 12, it is evident that 50% of the samples (20 out of 40) fall into the intrusion category, characterized by a predominant Ca–Cl water type. This suggests significant interaction between groundwater and saline waters. The anion triangle plot indicates that several samples are closer to the chloride apex. However, two distinct trends emerge: one group of samples is located between the bicarbonate and chloride poles, while the other group falls between the sulfate and chloride poles. These trends can be attributed to the dissolution of geological formations and the impact of seawater intrusion, which intensifies base exchange processes.

In the cation plot, most data points are concentrated around the ($Na^+ + K^+$) zone, with the exception of point W28, which lacks a dominant cation signature. Additionally, the majority of the points are positioned above the theoretical mixing line, indicating that many of the samples have been influenced by saltwater intrusion. The water types in the study area appear to be shaped by a combination of factors, including anthropogenic activities (e.g., agriculture and groundwater overuse), surface water-aquifer interactions, and the mixing of groundwater with marine waters.

V-3-2 Gibbs Diagram

The Gibbs diagrams (Figure 13) reveal that most samples are located within the evaporation domain and the rock weathering zone, trending towards the seawater influence field. Notably, sample W31 is positioned within the rock dominance region. This pattern indicates that the groundwater chemistry in the studied aquifer is predominantly influenced by rock weathering processes, which are associated with seawater intrusion. It's important to note that nearly all samples extend beyond the Gibbs diagram boundaries, especially in terms of the $Na^+ / (Na^+ + Ca^{2+})$ ratio. This occurrence is likely due to an excess of Ca^{2+} resulting from water-rock interactions and ion exchange processes influenced by seawater intrusion (Chen et al., 2020; Liu et al., 2020). Furthermore, the $Cl^- / (Cl^- + HCO_3^-)$ ratio shows variation with increasing total dissolved solids (TDS), which may be attributed to higher Cl^- concentrations resulting from evaporation-crystallization processes (Liu et al., 2020b), sea spray, or seawater intrusion.

Figure 13. Gibbs diagram representing the ratio of $Na^+/(Na^+ + Ca^{2+})$ and $Cl^-/(Cl^- + HCO_3^-)$ as a function of TDS in Akermoud coastal aquifers.

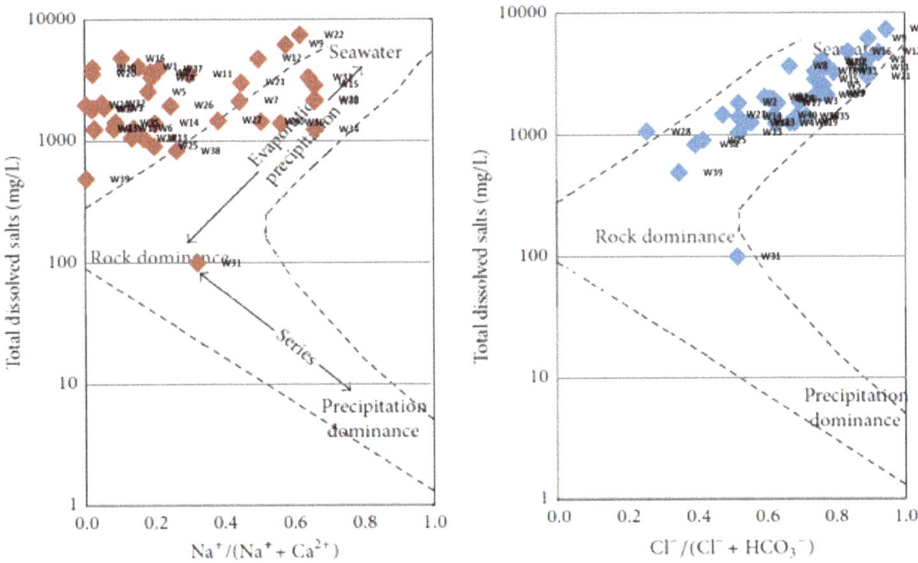

V-3-3 Sodium Absorption Ratio (SAR) and Sodium Percentage (Na^+ %)

It is important to note that, for the Wilcox graphs, we have once again categorized our samples into four groups based on their quality:

- Group 1: Includes samples W5, W7, W8, W9, W10, W11, W12, W15, W16, W18, W20, W21, W22, W24, W26, W29, W30, W32, W33, W34, W36, and W37.
- Group 2: Includes samples W2, W3, W14, W17, W27, W35, and W40.
- Group 3: Includes samples W4, W6, W13, W19, W23, W25, W28, W38, and W39.
- Group 4: Contains only sample W31.

In Figure 14, the SAR (Sodium Adsorption Ratio) values of the groundwater in the Akermoud aquifer are depicted, ranging from 0 to 5 with an average of 2. Analyzing the SAR diagrams in relation to electrical conductivity, we find that the samples fall into the following classes: C3S1, C4S1, C5S1, and C5S2 (Figure 14). These classes indicate high to very high salinity and low to medium alkalinity, suggesting that the water is unsuitable for irrigation purposes.

Figure 14. Classification of Akermoud aquifer and the groundwater suitability for irrigation purpose based on (EC and SAR).

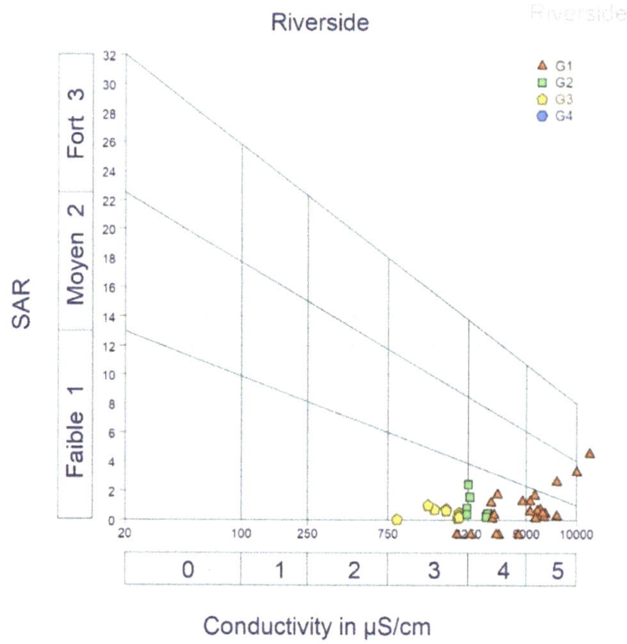

(Given that: $\boldsymbol{SAR} = \frac{\boldsymbol{Na}}{\sqrt{\frac{1}{2}(\boldsymbol{Ca} + \boldsymbol{Mg})}}$)

Additionally, in Figure 15, the Wilcox (1955) diagram presents the Na^+ % projection for the Akermoud coastal aquifers, showing a wide range of Na^+ % values from 0% to over 100%. Based on this projection, we can identify four main groups that were previously mentioned in this context. Group 1 corresponds to water of poor quality, Group 2 represents water of mediocre quality, Group 3 indicates good quality water, and Group 4 signifies excellent quality water. The majority of the water samples fall within the range of poor to mediocre quality. This situation can be attributed to an excess of sodium caused by the intrusion of seawater and water-rock interactions involving the dissolution of halite and evaporitic formations.

Figure 15. Classification of Akermoud coastal aquifer and groundwater suitability for irrigation purpose based on (EC and Na^+ %).

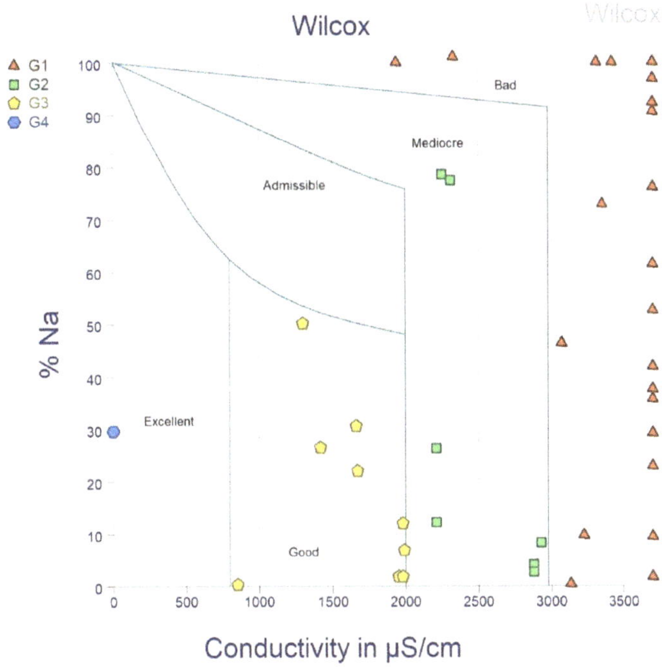

(Given that: $Na\% = \frac{100*(Na + K)}{(Ca + Mg + Na + K)}$)

VI. CONCLUSION

In summary, effectively managing the coastal aquifer in the Akermoud district necessitates a thorough understanding of the factors influencing groundwater salinity and geochemical changes. To this end, a combined approach of modeling and hydro-geochemical analysis was utilized to evaluate the extent of the aquifer's degradation due to climate change and human activities.

The GALDIT model was utilized to understand the encroachment and extent of seawater susceptibility into the groundwater system. The model indicated high to very high susceptibility index zones, primarily located parallel to the coast and in areas within a 3000 m inland band. Notably, the Tensift Outlet exhibited significant salinization of nearby wells due to the interaction between marine and continental waters. These findings were further validated through hydro-geochemical analysis to ensure reliability and consistency.

A comprehensive study was conducted, involving the collection of 40 groundwater samples from different types of tubewells. These samples were then subjected to analysis using the GALDIT Model, and hydro-geochemical techniques. The combined results of these analysis revealed the presence of marine intrusion in the study area.

Notably, the groundwater samples obtained from wells in close proximity to the coast, Tensift Oued, and certain areas in the eastern part of the study area exhibited significantly high values of electrical conductivity (EC) and total dissolved solids (TDS). This indicates a clear influence of seawater on the groundwater quality in these locations. The intrusion of marine water and the presence of saline lithological formations contribute to the observed geochemical variations.

The hydro-geochemical analysis focused on major ions and total dissolved solids (TDS). The results showed that a significant portion of the samples exhibited a dominant Ca-Cl water type, indicating groundwater interaction with saline waters. The analysis also revealed the dissolution of evaporitic formations and the impact of seawater intrusion, which led to the exchange of bases. Furthermore, the cation analysis demonstrated a tendency towards the domain of Na^{++} and K^{+} for most samples, suggesting the influence of seawater intrusion. Overall, the water types in the aquifer were affected by multiple factors such as anthropogenic activities, surface water interaction, and mixing with marine waters.

Overall, this multi-approach study provided valuable insights into the state of the Akermoud aquifer, including the susceptibility to seawater intrusion, hydro-geochemical characteristics, and subsurface features. The findings emphasized the need for effective management strategies to address the current groundwater quality and potential future impacts, particularly due to overexploitation and climate change.

The consequences of such geochemical variations are profound. The quality of irrigation water, as indicated by these samples, was found to be extremely poor in approximately half of the cases. Moreover, the quality of potable water has deteriorated to a significant extent due to the effects of marine intrusion and the contribution of saline lithological formations. This poses a serious concern for both agricultural activities and access to safe drinking water.

The impact of saline irrigation water on soil structure is particularly noteworthy. The higher the salinity of the irrigation water, the greater the adverse effects on soil quality and structure. Continuous irrigation with saline water leads to the accumulation of sodium in the soil over time, resulting in diminished soil fertility and decreased crop yields. This highlights the urgent need for improved water management practices to mitigate the detrimental effects on soil and ensure sustainable agricultural productivity.

Due to the scarcity of soil resources in these regions, it is vital to encourage societal adaptation and promote behavioral changes. Moreover, the formulation of suitable management strategies is necessary to support a shift towards sustainable hydrological conditions. By embracing sustainable practices and implementing effective water management approaches, the impacts of marine intrusion can be minimized, ensuring the long-term sustainability of the hydrological system in these areas.

REFERENCES

Alabjah, B., Amraoui, F., Chibout, M., & Slimani, M. (2018). Assessment of saltwater contamination extent in the coastal aquifers of Chaouia (Morocco) using the electric recognition. *Journal of Hydrology (Amsterdam)*, 566, 363–376. DOI: 10.1016/j.jhydrol.2018.09.003

Amiri, V., Nakhaei, M., Lak, R., & Kholghi, M. (2016). Geophysical, isotopic, and hydrogeochemical tools to identify potential impacts on coastal groundwater resources from Urmia hypersaline Lake, NW Iran. *Environmental Science and Pollution Research International*, 23(16), 16738–16760. DOI: 10.1007/s11356-016-6859-y PMID: 27184149

Badon-Ghyben, W. (1889). Nota in verband met de voogenomen put boring nabij Amsterdam (Notes on the probable results of the proposed well drilling near Amsterdam). *Tijdschr. Kon. Inst. Ing.*, 8-22.

Bahir, M., Mennani, A., Jalal, M., & Youbi, N. (2000). Ressources hydriques du bassin synclinal d'Essaouira (Maroc). *Estudios Geológicos (Madrid)*, 56(3-4), 185–195. DOI: 10.3989/egeol.00563-4150

Bahir, M., Ouhamdouch, S., Ouazar, D., & El Moçayd, N. (2020). Climate change effect on groundwater characteristics within semi-arid zones from western Morocco. *Groundwater for Sustainable Development*, 11, 100380. DOI: 10.1016/j.gsd.2020.100380

Bear, J., Cheng, A. H. D., Sorek, S., Ouazar, D., & Herrera, I. (Eds.). (1999). *Seawater intrusion in coastal aquifers: concepts, methods and practices* (Vol. 14). Springer Science & Business Media. DOI: 10.1007/978-94-017-2969-7

Benini, L., Antonellini, M., Laghi, M., & Mollema, P. N. (2016). Assessment of water resources availability and groundwater salinization in future climate and land use change scenarios: A case study from a coastal drainage basin in Italy. *Water Resources Management*, 30(2), 731–745. DOI: 10.1007/s11269-015-1187-4

Chachadi A.G., Lobo-Ferreira J.P. (2001). Sea water intrusion vulnerability mapping of aquifers using GALDIT method. Proceedings of the workshop on modelling in hydrogeology, 143–156.

Chachadi A.G., & Lobo-Ferreira J.P. (2005). Assessing aquifer vulnerability to seawater intrusion using GALDIT method: part 2- GALDIT Indicator Descriptions. In IAHS and LNEC, proceedings of the 4 the fourth interceltic colloquium on hydrology and management of water resources. University Minho.

Chachadi, A.G., & Lobo-Ferreira, J.P. (2007). Assessing aquifer vulnerability to seawater intrusion using GALDIT method: Part 2, GALDIT indicators description. *Water Celt Countries Quant Qual Clim Var*, 310, 172–180.

Custodio, E. (2002). Aquifer overexploitation: What does it mean? *Hydrogeology Journal*, 10(2), 254–277. DOI: 10.1007/s10040-002-0188-6

Custodio, E. (2010). Coastal aquifers of Europe: An overview. *Hydrogeology Journal*, 18(1), 269–280. DOI: 10.1007/s10040-009-0496-1

Derradji, F. (2015). *Intrusion marine dans les nappes côtières: cas de la nappe des graviers d'Annaba*. Extrême Nord-Est d'Algérie.

Erostate, M., Huneau, F., Garel, E., Ghiotti, S., Vystavna, Y., Garrido, M., & Pasqualini, V. (2020). Groundwater dependent ecosystems in coastal Mediterranean regions: Characterization, challenges and management for their protection. *Water Research*, 172, 115461. DOI: 10.1016/j.watres.2019.115461 PMID: 31951946

Ez-zaouy, Y., Bouchaou, L., Saad, A., Hssaisoune, M., Brouziyne, Y., Dhiba, D., & Chehbouni, A. (2022). Morocco's coastal aquifers: Recent observations, evolution and perspectives towards sustainability. *Environmental Pollution*, 293, 118498. DOI: 10.1016/j.envpol.2021.118498 PMID: 34798220

Fadili, A. (2014). *Etude hydrogéologique et géophysique de l'extension de l'intrusion marine dans le Sahel de l'Oualidia (Maroc): analyse statistique, hydrochimie et prospection électrique* (Doctoral dissertation, Univérsité de Chouaïb Doukkali Faculté Des Sciences El Jadida).

Gholami, V., Yousefi, Z., & Zabardast Rostami, H. (2010). Modeling of ground water salinity on the Caspian southern coasts. *Water Resources Management*, 24(7), 1415–1424. DOI: 10.1007/s11269-009-9506-2

Han, D., Kohfahl, C., Song, X., Xiao, G., & Yang, J. (2011). Geochemical and isotopic evidence for palaeo-seawater intrusion into the south coast aquifer of Laizhou Bay, China. *Applied Geochemistry*, 26(5), 863–883. DOI: 10.1016/j.apgeochem.2011.02.007

Herzberg, A. (1901). Die wasserversorgung einiger Nordseebader. *J. Gasbeleucht. Wasserversorg.*, 44, 815–819.

Hanich, L. (2001) Structure et fonctionnement d'un aquifère multicouche carbonaté. Exemple du bassin d'Essaouira. Guide méthodologique. Thèse d'Etat ès Sciences Naturelles, Université Cadi Ayyad, Faculté des Sciences et Techniques de Marrakech, Maroc.

Hermans, T., & Paepen, M. (2020). Combined inversion of land and marine electrical resistivity tomography for submarine groundwater discharge and saltwater intrusion characterization. Geophysical Research Letters, 47(3). DOI: 10.1029/2019GL085877

Hu, J., Dong, H., Xu, Q., Ling, W., Qu, J., & Qiang, Z. (2018). Impacts of water quality on the corrosion of cast iron pipes for water distribution and proposed source water switch strategy. *Water Research*, 129, 428–435. DOI: 10.1016/j.watres.2017.10.065 PMID: 29179122

Humphreys, E. R., Lafleur, P. M., Flanagan, L. B., Hedstrom, N., Syed, K. H., Glenn, A. J., & Granger, R. (2006). Summer carbon dioxide and water vapor fluxes across a range of northern peatlands. *Journal of Geophysical Research*, 111(G4), 2005JG000111. Advance online publication. DOI: 10.1029/2005JG000111

Khouz, A., Trindade, J., Santos, P. P., Oliveira, S. C., El Bchari, F., Bougadir, B., Garcia, R. A. C., Reis, E., Jadoud, M., Saouabe, T., & Rachidi, S. (2023). Flood susceptibility assessment through statistical models and HEC-RAS analysis for sustainable management in Essaouira Province, Morocco. *Geosciences*, 13(12), 382. DOI: 10.3390/geosciences13120382

Laouina, A. (2006). Le litoral marocain, milieux côtier et marin. *Environnement et Territoires, Cadre Natural*, 187-216.

Ma, C., Li, Y., Li, X., & Gao, L. (2020). Evaluation of groundwater sustainable development considering seawater intrusion in Beihai City, China. *Environmental Science and Pollution Research International*, 27(5), 4927–4943. DOI: 10.1007/s11356-019-07311-3 PMID: 31840220

Mdiker, N. A. B. I. L., El Achheb, A., Mandour, A., Younsi, A., El Maliki, S., & Outeyeb, B. (2009). Contribution à l'étude de la salinisation de la nappe côtière de sahel El Haouzia région d'El Jadida au Maroc. *Afrique Science: Revue Internationale des Sciences et Technologie, 5*(2).

Mirzavand, M., Ghasemieh, H., Sadatinejad, S. J., & Bagheri, R. (2020). An overview on source, mechanism and investigation approaches in groundwater salinization studies. *International Journal of Environmental Science and Technology*, 17(4), 2463–2476. DOI: 10.1007/s13762-020-02647-7

Monition, L. (1953). *Sur l'intérêt de sondages dans la plaine d'Akermoud*. Rap. Inéd. MTPC/DH/DRE.

Moujabber, M. E., Samra, B. B., Darwish, T., & Atallah, T. (2006). Comparison of different indicators for groundwater contamination by seawater intrusion on the Lebanese coast. *Water Resources Management*, 20(2), 161–180. DOI: 10.1007/s11269-006-7376-4

Moumane, A., El Ghazali, F. E., Al Karkouri, J., Delorme, J., Batchi, M., Chafiki, D., & Karmaoui, A. (2021). Monitoring spatiotemporal variation of groundwater level and salinity under land use change using integrated field measurements, GIS, geostatistical, and remote-sensing approach: Case study of the Feija aquifer, Middle Draa watershed, Moroccan Sahara. *Environmental Monitoring and Assessment*, 193(12), 769. DOI: 10.1007/s10661-021-09581-2 PMID: 34735624

Nadjla, B., Abdellatif, D., & Assia, S. (2021). Mapping of the groundwater vulnerability to saline intrusion using the modified GALDIT model (Case: The Ain Temouchent coastal aquifer,(North-Western Algeria)). *Environmental Earth Sciences*, 80(8), 319. DOI: 10.1007/s12665-021-09614-6

Parizi, E., Hosseini, S. M., Ataie-Ashtiani, B., & Simmons, C. T. (2019). Vulnerability mapping of coastal aquifers to seawater intrusion: Review, development and application. *Journal of Hydrology (Amsterdam)*, 570, 555–557. DOI: 10.1016/j.jhydrol.2018.12.021

Pinder, G. F. (1973). A Galerkin-finite element simulation of groundwater contamination on Long Island, New York. *Water Resources Research*, 9(6), 1657–1669. DOI: 10.1029/WR009i006p01657

Pulido-Leboeuf, P. (2004). Seawater intrusion and associated processes in a small coastal complex aquifer (Castell de Ferro, Spain). *Applied Geochemistry*, 19(10), 1517–1527. DOI: 10.1016/j.apgeochem.2004.02.004

Rodier, J., Legube, B., & Merlet, N. (2009). *L'Analyse de l'eau 9e édition. Entièrement mise à jour*. Dunod.

Salvan, H.M. (1984). Les formations évaporitiques du Trias marocain. Problèmes stratigraphiques, paléogéographiques et paléoclimatologiques. Quelques réflexions. *Revue de Géologie Dynamique et de Géographie Physique*, 25, 87–203.

Seddique, A. A., Masuda, H., Anma, R., Bhattacharya, P., Yokoo, Y., & Shimizu, Y. (2019). Hydrogeochemical and isotopic signatures for the identification of seawater intrusion in the paleobeach aquifer of Cox's Bazar city and its surrounding area, south-east Bangladesh. *Groundwater for Sustainable Development*, 9, 100215. DOI: 10.1016/j.gsd.2019.100215

Sherif, M. M., & Hamza, K. I. (2001). Mitigation of seawater intrusion by pumping brackish water. *Transport in Porous Media*, 43(1), 29–44. DOI: 10.1023/A:1010601208708

Talabi, A. O., Abdu-Raheem, Y. A., Afolagboye, L. O., Oguntuase, M. A., & Akinola, O. O. (2020). Hydrogeochemistry of shallow groundwater in Ado-Ekiti Area, Southwestern Nigeria. *Groundwater for Sustainable Development*, 11, 100386. DOI: 10.1016/j.gsd.2020.100386

Werner, A. D., & Simmons, C. T. (2009). Impact of sea-level rise on sea water intrusion in coastal aquifers. *Groundwater*, 47(2), 197–204. DOI: 10.1111/j.1745-6584.2008.00535.x PMID: 19191886

Xue, H., & Sigg, L. (1999). Comparison of the complexation of Cu and Cd by humic or fulvic acids and by ligands observed in lake waters. *Aquatic Geochemistry*, 5(4), 313–335. DOI: 10.1023/A:1009679819002

Zalidis, G., Stamatiadis, S., Takavakoglou, V., Eskridge, K., & Misopolinos, N. (2002). Impacts of agricultural practices on soil and water quality in the Mediterranean region and proposed assessment methodology. *Agriculture, Ecosystems & Environment*, 88(2), 137–146. DOI: 10.1016/S0167-8809(01)00249-3

Zghibi, A., Merzougui, A., Mansaray, A. S., Mirchi, A., Zouhri, L., Chekirbane, A., Msaddek, M. H., Souissi, D., Mabrouk-El-Asmi, A., & Boufekane, A. (2022). Vulnerability of a Tunisian coastal aquifer to seawater intrusion: Insights from the GALDIT Model. *Water (Basel)*, 14(7), 1177. DOI: 10.3390/w14071177

Chapter 3
A Review on Estimation of Climate and Land Use Change Impacts on Tropical Streamflow Using Remote Sensing and GIS:
CC and Land Use Impact on Thailand's Tropical Streamflow

Phyo Thandar Hlaing
https://orcid.org/0009-0007-7708-6446
King Mongkut's University of Technology, Thonburi, Bangkok, Thailand

Muhammad Waqas
https://orcid.org/0000-0002-9069-2762
King Mongkut's University of Technology, Thonburi, Bangkok, Thailand

Usa Wannasingha Humphries
King Mongkut's University of Technology, Thonburi, Bangkok, Thailand

ABSTRACT

The tropical climate is subjected to extreme hydrological events, exacerbated by climate change (CC) and changes in land use/land cover (LULC). Evaluating CC and LULC impacts on streamflow in tropical regions is vital for understanding the risks to vulnerable ecosystems. This chapter presents a meta-analysis of 30 system-

DOI: 10.4018/979-8-3693-9651-3.ch003

atically selected case studies that examine the spatiotemporal effects of CC and LULC on streamflow. The findings highlight the complex interactions among CC, LULC, and hydrological dynamics, underscoring the critical role of hydrological models, meteorological inputs, and scenario-based analyses. Models like SWAT and HEC-HMS, coupled with climate projections from GCMs (CMIP5 and CMIP6), are essential for projecting future streamflow changes. Integrating remote sensing (RS) and geographical information systems (GIS) enhances spatial accuracy and reliability. Incorporating diverse climate and LULC scenarios, including representative concentration pathways (RCPs), provides insights into adaptive water resource management in tropical regions.

1. INTRODUCTION

According to the IPCC-AR6 (2021), Climate Change (CC) is a prominent contemporary challenge in hydrological research. CC is projected to intensify these shifts in rainfall and temperature patterns, which will cause extreme hydrologic events (Ming et al. 2021; Adopted 2014). Addressing the challenges associated with these natural disasters by adapting water management practices is set to become a significant undertaking in the present century (Dlamini et al. 2017). Between 1960 and 2019, about 17% of Earth's terrestrial surface changed, including the loss of 0.8 million km^2 of forests and the expansion of agricultural land by roughly 1.0 million km^2, among other transformations (Winkler et al. 2021). The extensive land use/land cover (LULC) changes, predominantly linked to human activities, impact watershed hydrologic systems differently (Koneti, Sunkara, and Roy 2018). Thus, it is critical to quantify and comprehend the spatiotemporal dynamics of LULC changes. This knowledge is crucial for informing and actively contributing to the ongoing discussions on CC (Change 2019).

Nevertheless, despite the significant societal relevance of CC and LULC changes, researchers face challenges in understanding the evolution of global LULC over time and across space. This limitation is primarily attributed to the lack of comprehensive data and substantial uncertainties inherent in existing LULC transformation records (Prestele et al. 2017). However, the advancements in remote sensing (RS) technology, such as satellite imagery and LiDAR, combined with the analytical capabilities of GIS, have significantly enhanced the ability to detect, map, and analyze LULC changes with greater accuracy and detail. Despite these advancements in technology, LULC statistics still face certain constraints. These limitations arise from fragmented material, varying scales, inadequate geographic or temporal detail, and inconsistent time series (Bayer et al. 2021). Advanced RS and Geoinformation Systematic methods are popular and effective technologies that provide comprehensive, high-resolution data and tools to monitor temporal and spatial changes, offering critical insights for informing and actively contributing to CC and LULCC (Werf. 2010). The impact of CC on hydrological regimes and river flows exhibits variability between the catchment and regional scales (Liu and Curry 2010). In addition to the effects of CC, changes in LULC can significantly influence hydrological processes (Laurance 2007) (Bradshaw et al. 2007). The effects of these CCs fluctuate among regions, showing diverse magnitudes and manifestations (Dang et al. 2016). Therefore, determination of their impact on streamflows has proven challenging (Zhang et al. 2019; Waqas et al. 2021). Researchers can capture and model these regional differences by utilizing RS and GIS technologies coupled with hydrological models and advanced machine learning (ML) algorithms, offering an understanding of how CC and LULC changes influence streamflow across various global scales. As a tropical country, Thailand relies significantly on rainfall for its agriculture and economy.

Consequently, The presence of either an excessive or insufficient amount of rainfall presents a significant hazard, resulting in the potential occurrence of floods or droughts that can cause severe damage to the national economy (Waqas, Humphries, et al. 2023; Waqas, Bonnet, et al. 2023). Thailand is currently dealing with this problem, with the severity of floods and droughts escalating annually and anticipated to exacerbate further within 25 years. The recorded history highlights the most significant flood disaster in 2011, substantially damaging central and northern Thailand (Arunrat et al. 2017).

Thailand is divided into 22 distinct river basins, and most agricultural practices rely entirely on river water or tributaries (ชุ่ม 2020). Assessing the impact of CC and LULC on river streamflow is crucial, with remote sensing (RS) and GIS technologies being key tools in Thailand. Although research has addressed these impacts, gaps remain in understanding the detailed interplay between CC, LULC changes, and streamflow. This study aims to review recent methodologies to better understand CC effects on hydrological extremes in Thailand.

2. DESIGN OF THE STUDY

This review study analyzed recent research across various domains, highlighting the importance of systematic reviews in identifying trends and synthesizing insights from scientific findings. (Waqas, Humphries, et al. 2023). Figure 1 presents a visual representation of the flowchart detailing the steps undertaken during the review process.

2.1. Research Questions

RQ-1: What hydrological models are used to study CC impacts on streamflow?

RQ-2: What data sources and meteorological variables are integrated into hydrological models to analyze CC impacts on streamflow?

RQ-3: Which climate and LULC scenarios are utilized to assess the impacts on streamflow?

RQ-4: Which downscaling and bias-correction (BC) techniques are used to improve the accuracy of climate models?

RQ-5: How can integrating CC and LULC change scenarios into hydrological models provide a more comprehensive understanding of streamflow alterations in a changing environment?

RQ-6: How do RS and GIS technologies enhance the monitoring and prediction of streamflow changes due to CC and LU alterations?

Figure 1. The overall design of methodology for systematic review.

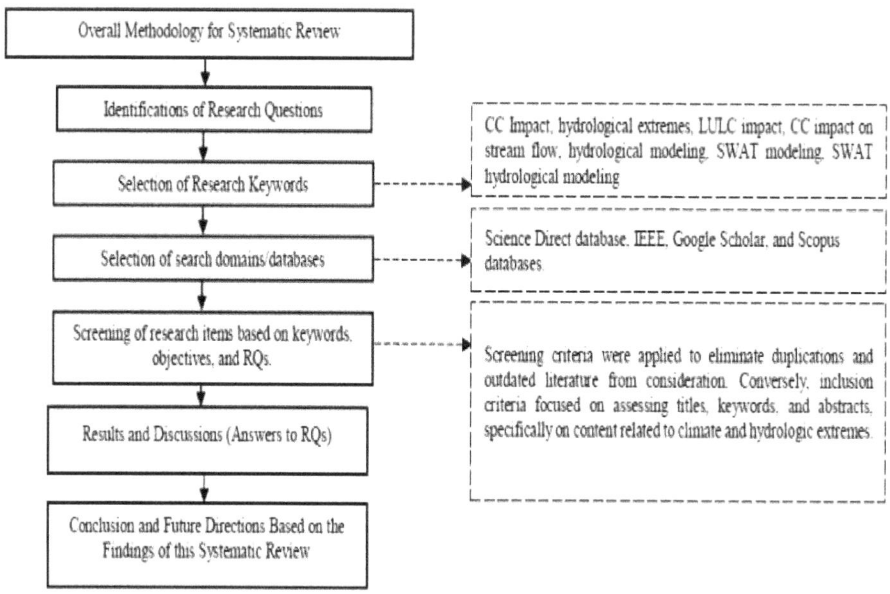

2.2. Keywords Selection

Selecting keywords is crucial to research (Waqas, Humphries, et al. 2023). This process is critical in enhancing the accuracy and effectiveness of the research. The following keywords used in this study are "CC," Impact of CC," "Land Use Land Cover Change," "Streamflow," "Hydrological Models," "Climate Scenarios," "SWAT," "Hydrological Model," "SWAT modeling," and "Remote Sensing and GIS."

2.3. Scientific Databases Selection

The research utilized specific keywords to identify peer-reviewed publications from online libraries, including ScienceDirect, IEEE, Google Scholar, and Scopus databases.

2.4. Screening Process

This study's screening process involved applying exclusion criteria to remove duplicate literature (Figure 2). Inclusion criteria were used to assess titles, keywords, and abstracts, focusing on content related to climate and hydrologic extremes. In the context of CC and LULC's impacts on streamflow for systematic review, we will include studies that utilize RS and GIS technologies for data collection, analysis, or modeling of hydrological changes. This selection process categorized the articles into three groups of ten: the first group examines the impacts of CC on streamflow, the second focuses on LULC change effects, and the third investigates the combined impacts of CC and LULC changes. A thorough analysis was performed on the selected articles, reviewing climate models, BC techniques, downscaling methods, hydrological models, and case studies. This study aimed to understand the outcomes reported in the literature comprehensively.

Figure 2. Overall literature screening process was used to select articles systematically

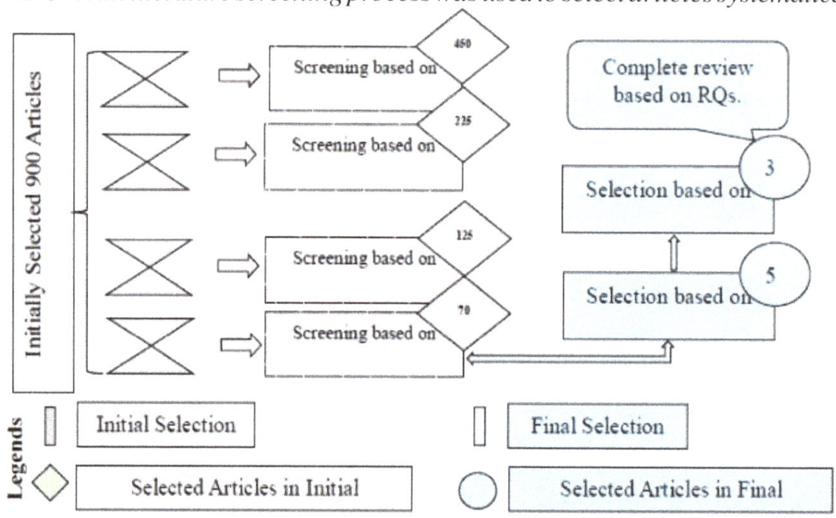

3. TROPICAL COUNTRIES/REGIONS AND CC

The tropics encompass the areas delineated by the Tropic of Cancer (23° 27′ N) and the Tropic of Capricorn (23° 27′ S), as illustrated in Figure 3 (Edelman et al. 2014). These regions experience high energy accumulation due to the elevated angles of the sun, resulting in a warm climate characterized by the absence of strong winters (Jahandideh-Tehrani et al. 2019). Tropical regions generally maintain a warm environment with average monthly temperatures above 18 °C year-round. Near the Equator, temperature variations are minimal, often under 4 °C, and in places like Quito, Ecuador, this difference can be less than 1 °C (Malhi and Wright 2004; Jahandideh-Tehrani et al. 2019). Global average air temperature has increased by approximately 0.5ºC over the past century. This study examines mean annual minimum temperatures to address the challenges of modeling subtle climate change signals, especially in tropical regions, over the last 50 years (Jáuregui 1992). A survey conducted by Eccles et al., (2019) highlighted that the vulnerability of tropical and subtropical regions to flooding is exacerbated by CC, leading to intensified precipitation and heightened global flood events (Eccles, Zhang, and Hamilton 2019).

Figure 3. Tropical regions around the world (Edelman et al. 2014)

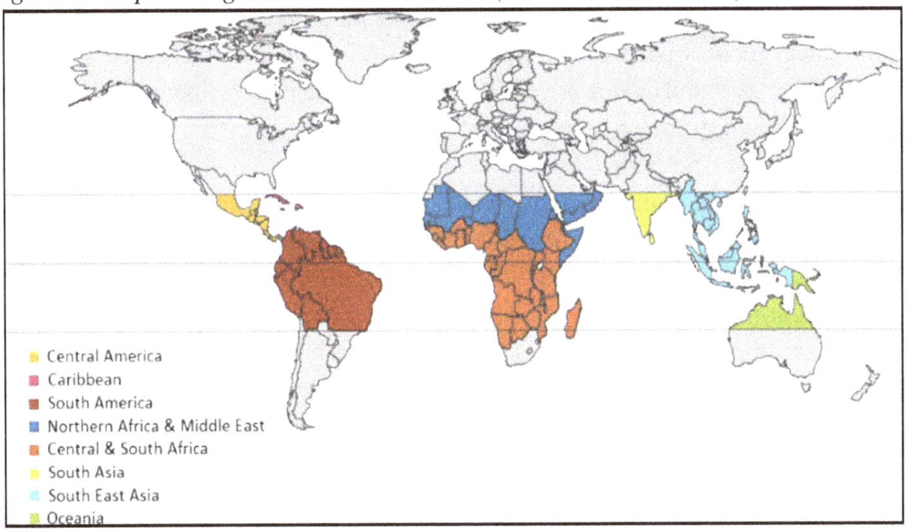

Eight of the world's ten largest rivers by water discharge are tropical, such as the Amazon, Congo, Orinoco, Brahmaputra, Paraná, and three Amazon tributaries (Negro, Madeira, and Japurá). These rivers show significant fluctuations in peak flows between wet and dry seasons, with some having two flood peaks each year (Latrubesse, Stevaux, and Sinha 2005). Discharge patterns can categorize tropical rivers: some show clear seasonal highs and lows, while others have two annual flood peaks. The Godavari River in India exemplifies this, with periods of very high discharge alternating with near-total dryness each year (Syvitski et al. 2014). From the literature reviewed, it is established that tropical regions are acknowledged as highly vulnerable to natural weather disasters, leading to substantial research on the impacts of CC in these areas. Marked by hot and humid conditions, frequent intense rainfall, cyclones, and elevated solar radiation, the tropics are particularly susceptible to climate-related changes.

4. HYDROLOGICAL MODELING TO ACCESS THE IMPACT OF CC AND LULC

Hydrological modeling involves a computational approach to simulate water dynamics in a specific catchment or river basin. This process utilizes mathematical and physical models to analyze, predict, and manage water resources within a defined geographical area (Te Chow, Maidment, and Mays 1988). RS and GIS enhance these models by providing detailed spatial data on land surface characteristics and temporal changes, which are critical for accurate parameterization and calibration. Hydrological models partition the study area into grids and cells, with each unit encompassing essential factors governing the hydrological cycle, including hydraulic conductivity, land surface roughness, soil storage capacity, and more. These parameters are utilized in the calibration process to simulate the model. Various hydrological models, ranging from simplified conceptual lumped models to more intricate physically distributed models, have been applied as analytical tools in assessing the impacts of CC in watershed studies (Singh and Marcy 2017). Integrating RS and GIS into these models allows for the utilization of up-to-date land use data and surface conditions, enhancing model accuracy and reliability. HEC-HMS, SWMM, MIKE-SHE, RRI, and SWAT are commonly used hydrological models. The basic working of all these hydrological models is presented in Figure 4.

Figure 4. Basic working principle of Hydrological Modeling to assess the impact of CC and LULC

5. CC IMPACTS ON STREAMFLOW

As per findings of IPCC-AR4 in 2007, numerous regions across the globe have been the subject of research about the effects of CC on water supplies, employing various hydrologic models (Arnell 1999; Fujihara et al. 2008; Mauser and Bach 2009). IPCC has prepared emission scenarios to evaluate the potential consequences of CC in the future. These scenarios are based on studies depicting potential trajectories of human growth across various dimensions (Nakicenovic et al. 2000). Emission scenarios in global coupled atmosphere-ocean general circulation models (GCMs) produce meteorological forecasts, revealing increased global temperatures and a more intense hydrological cycle. Downscaling techniques are then used to link these large-scale models with localized hydrologic models, improving the accuracy of water management strategies (Diaz-Nieto and Wilby 2005; Fowler, Blenkinsop, and Tebaldi 2007; Teutschbein and Seibert 2010). In a study by van Vliet et al. (2013), an increase in mean annual river discharge is expected for high northern latitudes and tropical regions, while a decrease is anticipated in mid-northern and southern latitudes. These changes are more pronounced in GCM experiments using the SRES A2 emissions scenario compared to the B1 scenario due to more significant projected shifts in precipitation and evaporation. Figure 5 shows these variations (van Vliet et al. 2013). In these studies, the application of RS and GIS has been the fundamentals in providing high-resolution data for LULC, topography, and soil type, which are integrated into hydrological models to calibrate and validate the spatial resolution of predictions.

Figure 5. Changes in worldwide river discharge in response to future CC (van Vliet et al. 2013)

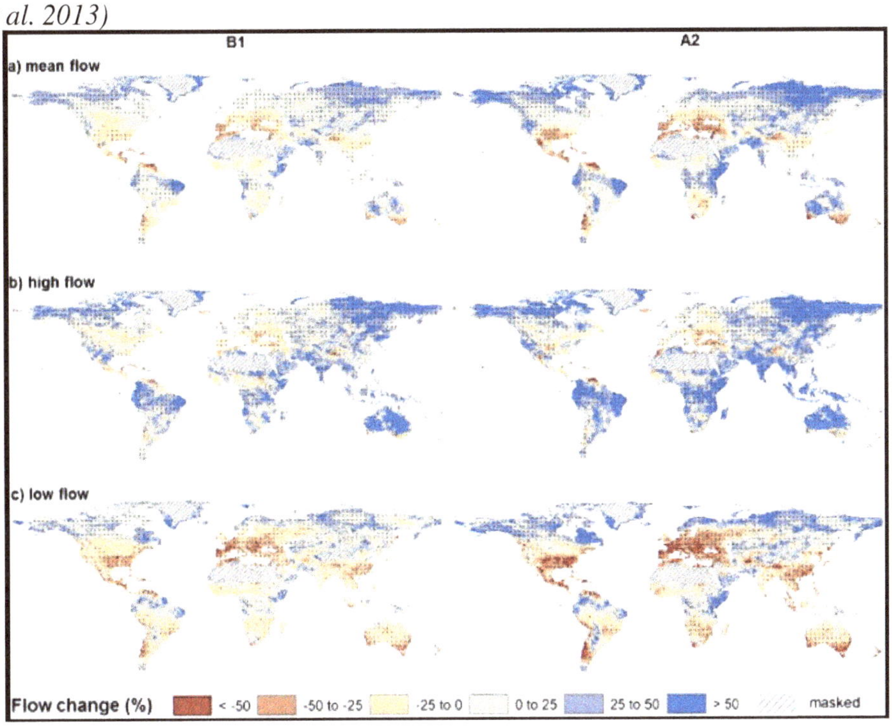

A detailed review of recent studies from 2018-2023 was conducted to understand these impacts and the role of various hydrological models. Table 2 summarizes global studies using hydrological models and climate scenarios to assess CC impacts on streamflow, regional variations' importance, and Thailand's adaptive strategies. These insights are critical for forecasting water availability shifts, flood risks, and sustainable water management in a changing climate.

Table 1. Overview of previous investigations on the impact of CC on streamflow in different regions.

County	Dataset	Climate Models	Climate Scenarios	Hydrological Models	Evaluation Methods	Outcomes/ Effectiveness of Study	Source
Upper Ayeyarwady River Basin, Myanmar	Precipitation, minimum and maximum temperature	CMIP5 models	RCP 2.6, RCP 4.5, RCP 8.5	SWAT	R^2, NSE, PBIAS, and RSR	CC scenarios predict yearly streamflow reductions by specific percentages across RCPs. The study shows that future scenarios show an increasing annual difference in high and low flow variations.	(Oo and Zin 2020)
Kelani River basin, Sri Lanka	Precipitation, minimum and maximum temperature	CanESM2	RCP 2.6, RCP4.5, RCP 8.5	HEC-HMS	Standardized descriptive indices are used to analyze changes	Findings suggested using an alternative hydrological model, such as ArcSWAT, to assess streamflow modifications attributed to CC. This approach is anticipated to offer more comprehensive insights into the associated impacts.	(Dissanayaka and Rajapakse 2019)
Kabul river sub-basin, Afghanistan	Precipitation, minimum and maximum temperature, solar radiation, relative humidity, wind speed, and discharge datasets.	Not applied	Not applied	SWAT-CUP	R^2, NSE, and PBIAS	The study highlights the SWAT model's success in predicting water resource dynamics, emphasizing its effective calibration and validation. It also reveals the significant impact of CC on precipitation, temperature, and streamflow.	(Aawar and Khare 2020)

continued on following page

Table 1. Continued

County	Dataset	Climate Models	Climate Scenarios	Hydrological Models	Evaluation Methods	Outcomes/ Effectiveness of Study	Source
Brahmaputra river basin, India	Precipitation, Min, and max temperature	CMIP5 GCMs (GFDL-ESM2M, HadGEM2-CC, and IPSL-CM5-1R)	RCP 4.5, RCP 8.5	Interpolation and Linear scaling BC method	Trend analysis and used RMSE and NSE for evaluation of GCMs	The study assesses CC effects on the Brahmaputra basin, projecting increased rainfall and temperatures under RCP8.5. Also, HadGEM2-CC was identified as the most appropriate GCM in the study area.	(Dutta et al. 2021)
Miyun Reservoir Basin (MRB), China	Streamflow, temperatures, and precipitation dataset	42 statistical downscaled CMIP5 GCMs	RCP 4.5, RCP 8.5	SWAT	The results are derived from the ensemble mean changes in precipitation and temperature projected for two future periods.	The study reveals future climate impacts on the Miyun Reservoir Basin, anticipating increased warmth, humidity, and precipitation. Projections indicate rising streamflow in summer, with higher probabilities in the shorter term (2021–2035).	(Yan et al. 2019)
Mun River, Thailand	APHRODITE daily precipitation, max and min Temp from GHCN, runoff data DWR	34 CMIP5 models	RCP 2.6, RCP 4.5, RCP 8.5	SWAT	Delta Downscaling	The study showed rising temperatures, especially in the dry season, and changing precipitation patterns that enhance streamflow.	(Li and Fang 2021)
Zarrine River Basin, Iran	Max & min Temp from IRIMO, runoff from Iran Ministry of Energy	CanESM, CCSM4, MRI-CGCM3, MIROC5, CNRM-CM5, CMCC-CM	RCP 4.5, RCP 8.5	SWAT	SDSM, Quantile Mapping	The analysis predicts the freshwater decline in the Basin for 2080–2098 due to CC.	(Emami and Koch 2018)

continued on following page

Table 1. Continued

County	Dataset	Climate Models	Climate Scenarios	Hydrological Models	Evaluation Methods	Outcomes/ Effectiveness of Study	Source
Thailand	Daily PCP from TMD, APHRODITE, Runoff data from RID	MIROC-K1, UKMO-HadCM3, CSIRO-Mk2.0, CGCM2.0, GFDL-R3.0	Not Applied	HEC-HMS	multi-day windows for correction of the coefficient of variation (CV)	Precipitation forecast bias indicated a growing trend in water availability for most catchments, which could increase flood risks.	(Deb, Babel, and Denis 2018)
Andasa watershed, Upper Blue Nile, Ethiopia	Rainfall, maximum temperature, minimum temperature, relative humidity, wind speed, sunshine hours, Ethiopian National Meteorological Services Agency (ENMSA).	GCMs (HadCm3, CanESM2)	H3A2a, H3B2a, canESM2_ rcp4.5, canESM2_ rcp8.5	SWAT	SDSM	Average annual precipitation decreases. While temperatures are expected to rise, precipitation may diminish and become more unpredictable.	(Tarekegn et al. 2021)

6. LULC CHANGES IMPACTS ON STREAMFLOW

Rapid socio-economic development significantly influences LULC changes, such as cropland conversion into urban areas due to increased urbanization (Stonestrom, Scanlon, and Zhang 2009). Researchers are leveraging advanced RS and GIS tools to analyze land use's spatial and temporal dynamics. Various global databases and algorithms have been developed to track changes in land use patterns (Fritz et al. 2011; See and Fritz 2006). Many investigations have been conducted by geospatial and statistical methodologies, including Markov Chain (MC) (Kamusoko et al. 2009), Cellular Automata (CA) (Han et al. 2009; Mitsova, Shuster, and Wang 2011), logistic regression models (Hu and Lo 2007; Wagner and Waske 2016), ML methods (Huang, Ding, and Zhou 2010), and integrated CA-MC models (Arsanjani et al. 2013). These methodologies rely on RS for temporal data and GIS for simulating and visualizing urban expansion and LULC transitions. Recently, Chen, Xu et al. (2019) Utilized freely accessible data and Google Earth Engine's programming environment and generated the FROM-GLC10 global land cover map at a 10-meter resolution based on Sentinel-2 data from 2017, shown in Figure 6 (Chen et al. 2019).

Figure 6. Global land cover map, known as FROM-GLC10, is derived from a 10-meter resolution Sentinel-2(2017) (Chen et al. 2019)

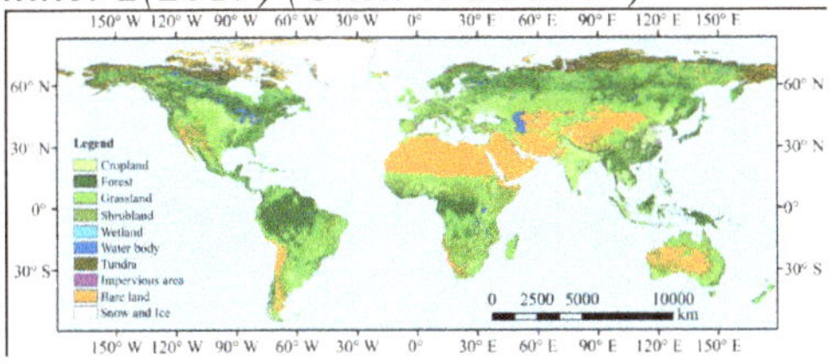

Multitemporal land use classifications are analyzed to track changes over specific periods, whether annually or decadal, due to variations in LULC classifications. Hydrologic models are crucial in assessing how these changes impact water resources historically and in future projections. Table 2 offers a detailed review of research on how LULC changes affect streamflow. Recent studies emphasize the complex interplay between LULC changes and streamflow globally, underscoring the need for integrated water resource management that accounts for these interactions. In Thailand, grasping these impacts is crucial due to its CC vulnerability and rising water demand, guiding the development of strategic policies for sustainable water management and resilience amid changing LULC and climate conditions.

Table 2. Overview of previous investigations on the impact of LULC on streamflow in different regions.

Country/ Region	Dataset	Land Use Classification	Hydrological Models	Land Use Prediction	Outcomes/Effectiveness of Study	Source
Central Chile	• The TanDEM-X DEM (12.5 m resolution) • Land cover classification utilized supervised Landsat OLI satellite data. • Soil dataset. • Daily temperature, precipitation, and discharge data from 2006 to 2019 were used.	The long-term national Forest Policy (FP) is evaluated against several scenarios: extreme afforestation (EX), adaptive plantation management (FM) with lower planting density, forced land displacement (FLD) to headwaters, and a pristine scenario (PR) preserving indigenous forests.	SWAT	Yes	In central Chile, afforestation with exotic pines intensifies water scarcity, whereas preserving and restoring native forests helps mitigate CC impacts on water resources. The study underscores the need to revise land management practices.	(Galleguillos et al. 2021)
Upper Gidabo Watershed in Ethiopia	• Satellite imageries • DEM • climate data (1983–2014) • River discharge records (1986–2006).	The survey identified seven land uses, with agroforestry comprising 64%. The findings indicate a shift towards expanded agricultural and urban areas, reducing forested and grassland areas.	SWAT-CUP algorithms using ArcSWAT	Yes	Changes in LULC affect hydrological responses by enhancing surface runoff and evapotranspiration. Wet season flows rise while dry season flows diminish, necessitating intervention in LULC management and water resource utilization.	(Belihu et al. 2020)
Upper Bhima Basin, India	• Object-based image analysis (OBIA) of satellite imageries for 1992, 2000, and 2009, DEM, climate data, and hydrological records.	OBIA is used for land use classification, identifying significant changes, including conversion from wastelands and agricultural land to built-up areas.	SWAT	Not Applied	This study underscores the importance of understanding the nature and extent of LULC changes for effective water resource management and highlights the need for future research to address emerging environmental challenges, including CC scenarios and model uncertainties.	(Samal and Gedam 2021)

continued on following page

Table 2. Continued

Country/ Region	Dataset	Land Use Classification	Hydrological Models	Land Use Prediction	Outcomes/Effectiveness of Study	Source
Upstream Area of the Yangtze River, China.	• Monthly streamflow • Meteorological data • LULC maps of 1990 and 2005.	LULC changes the transformation matrix	SWAT	Yes	The study reveals an increase in average annual runoff at Zhimenda station, with 98% attributed to CC and 2% to LULC change. Significant impacts on evapotranspiration and runoff are observed, emphasizing the influence of CC on hydrological processes.	(Ahmed et al. 2022)
Naoli River Basin, Northeast China.	• Catchment boundary meteorological data • DEM data • Discharge records • LULC data (1961-2014)	The Naoli River Basin experienced significant wetland loss, with wetland coverage decreasing from 45.8% in 1954 to 9.8% in 2010 and a corresponding increase in agricultural farmland.	NAM (Nedbor Afstromnings Model)	Yes	The findings indicate that hydrological droughts in the marshy Naoli River have increased in frequency and intensity since 1967. The study identifies CC, wetland loss (LULC change), and river dam construction as contributors to streamflow decline, with dam construction having the most significant impact.	(Qi, Xu, and Wang 2020)
Ireland, the United Kingdom, France, Spain, and Portugal (Northeast Atlantic)	• Streamflow data. • Climate, environmental, and Terrain Data.	Mann-Kendall trend test and employed autocorrelation in trend analysis	Regression-based techniques to analyze trends in annual streamflow and assess the significance and magnitude of these trends using the Mann-Kendall statistic and rainfall-runoff models	Not Applied	The study reveals that climate factors, especially changes in land use, explain the observed trends in yearly streamflow in northwest Europe and play a more significant role.	(Vicente-Serrano et al. 2019)

continued on following page

Table 2. Continued

Country/ Region	Dataset	Land Use Classification	Hydrological Models	Land Use Prediction	Outcomes/Effectiveness of Study	Source
Andasa watershed, Upper Blue Nile, Ethiopia	• Landsat 5 TM (1985) • Landsat-7 • ETM+ (2000) • Landsat-8 OLI_ TIRS (2015)	Hybrid classification (unsupervised and supervised), CA-Markov for LU prediction	SWAT	Yes	Historic LU changes, wet season, surface runoff, and water production increased yearly. Between 2030 and 2045, there is projected to be an increase in annual surface runoff and water yield during the wet season, while a decrease is expected in the dry season, groundwater, lateral flow, and evapotranspiration.	(Gashaw et al. 2018)
Krishna River basin, India.	• Meteorological data (1980-2013) • Climate model data (1970-2005) CORDEX-South Asia database. Streamflow data • Details of significant water storage structures	LULC changes were predicted (2025, 2055, and 2085) using the CA Markov Chain model.	SWAT	Yes	The study evaluates the combined influence of climate, LULC changes, and the presence of water storage structures on water balance components. It suggests that LULC changes significantly influence model simulations.	(Chanapathi and Thatikonda 2020)
Chi River Basin, Thailand	• LULC and Soil (LDD), • Observed Discharge (RID)	Land Use Scenarios	SWAT	Based on Scenarios	The study revealed that diverse land use scenarios had distinct impacts on yearly and seasonal water yield and evapotranspiration.	(Homdee, Pongput, and Kanae 2011)
Yom River Basin, Thailand	• Landsat 5 and 7 satellite images (1990, 1994, 1998, 2002, 2006)	Supervised Classification & Visual Interpretation (manually)	MIKE 11/ NAM	Not Applied	Increased urbanization and agricultural areas and decreased forested areas impact discharge (mean and extreme flows).	(Petchprayoon et al. 2010)

7. INTEGRATION OF CC AND LULC CHANGE IMPACTS INTO HYDROLOGICAL MODELING

Hydrological modeling shows that the interaction between climate trends and land use changes significantly affects watershed and river basin dynamics. Climate variability influences temperature, precipitation patterns, and extreme weather events, impacting water availability and hydrological processes (Li et al. 2022). Deforestation, infrastructure development, agriculture, and urbanization modify land use and cover. RS tools, such as the Normalized Difference Vegetation Index (NDVI), can quantify deforestation rates, while GIS correlates these changes with shifts in infiltration rates and streamflow patterns. These alterations impact the environment, potentially affecting evapotranspiration, infiltration, and surface discharge

(Dile, Berndtsson, and Setegn 2013). Hydrological models often integrate historical climate data, climatic projections, and land cover databases. Researchers can better assess streamflow, groundwater replenishment, and water quality by considering current and future climate scenarios and land use changes (Ayalew et al. 2022). In practice, GIS tools are used to model the hydrological impacts of these changes by incorporating topography data and LULC data available from RS to simulate water flow paths and identify potential flooding zones under various LULC scenarios.

Figure 7. Integration of Climate and LULC Changes Scenarios into Hydrological Modeling

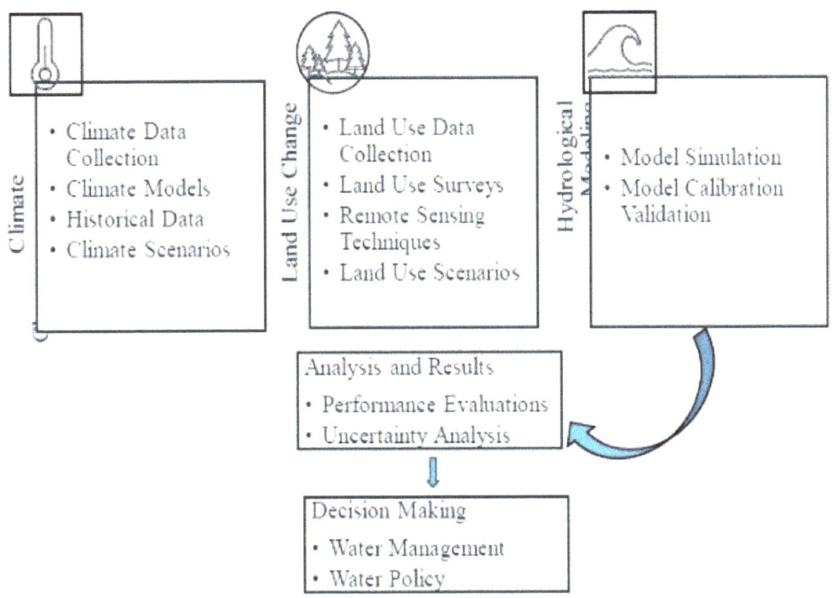

Table 3 summarizes research from tropical regions, including Thailand, Tanzania, Vietnam, Cambodia, Ethiopia, and South Korea, examining the combined effects of CC and LULC on streamflow. These studies employ diverse data sources, climate models, and hydrological models to evaluate their impacts on water resources. In Thailand, RS has been used to track annual forest cover changes, while GIS analysis has assessed the impact of these changes on regional water cycles, including streamflow and water storage (Akpejiori 2018). In Vietnam, RS has identified changes from forested to urban areas, while GIS modeling has been applied to evaluate the impact of these transitions on local hydrology and water resource management systems

(Nguyen 2020). Similarly, in Cambodia, studies have integrated RS-derived LULC data with GIS to explore the hydrological consequences of urban transitions on the Mekong River basin's flood patterns and water quality (Lauri 2012). The findings highlight the importance of understanding the interactions between CC and LULC changes in streamflow, essential for effective global water resource management. RS and GIS offer detailed insights into the spatial and temporal dynamics of environmental changes, supporting the development of region-specific water management strategies. These insights emphasize the need for tailored approaches in different regions, including Thailand, to address these complex dynamics.

Table 3. Previous Studies of Integrating Both CC and Land Use Land Cover Change Impacts on Streamflow

Country/ Region	Data	Climate Model	Climate Scenarios	Hydrological Model	LULC	Downscaling, BC, Model Performance	Effectiveness of Study	Source
Chi River Basin, Thailand	• Primary (runoff) and secondary (rainfall and socio-economic) data. • GIS data • Remote Sensing data	Not Applied	Not Applied	The CLUE-S model is employed for (LULC) change and Hydrological modeling.	CLUE-S model	Not Applied	This study identifies the most economically beneficial land use and cover allocation for flood prevention by 2049 under normal conditions.	(Phinyoyang and Ongsomwang 2021)
Morogoro region in south-central Tanzania	• DEM, soil maps, land use maps • Discharge data	CNRM-CM5 and MIROC5	RCP4.5 and RCP8.5	SWAT	Land Change Modeler (LCM) software	Applied BC for temperature and precipitation in future climate scenarios.	The study reveals a 6–8% decrease in low flows and up to an 84% increase in high flows due to combined LULC and CC scenarios, with CC having a more significant impact.	(Näschen et al. 2019)
Songkhram River Basin (SRB), Thailand	• Max and min temperatures • Observed streamflow. • Sediment data • Topographical data • Land use maps (2002, 2007, 2009, and 2014) • Soil map at the scale of 1:5,000,000	RCMs CMIP5, (ACCESS-CSIRO-CCAM, CNRM-CM5-CSIRO-CCAM, and MPI-ESM-LR-CSIRO-CCAM)	RCP 4.5 and RCP 8.5	SWAT	Two land use change scenarios were developed using the land use change model, Dyna-CLUE.	Not Applied	The study finds that future climate scenarios and LULC changes lead to decreased flow and sediment loading in the SRB.	(Shrestha et al. 2022)

continued on following page

Table 3. Continued

Country/ Region	Data	Climate Model	Climate Scenarios	Hydrological Model	LULC	Downscaling, BC, Model Performance	Effectiveness of Study	Source
Srepok River Basin (SRB) in Vietnam's Central Highlands	• Terrain Data (DEM) • Land-use Land use/cover classification data • Soil (FAO) • Meteorology Data	5 GCMs from IPCC-AR5 (EC-EARTH, GFDL-CM3, HadGEM2-ES, MIROC5, and MPI-ESM-MR)	RCP 4.5	SWAT	The LUCC scenario for 2050 was developed using the SWAT.	Not Applied	Integrating CC with LULC changes increases streamflow by 0.2% to 2.8%, surface runoff by 1.6% to 5.6%, and evapotranspiration by 1.0% to 3.1% while reducing groundwater discharge by 1.5% to 2.7%, offering insights for managing freshwater resources and developing adaptation strategies.	(Nhi et al. 2022)
Siem Reap River Basin, Cambodia	• Min &Max Temp, Precipitation • Daily discharge and water level data • Soil (FAO) • Climatic data • LULC classification data	CanESM2	RCP 2.6, RCP 4.5, and RCP 8.5.	SWAT	MaxLik Supervised Classification and Land Change Modeler (LCM) is used for LULC maps and change projections.	Artificial Neural Network (ANN) downscaling for precipitation and temperature.	The study predicts reduced streamflow and potential water scarcity during the dry season across all scenarios, highlighting the need to consider groundwater levels and their impact on water management for the Angkor Temple Complex.	(Chim et al. 2021)
Chao Phraya Basin, Thailand	• Min &Max Temp, Precipitation, RH, wind speed, sunshine hours, • 96 rain stations, • 14 runoff stations, • RID	ACCESS-ESMI-5 GFDL-ESM4 MIROC6 MPI-ESMI-2-LR MRI-ESM2-0	SSP-RCP SSP126 SSP245 SSP370 SSP585	GBHM	CA-Markov, Biome-BGC	Statistical QM, ML	The study showed that rising precipitation and temperature, a mid-21st century water scarcity issue, increased flood and drought risk.	(Yang et al. 2023)
Rift Valley Lakes Basin, Ethiopia	• Daily Climate Data (NMA), • Streamflow (MoWIE)	Not Applied	Not Applied	SWAT+	Random Forest on R environment	NSE, PBIAS, KGE, RSR	Results show that LULC modification improves surface runoff but decreases infiltration and ET. Climate dynamics benefited ET in all watersheds but hurt surface runoff and infiltration.	(Ayalew et al. 2023)
Gapcheon Watershed, South Korea	• Min &Max Temp, Precipitation, RH, wind speed, sunshine hours	Scenario Based	Not Applied	SWAT	1990, 2018	NSE, R2, KGE, RMSE	The total streamflow increased due to land use change, while the base flow decreased. Based flow is significant in the dry and wet seasons.	(Lee et al. 2023)
Lam Chi watershed, Northeast Thailand	• WATCH, • GSWP2, EFDEI	MIROC5, IPSL, MRI-CGCM3	CMIP5 (2022-2031)	H08 global hydrological model	CLUE	NSE	LULC change may have affected river discharge more than climatic change from historical to modern times. CC is expected to affect river discharge more than land use change shortly.	(Chacuttrikul, Kiguchi, and Oki 2018)

continued on following page

Table 3. Continued

Country/ Region	Data	Climate Model	Climate Scenarios	Hydrological Model	LULC	Downscaling, BC, Model Performance	Effectiveness of Study	Source
Lamtakhong, Northeastern Thailand	• Climate Data (1991-2009) from TMD, RID, DWR	ECHAM5 GCM, PRECIS RCM	Not Applied	SWAT	2002, 2011	SWAT-CUP, NSE, R2	LULC changes had little impact on Lamtakhong Basin hydrologic system yields and flows from 2002 to 2011. CC did. The 56-year CC scenario water flow estimates predict a river basin water shortage.	(Singkran et al. 2015)

8. RESULTS AND DISCUSSION

Based on the selected 30 articles and thorough review, this study finds out the answers of RQs as follows:

RQ-1: Hydrological Models Used to Study CC Impacts on Streamflow.

Various hydrological models used to study the impact of climate change on streamflow, including SWAT, HEC-HMS, and GCMs like CMIP5 and CMIP6, which are applied in regions such as the Ayeyarwady River Basin in Myanmar (Oo and Zin 2020), the Kelani River Basin in Sri Lanka(Dissanayaka and Rajapakse 2019), the Kabul River sub-basin in Afghanistan (Aawar and Khare 2020), and the Brahmaputra River basin in India(Dutta et al. 2021). Each of these models has its strengths and limitations, and the choice often depends on the specific characteristics and requirements of the study area.

RQ-2: Data Sources and Meteorological Variables Integrated into Hydrological Models for CC Impact Analysis.

To analyze the impacts of CC on streamflow, researchers integrate various data sources and meteorological variables into hydrological models. The studies mentioned utilize data from CMIP5 models, SWAT models, GIS, RS, and multiple climate models. Meteorological variables include precipitation, temperature, humidity, and other climate-related parameters. For instance, (Dutta et al. 2021) employed CMIP5 GCMs and a BC method to assess the Brahmaputra River basin in India, focusing on rainfall and temperature projections. These data sources and variables are crucial for accurately simulating the hydrological response to changing climate conditions.

RQ-3: Climate and Land Use Scenarios Utilized to Assess Streamflow Impacts

Climate scenarios, based on emission pathways such as SSPs and RCPs, are commonly utilized to assess the impacts of CC and LULC on streamflow. The reviewed studies provide examples of various river basins, such as in China (Yan et al. 2019), India (Samal and Gedam 2021), and Thailand (Chacuttrikul, Kiguchi, and Oki 2018), where researchers use different scenarios to project changes in streamflow.

These scenarios help understand potential hydrological change trajectories under different future conditions.

RQ-4: Downscaling and Bias-Correction Methods to Enhance Climate Model Accuracy

Downscaling approaches are employed to enhance the accuracy of climate models. The text mentions various downscaling methods, including Delta Downscaling (Li and Fang 2021) and statistical downscaled CMIP5 GCMs (Yan et al. 2019). BC methods are also applied, as seen in the study on the Brahmaputra River basin (Dutta et al. 2021). These methods are crucial for bridging the gap between macroscopic climate models and more localized hydrological models, ensuring that the projections are reliable and applicable at more minor scales.

RQ-5: Integration of CC and Land Use Change Scenarios into Hydrological Models

Integrating CC and land use change scenarios into hydrological models provides a more comprehensive understanding of streamflow alterations in a changing environment. The reviewed studies emphasize the intricate relationship between climate, land use, and hydrological processes. For example, in Thailand, where the vulnerability to CC is high, the information gained from these integrated models is vital for developing effective policies and management strategies. The studies in different regions, including Thailand, Tanzania, Vietnam, Cambodia, Ethiopia, and South Korea, showcase the need for tailored and integrated water resource management strategies that account for the complex interplay between land use changes and climate dynamics. These studies discussed in the text employ a variety of hydrological models, data sources, meteorological variables, CC and LULC change scenarios, and downscaling and bias-correction methods to analyze the impacts of CC and LULC comprehensive changes on streamflow. Integrating these factors into hydrological models contributes to a nuanced understanding of the complex interactions in a changing environment.

RQ-6: Monitoring and Predicting Streamflow Changes due to CC and LU with RS and GIS

RS and GIS significantly enhance the monitoring and prediction of streamflow changes resulting from CC and LU changes. RS technologies, including satellite imagery and aerial sensors, provide critical data on terrain, vegetation, and surface cover changes with high temporal resolution. This data is vital for identifying deforestation, urbanization, and agricultural expansion trends that influence hydrological processes and streamflow patterns (Tarekegn et al. 2021). GIS integrates these diverse spatial datasets with hydrological models, enabling detailed analysis and visualization of how CC and LU changes impact river systems. For instance, studies in the Amazon basin have applied RS data within GIS frameworks to predict the impacts of deforestation and climate variability on streamflow, highlighting significant reductions in river discharge (de Oliveira 2021). By offering a robust means to assess and predict environmental impacts on water resources, RS and GIS are critical fundamental applications in devising adaptive strategies to mitigate the adverse effects of CC and LU changes on global water dynamics.

9. CONCLUSION

This review investigates the impacts of CC and LULC on streamflow and reveals a complex and interdependent relationship between climate dynamics and land use alterations. This interaction highlights the importance of understanding these factors for effective water resource management in regions highly vulnerable to CC. The hydrological models employed, such as SWAT and HEC-HMS, have been invaluable in simulating and predicting streamflow variations, while GCMs from CMIP5 and CMIP6 provide essential climate projections that underpin these hydrological assessments. The integration of various scenarios, including RCPs and LULC change scenarios, offers critical insights into future water resource challenges. Such scenario-based analyses emphasize the importance of developing adaptive and region-specific strategies to manage water resources sustainably, considering the diverse impacts of CC on precipitation, temperature, and hydrological patterns across different geographical areas. The observed variability in streamflow across river basins further reinforces the need for tailored management approaches that address each region's unique environmental and socio-economic contexts. RS and GIS have emerged as indispensable tools in enhancing the precision and scope of hydrological studies. RS provides high-resolution spatial and temporal data, which is crucial for accurately capturing changes in real-time land use and environmental conditions. GIS, in turn, facilitates the integration of these data into hydrological models, improving the analysis and visualization of CC and LULC impacts across multiple scales. This integration allows for more refined predictions and enables adaptive management strategies responsive to evolving land and climate conditions.

Future research should focus on hydrological models to better represent the complexities of CC and LULC impacts on streamflow. Incorporating ML and DL techniques holds promise for capturing nonlinear relationships within hydrological systems.

REFERENCES

Aawar, T., & Khare, D. (2020). Assessment of climate change impacts on streamflow through hydrological model using SWAT model: A case study of Afghanistan. *Modeling Earth Systems and Environment*, 6(3), 1427–1437. DOI: 10.1007/s40808-020-00759-0

Ahmed, N., Wang, G., Booij, M. J., Xiangyang, S., Hussain, F., & Nabi, G. (2022). Separation of the impact of landuse/landcover change and climate change on runoff in the upstream area of the Yangtze River, China. *Water Resources Management*, 36(1), 1–21. DOI: 10.1007/s11269-021-03021-z

Akpejiori, I. J. (2018). Assessing the impacts of land management practices on surface runoff and soil erosion in Ham-Chun Watershed, Northern Thailand (Master's thesis, University of Twente). 2018.

Arnell, N. W. (1999). Climate change and global water resources. *Global Environmental Change*, 9, S31–S49. DOI: 10.1016/S0959-3780(99)00017-5

Arsanjani, J. J., Helbich, M., Kainz, W., & Boloorani, A. D. (2013). Integration of logistic regression, Markov chain and cellular automata models to simulate urban expansion. *International Journal of Applied Earth Observation and Geoinformation*, 21, 265–275. DOI: 10.1016/j.jag.2011.12.014

Arunrat, N., Wang, C., Pumijumnong, N., Sereenonchai, S., & Cai, W. (2017). 'Farmers' intention and decision to adapt to climate change: A case study in the Yom and Nan basins, Phichit province of Thailand'. *Journal of Cleaner Production*, 143, 672–685. DOI: 10.1016/j.jclepro.2016.12.058

Ayalew, A. D., Wagner, P. D., Sahlu, D., & Fohrer, N. (2022). Land use change and climate dynamics in the Rift Valley Lake Basin, Ethiopia. *Environmental Monitoring and Assessment*, 194(11), 791. DOI: 10.1007/s10661-022-10393-1 PMID: 36107274

Ayalew, A. D., Wagner, P. D., Tigabu, T. B., Sahlu, D., & Fohrer, N. (2023). Hydrological responses to land use and land cover change and climate dynamics in the Rift Valley Lakes Basin, Ethiopia. *Journal of Water and Climate Change*, 14(8), 2788–2807. DOI: 10.2166/wcc.2023.138

Bayer, A. D., Fuchs, R., Mey, R., Krause, A., Verburg, P. H., Anthoni, P., & Arneth, A. (2021). Diverging land-use projections cause large variability in their impacts on ecosystems and related indicators for ecosystem services. *Earth System Dynamics: ESD*, 12(1), 327–351. DOI: 10.5194/esd-12-327-2021

Belihu, M., Tekleab, S., Abate, B., & Bewket, W. (2020). Hydrologic response to land use land cover change in the Upper Gidabo Watershed, Rift Valley Lakes Basin, Ethiopia. *HydroResearch*, 3, 85–94. DOI: 10.1016/j.hydres.2020.07.001

Bradshaw, Sodhi, Peh, & Brook. (2007). Global evidence that deforestation amplifies flood risk and severity in the developing world. *Global Change Biology*, *13*, 2379-95.

Chacuttrikul, P., Kiguchi, M., & Oki, T. (2018). Impacts of climate and land use changes on river discharge in a small watershed: A case study of the Lam Chi sub-watershed, northeast Thailand. *Hydrological Research Letters*, 12(2), 7–13. DOI: 10.3178/hrl.12.7

Chanapathi, T., & Thatikonda, S. (2020). Investigating the impact of climate and land-use land cover changes on hydrological predictions over the Krishna river basin under present and future scenarios. *The Science of the Total Environment*, 721, 137736. DOI: 10.1016/j.scitotenv.2020.137736 PMID: 32169648

Change, IPCC Climate. (2019). Land: An IPCC Special Report on Climate Change. *Desertification, Land Degradation, Sustainable Land Management, Food Security, and Greenhouse Gas Fluxes in Terrestrial Ecosystems.*

Chen, , Xu, Zhu, Yuan, Suen, Guo, Xu, Li, Zhao, & Yang. (2019). Stable classification with limited sample: Transferring a 30-m resolution sample set collected in 2015 to mapping 10-m resolution global land cover in 2017. *Science Bulletin*, 64, 3. PMID: 36659725

Chim, K., Tunnicliffe, J., Shamseldin, A. Y., & Bun, H. (2021). Assessment of land use and climate change effects on hydrology in the upper Siem Reap River and Angkor Temple Complex, Cambodia. *Environmental Development*, 39, 100615. DOI: 10.1016/j.envdev.2021.100615

Dang, T. D., Cochrane, T. A., Arias, M. E., Van, P. D. T., & de Vries, T. T. (2016). Hydrological alterations from water infrastructure development in the Mekong floodplains. *Hydrological Processes*, 30(21), 3824–3838. DOI: 10.1002/hyp.10894

de Oliveira, R. G., Júnior, L. C. G. V., da Silva, J. B., Espíndola, D. A., Lopes, R. D., Nogueira, J. S., & Rodrigues, T. R. (2021). Temporal trend changes in reference evapotranspiration contrasting different land uses in southern Amazon basin. Agricultural Water Management, 250.

Deb, P., Babel, M. S., & Denis, A. F. (2018). Multi-GCMs approach for assessing climate change impact on water resources in Thailand. *Modeling Earth Systems and Environment*, 4(2), 825–839. DOI: 10.1007/s40808-018-0428-y

Diaz-Nieto, J., & Wilby, R. L. (2005). A comparison of statistical downscaling and climate change factor methods: Impacts on low flows in the River Thames, United Kingdom. *Climatic Change*, 69(2-3), 245–268. DOI: 10.1007/s10584-005-1157-6

Dile, Y. T., Berndtsson, R., & Setegn, S. G. (2013). Hydrological response to climate change for gilgel abay river, in the lake tana basin-upper blue Nile basin of Ethiopia. *PLoS One*, 8(10), e79296. DOI: 10.1371/journal.pone.0079296 PMID: 24250755

Dissanayaka, K. D. C. R., & Rajapakse, R. L. H. L. (2019). Long-term precipitation trends and climate extremes in the Kelani River basin, Sri Lanka, and their impact on streamflow variability under climate change. *Paddy and Water Environment*, 17(2), 281–289. DOI: 10.1007/s10333-019-00721-6

Dlamini, N. S., Kamal, M. R., Soom, M. A. B. M., Mohd, M. S. F., Abdullah, A. F. B., & Hin, L. S. (2017). Modeling potential impacts of climate change on streamflow using projections of the 5th assessment report for the Bernam River Basin, Malaysia. *Water (Basel)*, 9(3), 226. DOI: 10.3390/w9030226

Dutta, P., Hinge, G., Marak, J. D. K., & Sarma, A. K. (2021). Future climate and its impact on streamflow: A case study of the Brahmaputra river basin. *Modeling Earth Systems and Environment*, 7(4), 2475–2490. DOI: 10.1007/s40808-020-01022-2

Eccles, R., Zhang, H., & Hamilton, D. (2019). A review of the effects of climate change on riverine flooding in subtropical and tropical regions. *Journal of Water and Climate Change*, 10(4), 687–707. DOI: 10.2166/wcc.2019.175

Edelman, Alexandra, Andrea Gelding, Elena Konovalov, Rodney McComiskie, Anne Penny, Nicholas Roberts, Shelley Templeman, Dennis Trewin, Mark Ziembicki, and Blair Trewin. 2014. 'State of the Tropics 2014 report'.

Emami, F., & Koch, M. (2018). Evaluation of statistical-downscaling/bias-correction methods to predict hydrologic responses to climate change in the zarrine river basin, Iran. *Climate (Basel)*, 6(2), 30. DOI: 10.3390/cli6020030

Fowler, H. J., Blenkinsop, S., & Tebaldi, C. (2007). Linking climate change modelling to impacts studies: Recent advances in downscaling techniques for hydrological modelling. *International Journal of Climatology*, 27(12), 1547–1578. DOI: 10.1002/joc.1556

Fritz, S., See, L., McCallum, I., Schill, C., Obersteiner, M., Van der Velde, M., Boettcher, H., Havlík, P., & Achard, F. (2011). Highlighting continued uncertainty in global land cover maps for the user community. *Environmental Research Letters*, 6(4), 044005. DOI: 10.1088/1748-9326/6/4/044005

Fujihara, Y., Tanaka, K., Watanabe, T., Nagano, T., & Kojiri, T. (2008). Assessing the impacts of climate change on the water resources of the Seyhan River Basin in Turkey: Use of dynamically downscaled data for hydrologic simulations. *Journal of Hydrology (Amsterdam)*, 353(1-2), 33–48. DOI: 10.1016/j.jhydrol.2008.01.024

Galleguillos, M., Gimeno, F., Puelma, C., Zambrano-Bigiarini, M., Lara, A., & Rojas, M. (2021). Disentangling the effect of future land use strategies and climate change on streamflow in a Mediterranean catchment dominated by tree plantations. *Journal of Hydrology (Amsterdam)*, 595, 126047. DOI: 10.1016/j.jhydrol.2021.126047

Gashaw, T., Tulu, T., Argaw, M., & Worqlul, A. W. (2018). Modeling the hydrological impacts of land use/land cover changes in the Andassa watershed, Blue Nile Basin, Ethiopia. *The Science of the Total Environment*, 619, 1394–1408. DOI: 10.1016/j.scitotenv.2017.11.191 PMID: 29734616

Han, J., Hayashi, Y., Cao, X., & Imura, H. (2009). Application of an integrated system dynamics and cellular automata model for urban growth assessment: A case study of Shanghai, China. *Landscape and Urban Planning*, 91(3), 133–141. DOI: 10.1016/j.landurbplan.2008.12.002

Homdee, Pongput, & Kanae. (2011). Impacts of land cover changes on hydrologic responses: A case study of Chi River Basin, Thailand. *Journal of Japan Society of Civil Engineers, Ser. B1 (Hydraulic Engineering)*, 67, I_31-I_36.

Hu, Z., & Lo, C. P. (2007). Modeling urban growth in Atlanta using logistic regression. *Computers, Environment and Urban Systems*, 31(6), 667–688. DOI: 10.1016/j.compenvurbsys.2006.11.001

Huang, G.-B., Ding, X., & Zhou, H. (2010). Optimization method based extreme learning machine for classification. *Neurocomputing*, 74(1-3), 155–163. DOI: 10.1016/j.neucom.2010.02.019

Jahandideh-Tehrani, M., Zhang, H., Helfer, F., & Yu, Y. (2019). Review of climate change impacts on predicted river streamflow in tropical rivers. *Environmental Monitoring and Assessment*, 191(12), 1–23. DOI: 10.1007/s10661-019-7841-1 PMID: 31732799

Jáuregui, E. (1992). Aspects of monitoring local/regional-climate change in a tropical region. *Atmosfera*, 5.

Kamusoko, C., Aniya, M., Adi, B., & Manjoro, M. (2009). Rural sustainability under threat in Zimbabwe–simulation of future land use/cover changes in the Bindura district based on the Markov-cellular automata model. *Applied Geography (Sevenoaks, England)*, 29(3), 435–447. DOI: 10.1016/j.apgeog.2008.10.002

Koneti, S., Sunkara, S. L., & Roy, P. S. (2018). Hydrological modeling with respect to impact of land-use and land-cover change on the runoff dynamics in Godavari River Basin using the HEC-HMS model. *ISPRS International Journal of Geo-Information*, 7(6), 206. DOI: 10.3390/ijgi7060206

Latrubesse, E. M., Stevaux, J. C., & Sinha, R. (2005). Tropical rivers. *Geomorphology*, 70(3-4), 187–206. DOI: 10.1016/j.geomorph.2005.02.005

Laurance, W. F. (2007). Forests and floods. *Nature*, 449(7161), 409–410. DOI: 10.1038/449409a PMID: 17898753

Lauri, H., de Moel, H., Ward, P. J., Räsänen, T. A., Keskinen, M., & Kummu, M. (2012). Future changes in Mekong River hydrology: impact of climate change and reservoir operation on discharge. Hydrology and Earth system sciences, 16(12), 4603-4619. DOI: 10.5194/hess-16-4603-2012

Lee, J., Park, M., Min, J.-H., & Na, E. H. (2023). Integrated Assessment of the Land Use Change and Climate Change Impact on Baseflow by Using Hydrologic Model. *Sustainability (Basel)*, 15(16), 12465. DOI: 10.3390/su151612465

Li, C., & Fang, H. (2021). Assessment of climate change impacts on the streamflow for the Mun River in the Mekong Basin, Southeast Asia: Using SWAT model. *Catena*, 201, 105199. DOI: 10.1016/j.catena.2021.105199

Li, X., Tan, L., Li, Y., Qi, J., Feng, P., Li, B., Liu, D. L., & Zhang, X. (2022). Effects of global climate change on the hydrological cycle and crop growth under heavily irrigated management–A comparison between CMIP5 and CMIP6. *Computers and Electronics in Agriculture*, 202, 107408. DOI: 10.1016/j.compag.2022.107408

Liu, & Curry. (2010). Accelerated warming of the Southern Ocean and its impacts on the hydrological cycle and sea ice. *Proceedings of the National Academy of Sciences*, 107, 14987-92. DOI: 10.1073/pnas.1003336107

Malhi, Y., & Wright, J. (2004). Spatial patterns and recent trends in the climate of tropical rainforest regions. *Philosophical Transactions of the Royal Society of London. Series B, Biological Sciences*, 359(1443), 311–329. DOI: 10.1098/rstb.2003.1433 PMID: 15212087

Mauser, W., & Bach, H. (2009). PROMET–Large scale distributed hydrological modelling to study the impact of climate change on the water flows of mountain watersheds. *Journal of Hydrology (Amsterdam)*, 376(3-4), 362–377. DOI: 10.1016/j.jhydrol.2009.07.046

Ming, Alison, Isobel Rowell, Sam Lewin, Robert Rouse, Thomas Aubry, and Emma Boland. 2021. 'Key messages from the IPCC AR6 climate science report'.

Mitsova, D., Shuster, W., & Wang, X. (2011). A cellular automata model of land cover change to integrate urban growth with open space conservation. *Landscape and Urban Planning*, 99(2), 141–153. DOI: 10.1016/j.landurbplan.2010.10.001

Näschen, K., Diekkrüger, B., Evers, M., Höllermann, B., Steinbach, S., & Thonfeld, F. (2019). The impact of land use/land cover change (LULCC) on water resources in a tropical catchment in Tanzania under different climate change scenarios. *Sustainability (Basel)*, 11(24), 7083. DOI: 10.3390/su11247083

Nguyen, T. T., Ngo, H. H., Guo, W., Nguyen, H. Q., Luu, C., Dang, K. B., & Zhang, X. (2020). New approach of water quantity vulnerability assessment using satellite images and GIS-based model: An application to a case study in Vietnam. Science of the Total Environment, 737, 139784.

Nhi, P. T. T., Khoi, D. N., Nguyen, T. T. T., Van Ty, T., & Fang, S. (2022). Hydrological impacts of future climate and land use/cover changes in the Lower Mekong Basin: A case study of the Srepok River Basin, Vietnam. *Environmental Monitoring and Assessment*, 194(S2), 768. DOI: 10.1007/s10661-022-10175-9 PMID: 36255530

Oo, H. T., Zin, W. W., & Thin Kyi, C. C. (2020). Analysis of streamflow response to changing climate conditions using SWAT model. *Civil Engineering Journal*, 6(2), 194–209. DOI: 10.28991/cej-2020-03091464

Petchprayoon, P., Blanken, P. D., Ekkawatpanit, C., & Hussein, K. (2010). Hydrological impacts of land use/land cover change in a large river basin in central–northern Thailand. *International Journal of Climatology*, 30(13), 1917–1930. DOI: 10.1002/joc.2131

Phinyoyang, A., & Ongsomwang, S. (2021). Optimizing Land Use and Land Cover Allocation for Flood Mitigation Using Land Use Change and Hydrological Models with Goal Programming, Chaiyaphum, Thailand. *Land (Basel)*, 10(12), 1317. DOI: 10.3390/land10121317

Prestele, R., Arneth, A., Bondeau, A., de Noblet-Ducoudré, N., Pugh, T. A. M., Sitch, S., Stehfest, E., & Verburg, P. H. (2017). Current challenges of implementing anthropogenic land-use and land-cover change in models contributing to climate change assessments. *Earth System Dynamics: ESD*, 8(2), 369–386. DOI: 10.5194/esd-8-369-2017

Qi, P., Xu, Y. J., & Wang, G. (2020). Quantifying the individual contributions of climate change, dam construction, and land use/land cover change to hydrological drought in a marshy river. *Sustainability (Basel)*, 12(9), 3777. DOI: 10.3390/su12093777

Samal, D. R., & Gedam, S. (2021). Assessing the impacts of land use and land cover change on water resources in the Upper Bhima river basin, India. *Environmental Challenges*, 5, 100251. DOI: 10.1016/j.envc.2021.100251

See, L. M., & Fritz, S. (2006). A method to compare and improve land cover datasets: Application to the GLC-2000 and MODIS land cover products. *IEEE Transactions on Geoscience and Remote Sensing*, 44(7), 1740–1746. DOI: 10.1109/TGRS.2006.874750

Shrestha, S., Bhatta, B., Talchabhadel, R., & Salvatore, G. P. V. (2022). Integrated assessment of the landuse change and climate change impacts on the sediment yield in the Songkhram River Basin, Thailand. *Catena*, 209, 105859. DOI: 10.1016/j.catena.2021.105859

Singh, S. K., & Marcy, N. (2017). Comparison of simple and complex hydrological models for predicting catchment discharge under climate change. *AIMS Geosciences*, 3(3), 467–497. DOI: 10.3934/geosci.2017.3.467

Singkran, N., Tosang, J., Waijaroen, D., Intharawichian, N., Vannarart, O., Anantawong, P., Kunta, K., Wisetsopa, P., Tipvong, T., Janjirawuttikul, N., Masthawee, F., Amornpatanawat, S., & Kirtsaeng, S. (2015). Influences of land use and climate changes on hydrologic system in the northeastern river basin of Thailand. *Journal of Water and Climate Change*, 6(2), 325–340. DOI: 10.2166/wcc.2014.127

Stonestrom, D. A., Scanlon, B. R., & Zhang, L. (2009). Introduction to special section on impacts of land use change on water resources. *Water Resources Research*, 45(7), 45. DOI: 10.1029/2009WR007937

Syvitski, Cohen, Kettner, & Brakenridge. (2014). How important and different are tropical rivers? An overview. *Geomorphology*, 227, 5-17.

Tarekegn, N., Abate, B., Muluneh, A., & Dile, Y. (2021). Modeling the impact of climate change on the hydrology of Andasa watershed. *Modeling Earth Systems and Environment*, 1–17.

Teutschbein, C., & Seibert, J. (2010). Regional climate models for hydrological impact studies at the catchment scale: A review of recent modeling strategies. *Geography Compass*, 4(7), 834–860. DOI: 10.1111/j.1749-8198.2010.00357.x

van Vliet, M. T. H., Franssen, W. H. P., Yearsley, J. R., Ludwig, F., Haddeland, I., Lettenmaier, D. P., & Kabat, P. (2013). Global river discharge and water temperature under climate change. *Global Environmental Change*, 23(2), 450–464. DOI: 10.1016/j.gloenvcha.2012.11.002

Vicente-Serrano, S. M., Peña-Gallardo, M., Hannaford, J., Murphy, C., Lorenzo-Lacruz, J., Dominguez-Castro, F., López-Moreno, J. I., Beguería, S., Noguera, I., Harrigan, S., & Vidal, J.-P. (2019). Climate, irrigation, and land cover change explain streamflow trends in countries bordering the northeast Atlantic. *Geophysical Research Letters*, 46(19), 10821–10833. DOI: 10.1029/2019GL084084

Wagner, P. D., & Waske, B. (2016). Importance of spatially distributed hydrologic variables for land use change modeling. *Environmental Modelling & Software*, 83, 245–254. DOI: 10.1016/j.envsoft.2016.06.005

Waqas, M., Bonnet, S., Wannasing, U. H., Hlaing, P. T., Lin, H. A., & Hashim, S. (2023). Assessment of Advanced Artificial Intelligence Techniques for Flood Forecasting. *2023 International Multi-disciplinary Conference in Emerging Research Trends (IMCERT)*, 1-6. DOI: 10.1109/IMCERT57083.2023.10075119

Waqas, M., Humphries, U. W., Wangwongchai, A., Dechpichai, P., & Ahmad, S. (2023). Potential of Artificial Intelligence-Based Techniques for Rainfall Forecasting in Thailand: A Comprehensive Review. *Water (Basel)*, 15(16), 2979. DOI: 10.3390/w15162979

Waqas, M., Shoaib, M., Saifullah, M., Naseem, A., Hashim, S., Ehsan, F., Ali, I., & Khan, A. (2021). Assessment of advanced artificial intelligence techniques for streamflow forecasting in Jhelum River Basin. *Pakistan Journal of Agricultural Research*, 34(3), 580. DOI: 10.17582/journal.pjar/2021/34.3.580.598

Werf, G. R., Randerson, J. T., Giglio, L., Collatz, G. J., Mu, M., Kasibhatla, P. S., Morton, D. C., DeFries, R. S., Jin, Y., & van Leeuwen, T. T. (2010). Global Fire emissions and the contribution of deforestation, savanna, forest, agricultural, and peat fires. *Atmospheric Chemistry and Physics*, 10.

Winkler, K., Fuchs, R., Rounsevell, M., & Herold, M. (2021). Global land use changes are four times greater than previously estimated. *Nature Communications*, 12(1), 2501. DOI: 10.1038/s41467-021-22702-2 PMID: 33976120

Yan, T., Bai, J., Arsenio, T., Liu, J., & Shen, Z. (2019). Future climate change impacts on streamflow and nitrogen exports based on CMIP5 projection in the Miyun Reservoir Basin, China. *Ecohydrology & Hydrobiology*, 19(2), 266–278. DOI: 10.1016/j.ecohyd.2018.09.001

Yang, S., Zhao, B., Yang, D., Wang, T., Yang, Y., Ma, T., & Santisirisomboon, J. (2023). Future changes in water resources, floods and droughts under the joint impact of climate and land-use changes in the Chao Phraya basin, Thailand. *Journal of Hydrology (Amsterdam)*, 620, 129454. DOI: 10.1016/j.jhydrol.2023.129454

Zhang, H., Wang, B., Liu, D. L., Zhang, M., Feng, P., Cheng, L., Yu, Q., & Eamus, D. (2019). Impacts of future climate change on water resource availability of eastern Australia: A case study of the Manning River basin. *Journal of Hydrology (Amsterdam)*, 573, 49–59. DOI: 10.1016/j.jhydrol.2019.03.067

Chapter 4
Detailed JavaScript API for Google Earth Engine to Enhance Monitoring of Surface Water Dynamics in the Drought-Affected Mediterranean Region

Adil Moumane
https://orcid.org/0000-0003-0296-2679
University Ibn Tofail, Morocco

Jamal Al Karkouri
University Ibn Tofail, Morocco

Batchi Mouhcine
https://orcid.org/0000-0002-0915-6592
University Ibn Tofail, Morocco

ABSTRACT

This study focuses on monitoring surface water dynamics in the Mediterranean region, a region increasingly grappling with water scarcity. The authors analyzed five dams—Al Massira and Mansour Eddahbi in Morocco, Sidi Salem in Tunisia, Sau Reservoir in Spain, and Cabril Dam in Portugal—using a custom JavaScript API developed for Google Earth Engine (GEE). This API enabled efficient processing of Landsat satellite imagery to calculate annual average water areas and extract representative water body images. The findings reveal significant reductions in water

DOI: 10.4018/979-8-3693-9651-3.ch004

area across all studied reservoirs, underscoring the impact of prolonged drought conditions on water availability in the region. The Al Massira Dam in Morocco experienced the most pronounced decline. This research emphasizes the critical role of advanced remote sensing tools in supporting water resource management and policymaking in the Mediterranean region.

1. INTRODUCTION

Water scarcity is a growing global challenge, with the Mediterranean region experiencing particularly acute pressures due to climate change, increasing population, and intensive agriculture (Achite et al., 2023; Elair et al., 2023; Hakam et al., 2023; Garrido-Perez et al., 2024; Moreno-de-las-Heras et al., 2023; Ramat et al., 2023; Rossi et al., 2023). The Mediterranean's semi-arid climate, combined with variable precipitation patterns and intense seasonal demands, exacerbates the vulnerability of its water resources. Accurate and timely monitoring of surface water bodies is crucial for understanding these dynamics and informing effective water resource management.

Traditional remote sensing methods, such as satellite imagery, have been used to study surface water changes, employing indices like the Normalized Difference Water Index (NDWI) (Gao, 1996) and the Modified Normalized Difference Water Index (MNDWI) (Xu, 2006). These indices help differentiate water bodies from other land cover types by leveraging the spectral properties of water in various bands. However, these methods face challenges in processing large volumes of data and providing timely information. The sheer size of satellite datasets and the computational power required to process them can be prohibitive, especially for continuous and large-scale monitoring efforts.

Recent advancements in cloud-based platforms, particularly Google Earth Engine (GEE), have revolutionized the analysis of satellite imagery, offering unparalleled efficiency and scalability (Tamiminia et al., 2020; Deng et al., 2019; Sur et al., 2021; Wang et al., 2022). GEE's robust computational infrastructure, coupled with its extensive data repository, empowers researchers to rapidly process and analyze massive volumes of geospatial data with precision. This capability is especially transformative for water resource management, enabling detailed monitoring of surface water dynamics across vast spatial extents and fine temporal resolutions (Garnier et al. 2024; Kazemi Garajeh et al. 2024). By leveraging GEE, researchers can now track changes in surface water availability, quality, and distribution with unprecedented accuracy, supporting more informed decision-making in the context of water conservation and management.

This study leverages GEE to develop a JavaScript API for calculating the annual average area of dams and reservoirs in the Mediterranean region. By focusing on this drought-affected region, we aim to provide insights into the temporal changes of surface water bodies and their relationship with hydrological factors. The API automates the extraction and processing of satellite imagery, applying water indices to delineate water bodies and calculate their areas over time. This automated approach ensures consistent and reproducible results, which are crucial for longitudinal studies and trend analyses.

Understanding these dynamics is crucial for supporting policymakers and water managers in developing effective water management strategies. The insights gained from this research can help identify periods of significant water loss or gain, correlate these changes with climatic and anthropogenic factors, and inform proactive measures to mitigate water scarcity. Moreover, this methodology can be adapted and applied to other regions facing similar challenges, contributing to a broader understanding of global water resource dynamics.

This research contributes to the growing body of knowledge on water resource management by offering a robust methodology for monitoring surface water changes and providing valuable insights into the complex interactions between climate, human activities, and water availability in the Mediterranean region. The integration of advanced remote sensing techniques with cloud-based processing platforms represents a significant step forward in the ability to monitor and manage water resources effectively, addressing one of the most pressing environmental issues of our time.

2. DROUGHT-AFFECTED MEDITERRANEAN REGION

Droughts, characterized by prolonged periods of significantly below-average water supply, have severe consequences for society, the economy, and the environment. To effectively manage water resources and reduce the negative impacts of drought, it's essential to closely track how and where drought conditions change over time. (Kesgin et al. 2024).

The Mediterranean region, which includes parts of Northern Africa and Southern Europe, is particularly vulnerable to severe and prolonged droughts (Derdous et al. 2021). Characterized by its unique climatic conditions, this area experiences a Mediterranean climate with hot, dry summers and mild, wet winters. However, recent years have seen increasingly severe and prolonged drought events, exacerbated by climate change and other environmental factors.

In recent years, Northern Africa has suffered from severe and persistent droughts, leading to critical water shortages. Countries like Egypt, Libya, and Morocco have faced diminished water supplies, negatively impacting their agriculture, industries, and daily lives (El Qorchi et al. 2023; Elfadli et al. 2024; Hadri et al. 2021; Tanarhte et al. 2024). Simultaneously, Southern European nations, including Spain, Italy, and Greece, have endured prolonged dry periods, resulting in severe water scarcity and increased wildfire risks (European Commission. Joint Research Centre. 2020; Gaitán et al. 2020 ; Moravec et al. 2021; Quirós and Fragoso-Campón 2024).

These droughts have profound implications for the region. Reduced water availability affects crop yields, threatens food security, and increases competition for water resources among various sectors. Additionally, prolonged droughts strain water infrastructure, leading to higher costs for water management and potentially increasing the frequency and severity of water-related conflicts.

The Mediterranean region, which includes parts of Northern Africa and Southern Europe, is increasingly experiencing severe and prolonged droughts. This area, characterized by its Mediterranean climate with hot, dry summers and mild, wet winters, is particularly vulnerable to the impacts of climate change. Recent trends suggest that regular droughts could become the Mediterranean's 'new normal,' exacerbating existing water scarcity issues.

In April 2023, the western Mediterranean region experienced an exceptional and unprecedented early heatwave, particularly affecting the Iberian Peninsula and northwestern Africa (Lemus-Canovas et al. 2024).

3. STUDY AREA

The study focuses on four dams located in the Mediterranean region, with two situated in the southern countries of Morocco and Tunisia, and two in the northern countries of Spain and Portugal. These dams have been selected due to their significant vulnerability to drought conditions.

• Al Massira Reservoir, Morocco

Al Massira Reservoir, situated on the Oum Er-Rbia River near the city of Settat, Morocco, is one of the country's largest reservoirs. Constructed in the 1970s, it serves multiple functions, including irrigation, potable water supply (Aïtelghazi et al. 2019; Mahajan and Martinez 2021), and hydroelectric power generation. The reservoir is critical for agriculture, providing water to extensive irrigation schemes in the surrounding regions. Al Massira also plays a role in flood control and supports local biodiversity. However, it faces challenges from sedimentation and fluctuating water levels due to variable rainfall and increasing water demand, which impact its capacity and sustainability.

• Mansour Eddahbi Reservoir, Morocco

The Mansour Eddahbi Reservoir, located on the Draa River in the Ouarzazate province of Morocco, is an essential water resource for the arid regions it serves. Completed in the early 1970s, the reservoir supports agricultural activities, provides drinking water, and helps regulate river flow. It is particularly vital for the Draa oasis regions downstream, which depend on its water for sustaining agriculture and local communities (Moumane et al. 2022). The reservoir is also a part of the region's hydroelectric power generation infrastructure. Like many reservoirs in semi-arid regions, Mansour Eddahbi is affected by issues such as sedimentation, water scarcity, and climate variability, which pose challenges to its long-term sustainability.

• Sidi Salem Dam, Tunisia

The Sidi Salem Dam, located on the Medjerda River near the town of Testour in Tunisia, is one of the country's most important water infrastructures. Constructed in the late 1970s and early 1980s, it plays a key role in irrigation, drinking water supply, and flood control. The reservoir is central to the agricultural productivity of the region, providing water to extensive irrigation systems that support a variety of crops (Bachta and Ben Nasr 2020). Additionally, Sidi Salem helps regulate water flow, mitigating the impacts of both droughts and floods. The dam faces challenges such as sedimentation and water quality issues, exacerbated by climate change and increasing water demand.

• Sau Reservoir (Pantà De Sau), Spain

The Sau Reservoir (Pantà De Sau), situated 80 kilometers (50 miles) inland from Barcelona, is the second-largest reservoir in the region. Established in 1966 by damming the Ter River and submerging the village of Santa Romà de Sau visible when water levels are low (Marcé and Armengol 2009), it provides essential water resources for numerous towns, including Barcelona.

• Cabril Dam, Portugal

Cabril Dam, situated on the Zêzere River in central Portugal, is an important component of the country's hydroelectric power infrastructure. Completed in the early 1950s (Cardoso De Matos 2022), the dam and its reservoir serve multiple purposes, including electricity generation, flood control, and water supply for irrigation. The reservoir is surrounded by scenic landscapes, making it a popular spot for tourism and recreational activities such as boating and fishing. Cabril Dam is also part of a broader network of water management systems in Portugal. The region faces challenges related to water resource management, including the impacts of climate change, which affect rainfall patterns and water availability.

4. METHODS

4.1 GEE JavaScript API for Water Area Calculation

This script is written using Google Earth Engine's JavaScript API and is designed to calculate the average water area within a specified Area of Interest (AOI) over a series of years using Landsat satellite data. Figure 1 summarizes the code as a flowchart. Below is an explanation of each part of the code:

- Area of Interest and Map Centering

```
// Define the area of interest
Map.centerObject(areaOfInterest);
Map.addLayer(areaOfInterest);
```

- This part of the code sets the map's view to center on the `areaOfInterest` and adds a visual layer of this area to the map. `areaOfInterest` is assumed to be a pre-defined variable that specifies the geographic region of interest.

- Date Range Definition

```
// Define the start and end date
var startDate = '1984-01-01';
var endDate = '2024-07-25';
```

- These variables define the temporal range for the analysis, with data being considered from January 1, 2014, to July 25, 2024.

- Function to Calculate Water Area Using NDWI

```
// Function to calculate water area for each image with correct band names
var calculateWaterArea = function(image) {
var ndwi;
// Check the sensor type
var sensor = image.get('SATELLITE');

// Apply correct bands for each Landsat satellite
```

```
if (sensor === 'LANDSAT_5') {
ndwi = image.normalizedDifference(['B2', 'B4']).rename('NDWI');
} else { // For Landsat 8 and 9
ndwi = image.normalizedDifference(['B3', 'B5']).rename('NDWI');
}

// Create NDWI mask
var ndwiThreshold = ndwi.gte(0.0);

// Mask out non-water areas
var waterMask = ndwiThreshold.gt(0);

// Compute the area of water pixels
var waterArea = waterMask.multiply(ee.Image.pixelArea()).rename('Water_Area');

// Return the water area image for this image
return waterArea;
};
```

- NDWI Calculation: The Normalized Difference Water Index (NDWI) is calculated using bands B3 and B5 of the Landsat imagery.
- Mask Creation: The NDWI values are thresholder to distinguish water from non-water areas.
- Water Area Calculation**: The area of pixels identified as water is computed.

• Function to Calculate Average Water Area for Each Year

```
var calculateAverageWaterAreaForYear = function(year, collection) {
var yearStartDate = ee.Date.fromYMD(year, 1, 1);
var yearEndDate = ee.Date.fromYMD(year, 12, 31);

var landsatImageCollection = collection
.filterBounds(areaOfInterest)
.filterDate(yearStartDate, yearEndDate)
.filter(ee.Filter.lt('CLOUD_COVER', 5));

print("Number of images for year" + year + "=", landsatImageCollection.size());

var waterAreaImages = landsatImageCollection.map(calculateWaterArea);
var waterAreaSum = waterAreaImages.sum();

var waterAreaTotal = waterAreaSum.reduceRegion({
```

```
reducer: ee.Reducer.sum(),
geometry: areaOfInterest,
scale: 30,
bestEffort: true
}).get('Water_Area');

var waterAreaTotalKm2 = ee.Number(waterAreaTotal).divide(1e6);
var count = landsatImageCollection.size();
var averageWaterArea = waterAreaTotalKm2.divide(count);

return ee.Feature(null, {
'year': year,
'averageWaterArea_km2': averageWaterArea
});
};
```

- Filtering Image Collection: Filters the Landsat image collection for the specified year, AOI, and cloud cover.
- Water Area Calculation: Applies the `calculateWaterArea` function to each image and sums the results.
- Average Calculation: Computes the average water area for the year.

• Processing Landsat Data for Each Year

```
var results = [];
var startYear = ee.Date(startDate).get('year').getInfo();
var endYear = ee.Date(endDate).get('year').getInfo();

// Loop for Landsat 5 Data (1984-2011)
for (var year = Math.max(startYear, 1984); year <= Math.min(endYear, 2011);
year++) {
var collection = ee.ImageCollection("LANDSAT/LT05/C02/T1_TOA");
var feature = calculateAverageWaterAreaForYear(year, collection);
results.push(feature);
}

// Loop for Landsat 8 and 9 Data (2013-2024)
for (var year = Math.max(startYear, 2013); year <= Math.min(endYear, 2024);
year++) {
var landsat8Collection = ee.ImageCollection("LANDSAT/LC08/C02/T1_
TOA");
```

```
var landsat9Collection = ee.ImageCollection("LANDSAT/LC09/C02/T1_TOA");
var collection = landsat8Collection.merge(landsat9Collection);
var feature = calculateAverageWaterAreaForYear(year, collection);
results.push(feature);
}
```

- Loop for Landsat 5 Data (1984-2011)

This loop iterates over each year from 1984 to 2011 (if those years are within the specified date range). It processes the Landsat 5 satellite data to calculate the average water area for each year. The loop runs from the startYear (which is the maximum of the startYear and 1984, to ensure the start is not before the availability of Landsat 5 data) to the endYear (which is the minimum of the endYear and 2011, as Landsat 5 data is only available until 2011). For each year, the Landsat 5 image collection is filtered for the AOI, date range, and cloud cover. The calculateAverageWaterAreaForYear function is called for each year, which calculates the average water area using the NDWI method. The computed average water area for each year is stored as a Feature in the results array.

- Loop for Landsat 8 and 9 Data (2013-2024)

This loop covers the years from 2013 to 2024, using data from Landsat 8 and 9 satellites to calculate the average water area for each year.

The loop starts from the startYear or 2013 (whichever is later) and goes up to the endYear or 2024 (whichever is earlier). This ensures data processing aligns with the availability of Landsat 8 and 9 data. For each year, data from both Landsat 8 and 9 is merged into a single collection to cover the time period.

Similar to the previous loop, the calculateAverageWaterAreaForYear function calculates the average water area for each year. The results are stored as Feature objects in the results array.

- Output and Visualization

```
var featureCollection = ee.FeatureCollection(results);
print('Feature Collection:', featureCollection);

Export.table.toDrive({
```

```
collection: featureCollection,
description: 'AverageWaterAreaByYear_Landsat5_8_9',
fileFormat: 'CSV'
});

var chart = ui.Chart.feature.byFeature(featureCollection, 'year', 'averageWaterArea_km2')
.setChartType('LineChart')
.setOptions({
title: 'Average Water Area by Year',
hAxis: {title: 'Year'},
vAxis: {title: 'Average Water Area (km^2)'},
lineWidth: 1,
pointSize: 3
});

print(chart);
```

- **Feature Collection**: Combines all calculated features into a collection.
- **Export**: Exports the results as a CSV file to Google Drive.
- **Chart Visualization**: Creates and prints a line chart of the average water area by year.

-Feature Collection: The `ee.FeatureCollection(results)` creates a collection from the variable `results`, which contains features related to water area measurements. This is printed to the console for inspection.

-Export: The `Export.table.toDrive` function exports the feature collection to Google Drive as a CSV file named `AverageWaterAreaByYear_Landsat5_8_9`. This allows you to download the data for further analysis.

-Chart Visualization: The `ui.Chart.feature.byFeature` function generates a line chart displaying the average water area (`averageWaterArea_km2`) over time (`year`). The chart is customized with titles for the axes and visual styles, and then printed to the console for viewing.

4.2 Average NDWI Calculation and Visualization Using GEE API

This API function in Google Earth Engine calculates the average Normalized Difference Water Index (NDWI) over a specified area and time period using Landsat data, then visualizes and exports the results.

- Area of Interest Definition

var areaOfInterest = /* your area of interest geometry */;

- This line defines a variable `areaOfInterest` that represents the geographical area you want to analyze. This area is specified as a geometry, such as a polygon or other shapes.

- Map Centering and Layer Addition

Map.centerObject(areaOfInterest);
Map.addLayer(areaOfInterest);

- Map.centerObject(areaOfInterest); centers the map view on the defined area of interest.
- Map.addLayer(areaOfInterest); adds the area of interest as a layer on the map for visualization.

- Date Range for Analysis

```
var startDate = '1984-09-01';
var endDate = '2024-08-30';
```

- These variables define the time period for which Landsat data will be analyzed, spanning from 1984 to 2024.

- Landsat Collection Selection

```
var getLandsatCollection = function(startDate, endDate) {
if (endDate <= '2013-08-30') {
return ee.ImageCollection('LANDSAT/LT05/C01/T1_TOA')
.filterDate(startDate, endDate)
.filterBounds(areaOfInterest);
} else if (startDate >= '2014-09-01' && endDate <= '2023-08-30') {
return ee.ImageCollection('LANDSAT/LC08/C02/T1_TOA')
.filterDate(startDate, endDate)
.filterBounds(areaOfInterest);
} else if (startDate >= '2023-09-01') {
return ee.ImageCollection('LANDSAT/LC09/C02/T1_TOA')
.filterDate(startDate, endDate)
.filterBounds(areaOfInterest);
} else {
var collection5 = ee.ImageCollection('LANDSAT/LT05/C01/T1_TOA')
.filterDate(startDate, '2013-08-30')
.filterBounds(areaOfInterest);
var collection8 = ee.ImageCollection('LANDSAT/LC08/C02/T1_TOA')
.filterDate('2014-09-01', '2023-08-30')
```

.filterBounds(areaOfInterest);
var collection9 = ee.ImageCollection('LANDSAT/LC09/C02/T1_TOA')
.filterDate('2023-09-01', endDate)
.filterBounds(areaOfInterest);
return collection5.merge(collection8).merge(collection9);
}
};

- This function selects the appropriate Landsat image collection based on the specified date range:

- Landsat 5 (`'LANDSAT/LT05/C01/T1_TOA'`) for dates up to August 30, 2013.

- Landsat 8 (`'LANDSAT/LC08/C02/T1_TOA'`) for dates between September 1, 2014, and August 30, 2023.

- Landsat 9 (`'LANDSAT/LC09/C02/T1_TOA'`) for dates from September 1, 2023, onwards.

- If the date range spans across different satellite missions, it merges the collections.

• .Load and Filter Landsat Data**

var landsatCollection = getLandsatCollection(startDate, endDate)
.filter(ee.Filter.lt('CLOUD_COVER', 5));

- Calls the `getLandsatCollection` function with the start and end dates to get the appropriate Landsat image collection.

- Applies a filter to only include images with less than 5% cloud cover.

• NDWI Calculation Function

```
var calculateNDWI = function(image) {
var ndwi;
var sensor = image.get('SPACECRAFT_ID');
if (sensor === 'LANDSAT_5') {
ndwi = image.normalizedDifference(['B2', 'B4']).rename('NDWI');
} else {
ndwi = image.normalizedDifference(['B3', 'B5']).rename('NDWI');
}
return ndwi;
};
```

- Defines a function `calculateNDWI` that computes the Normalized Difference Water Index (NDWI) for an image.

- For Landsat 5, it uses bands B2 (green) and B4 (near-infrared).

- For Landsat 8 and 9, it uses bands B3 (green) and B5 (near-infrared).

- The NDWI highlights water bodies by using the normalized difference between the green and near-infrared bands.

• NDWI Calculation for the Collection

var ndwiImages = landsatCollection.map(calculateNDWI);

- Applies the `calculateNDWI` function to each image in the Landsat collection, resulting in a new collection of NDWI images.

- Average NDWI Calculation

```
var averageNDWI = ndwiImages.reduce(ee.Reducer.mean());
```

- Computes the average NDWI value across all images in the NDWI image collection.

- Visualization and Export Parameters

```
var ndwiVis = {
min: -1,
max: 1,
palette: ['red', 'yellow', 'green', 'cyan', 'blue']
};
```

- Defines visualization parameters for displaying the NDWI images, specifying the value range and color palette.

- Adding NDWI Layer to the Map

```
Map.addLayer(averageNDWI, ndwiVis, 'Average NDWI');
```

- Adds the average NDWI image to the map for visualization.

- Exporting NDWI Image

```
Export.image.toDrive({
image: averageNDWI,
description: 'average_ndwi_image',
region: areaOfInterest,
scale: 30,
fileFormat: 'GeoTIFF',
});
```

- Exports the average NDWI image to Google Drive as a GeoTIFF file.

- Printing NDWI Image Information

```
print('Average NDWI image:', averageNDWI);
```

- Prints the average NDWI image object to the console, providing additional metadata and information.

Figure 1. Flowchart: (a) Water Area Calculation Process, (b) NDWI Calculation and Visualization Process

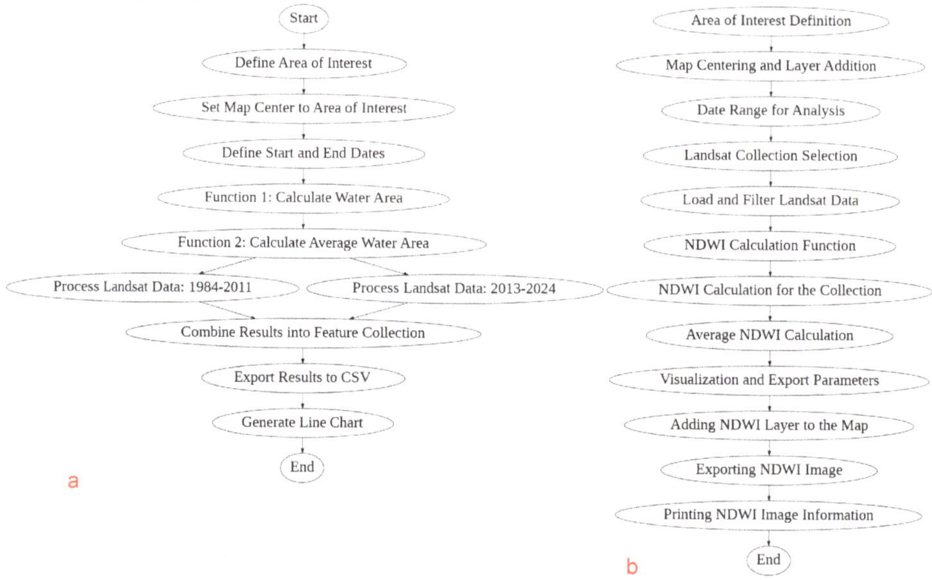

4. RESULTS AND DISCUSSION

4.1 Analysis of Declining Reservoir Levels in Mediterranean Dams

The Al Massira Dam in Morocco has experienced a significant decline in its average water area over the past few decades. Starting with values around 120 km^2 in the mid-1990s, the water area has seen a sharp decrease, especially noticeable after 2010, with the latest figures dropping below 40 km^2 (Figure 2) . This downward trend is indicative of potential challenges such as prolonged drought periods, increased water consumption, or changes in reservoir management practices. The trendline in the graph emphasizes this decline, suggesting that the dam's water storage capacity is being increasingly compromised, which could have serious implications for water supply and agriculture in the region.

The Sidi Salem Dam in Tunisia shows a slight decreasing trend in its average water area, though the fluctuations are less pronounced compared to the Al Massira Dam. The water area has generally ranged between 25 and 30 km^2, with occasional peaks and troughs. The trendline suggests a modest decline, reflecting possibly stable but slightly decreasing water inflows or increased water demand. The average volume remains around 25.69 km^2, indicating a relatively stable but potentially vulnerable water resource, essential for local agriculture, drinking water, and other uses. Since 2021, the Sidi Salem Dam in Tunisia has averaged about 15 km^2, reflecting a decrease of approximately 41.6% from the previous average of 25.69 km^2. This decline can be attributed to the severe drought conditions that have affected Tunisia in recent years. Several authors, such as Sawassi et al. (2024) and Berhail and Katipoğlu (2023), have documented this trend in the region.

The Sau Reservoir in Spain has generally maintained a stable average water area, typically ranging between 2 and 6 km^2. However, there is a slight decreasing trend, suggesting subtle changes in precipitation patterns, water management policies, or consumption rates. The average volume of 3.68 km^2 indicates consistent capacity, but since 2021, the average water area has recorded a significant decline due to consecutive years of record low precipitation, severely impacting northeastern Spain's Catalonia region. According to the Meteorological Service of Catalonia (2024), the year 2023 was the second-driest year on record, only surpassed by 2022. The past three years have seen the lowest rainfall since 1914, with the 2023 drought described as the "worst on record." Conditions remained dire into 2024, with parts of Catalonia receiving less than half of the average rainfall from January through early March, leading to dried-up crops and reservoirs. On February 1, 2024, the Catalan government declared a drought emergency, imposing water restrictions on the region's 6 million residents, businesses, and farmers. The Sau Reservoir, the region's second-largest, has been particularly affected.

The Cabril Dam in Portugal has experienced a decreasing trend in its average water area, similar to the other reservoirs observed. The water area has shown significant variability, peaking around 9-10 km^2 and occasionally dropping below 2 km^2. This variability and the declining trendline suggest potential challenges in water management, possibly linked to factors like changing rainfall patterns, evaporation rates, or increasing water usage. The average volume of 5.82 km^2 indicates a substantial water resource, but the decreasing trend raises concerns about the dam's long-term water storage capability and the sustainability of water supply in the region.

The three reservoirs—Al Massira Dam, Sidi Salem Dam, and Sau Reservoir—exhibit a common trend of declining water areas, though the severity and nature of the decline vary. Al Massira Dam shows a pronounced and worrying decrease, due to factors such as drought and increased water extraction (Mohammed Bounif et al. 2023). These trends underline the importance of sustainable water management practices, regional cooperation, and adaptive strategies to address the challenges posed by changing environmental conditions and growing water demands.

Figure 2. The average water area (in square kilometers) over time for four different dams or reservoirs: Al Massira Dam in Morocco, Sidi Salem Dam in Tunisia, Sau Reservoir in Spain, and Cabril Dam in Portugal. Each graph includes a blue line representing the average water area, a red dashed trendline, and a green line indicating the average volume over the period

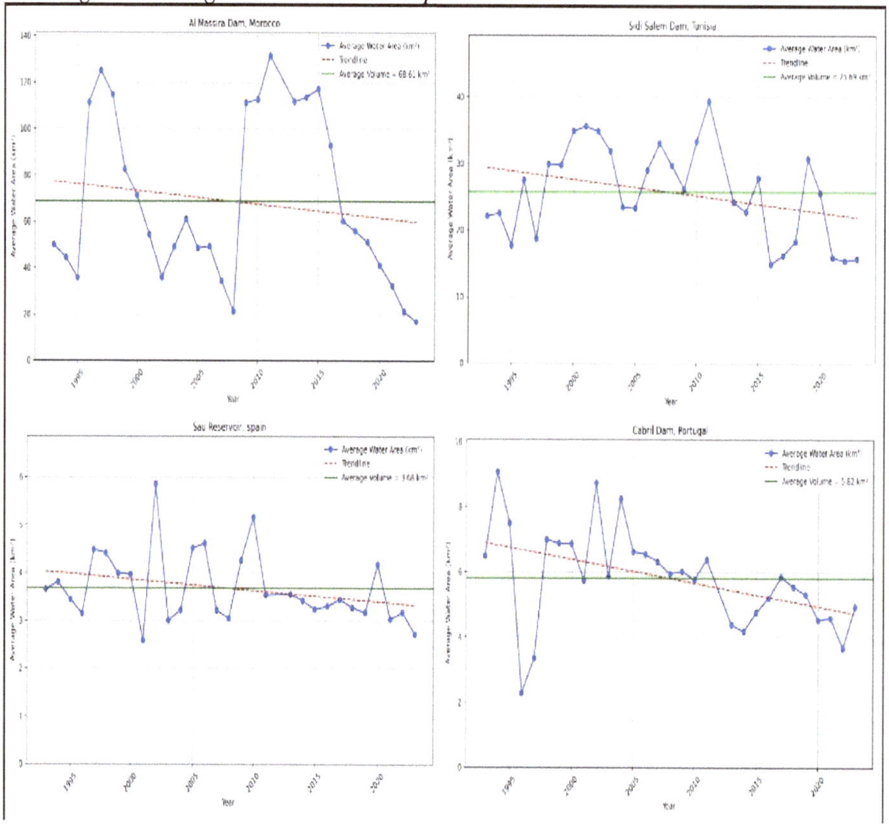

4.2 Comparative Analysis of Water Area Changes Through Mapping

Figure 3 represents the visual overlay of two datasets (average water area from 1993 to 2023 in blue and 2023 in red) on the maps, highlighting changes in water coverage over time. This overlay facilitates a direct comparison, emphasizing areas where water bodies have either shrunk or remained stable.

Al Massira Dam, Morocco: The map shows that the water area in 2023 has significantly decreased compared to the historical average. The blue areas represent the historical water extent, while the red areas show the water extent for 2023. The red overlay on the blue areas indicates a substantial reduction in water levels.

Sidi Salem Dam, Tunisia: - Similar to Al Massira Dam, the map shows a noticeable reduction in water area for 2023 compared to the historical average. The extent of water has decreased, with the 2023 extent (red) being much smaller than the average from previous years (blue).

Sau Reservoir, Spain: The superposition shows a drastic reduction in water area, with the 2023 extent (red) significantly smaller than the average extent (blue). This suggests a severe reduction in water levels in 2023.

Cabril Dam, Portugal: - The map highlights a dramatic decrease in water area in 2023. The red areas are predominant, indicating that the current water levels are much lower than the historical average, as represented by the blue areas.

The superposition of the two datasets visually underscores the trend of decreasing water areas in 2023 across all four locations compared to the historical average from 1993 to 2023. This decline is significant and raises concerns about water resource management, climate change impacts, and environmental sustainability. The visual data clearly communicates the extent of the changes, providing a compelling illustration of the challenges faced by these water bodies.

Figure 3. The visual superposition of the two datasets (average water area from 1993 to 2023 in blue and 2023 in red)

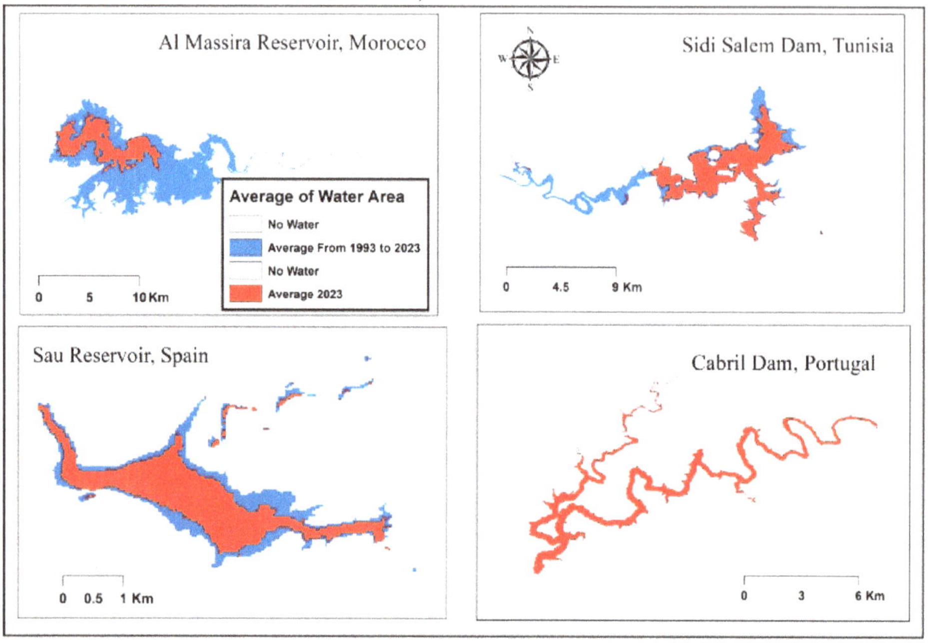

4.3 Drought Impact on Reservoirs: A Visual Perspective

These images visually confirm the trends observed in the earlier water area graphs, highlighting a decline in water surface areas over time for both reservoirs (Figure 4 and 5).

The provided image shows true-color Sentinel-2 satellite images of the Al Massira Dam in Morocco (Images a and b) and the Sidi Salem Dam in Tunisia (Images c and d), comparing the water surface areas between April 2017 and April 2023.

April 2017 (Image a): The image shows a substantial water surface area, indicating relatively high water levels in the reservoir. The water body is extensive, with clear demarcations of the shoreline. In April 2023 (Image b): The water surface area has noticeably reduced, revealing more land and sediment. The decrease in water levels is apparent, with previously submerged areas now exposed.

April 2017 (Image c): The reservoir appears relatively full, with a significant expanse of water visible. The surrounding landscape includes agricultural areas, suggesting the importance of the reservoir for irrigation. In April 2023 (Image d): Similar to the Al Massira Dam, the water surface area in the Sidi Salem Dam has decreased. There are visible reductions in the water level, exposing more of the reservoir bed and reducing the overall water volume.

Figure 4. Declining Water Levels in North African Reservoirs: A Satellite Imagery and Temporal Analysis of Al Massira Dam and Sidi Salem Dam (2017-2023). Image a (April 2017): Al Massira Dam, Morocco: High Water Levels; Image b (April 2023): Al Massira Dam, Morocco: Reduced Water Surface Area; Image c (April 2017): Sidi Salem Dam, Tunisia -: Full Reservoir; Image d (April 2023): Sidi Salem Dam, Tunisia; Declining Water Levels

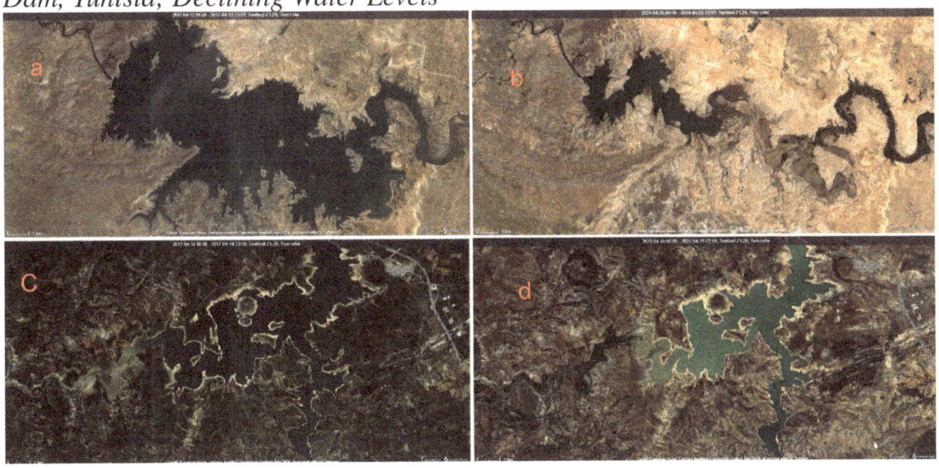

Figure 5. Declining Water Levels in Southern African Reservoirs: A Satellite Imagery and Temporal Analysis of Sau Reservoir and Cabril Dam, Portugal (2017-2023). Image a (April 2017):Sau Reservoir, Spain: Full Reservoir. Image b (April 2023): Sau Reservoir, Spain: Notable Water Decline; Image c (April 2017): Cabril Dam, Portugal - April 2017: Ample Water Levels; Image d (April 2023):Cabril Dam, Portugal - April 2023: Reduced Reservoir Capacity

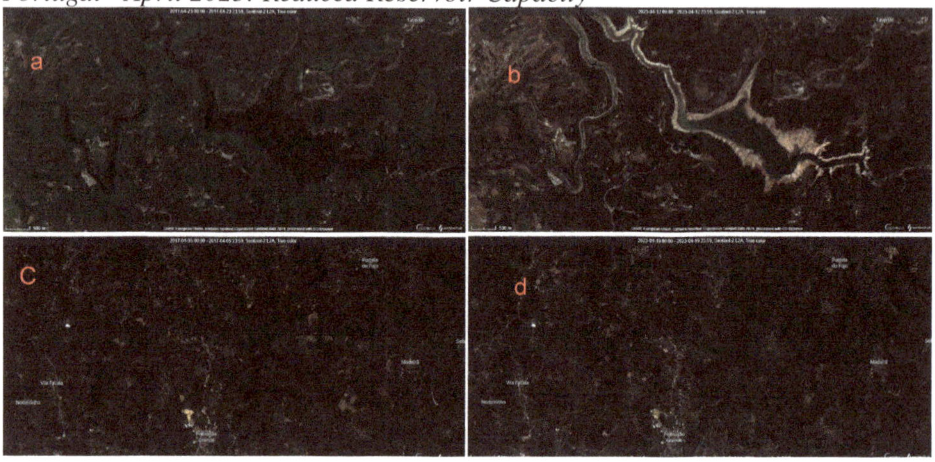

4.4. Validating GEE Water Area Estimates With Inflow Data

To validate the output images generated by the Google Earth Engine (GEE) JavaScript API, we analyzed the correlation between surface water areas derived from GEE and inflow data for the Mansour Dahbi Dam. The inflow data is recorded for each agricultural season, spanning from September of one year to July of the next. Consequently, we adjusted our script to calculate the water area for the corresponding periods of each agricultural season.

The figure 6 represents variation in surface water area (km^2) of Mansour Dahbi Dam, Morocco from 2013/2014 to 2023/2024.

Figure 6 includes two graphical representations: a time series plot (Figure 7a) illustrating the temporal evolution of inflow into and surface water area of the Mansour Dahbi Dam, and a scatter plot (Figure 7b) examining the correlation between these variables. The time series plot reveals fluctuations in both inflow and area over time.

Correlation analysis between inflow and surface water area at the Mansour Dahbi Dam in Morocco demonstrates a complex relationship. While a positive correlation exists, it is not strictly linear, indicating the influence of factors beyond inflow on reservoir levels. The scatter plot, with a correlation coefficient of 0.62, indicates a moderate positive relationship between the two variables. This suggests that as inflow increases, there is a general tendency for the surface water area to expand.

However, the scattered data points in the scatter plot indicate the influence of additional factors beyond inflow on the surface water area. While a general correlation exists, the relationship is not strictly linear, suggesting other influencing factors such as evaporation, seepage, and water extraction. Notably, changes in surface water area often lag behind inflow variations, indicating the reservoir's gradual response. The considerable data variability seen in the scatter plot limits the predictive power of the regression line, highlighting that correlation does not imply causation. Furthermore, the scarcity of comprehensive official data on the studied dams limited the depth of our correlation analysis.

Figure 6. The Variation in Surface Water Area (km^2) of Mansour Dahbi Dam, Morocco from 2013/2014 to 2023/2024

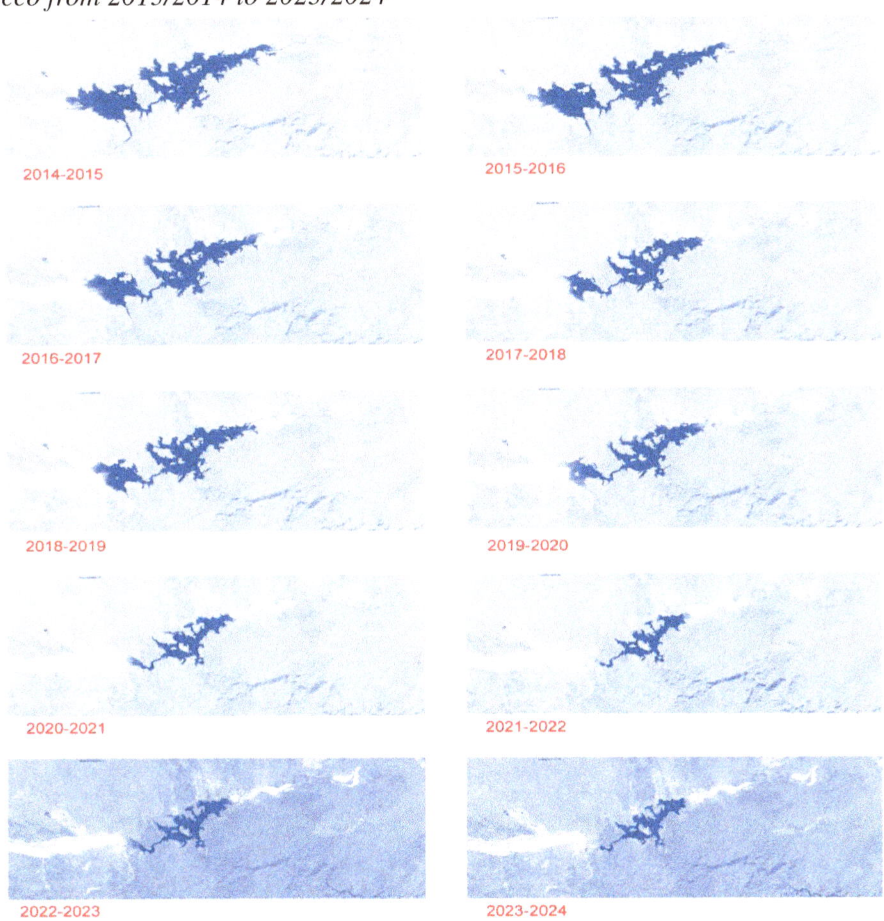

Figure 7. (a) Inflow and Surface Water Area of Mansour Dahbi Dam from 1984/1985 to 2020/2021, Derived from Landsat Imagery. (Figure 6b) the correlation between these variables.

5. CONCLUSION

This study highlights the critical importance of monitoring surface water dynamics in the drought-affected Mediterranean region. Utilizing advanced tools like Google Earth Engine's JavaScript API, we were able to efficiently analyze large-scale satellite data, providing detailed insights into the temporal changes in water bodies. Our analysis reveals a significant decline in water levels across multiple reservoirs in North Africa and Iberia, with the Al Massira Dam in Morocco experiencing the most pronounced decrease.

Correlation analysis between inflow and surface water area at the Mansour Dahbi Dam in Morocco demonstrates a complex relationship. While a positive correlation exists, it is not strictly linear, indicating the influence of factors beyond inflow on reservoir levels. However, the scarcity of comprehensive official data on the studied dams limited the depth of our correlation analysis.

These findings underscore the urgent need for comprehensive water resource management strategies to address the challenges posed by climate change, increasing water demand, and population growth in the region. Implementing sustainable water use practices, investing in water conservation technologies, and exploring alternative water sources are crucial steps to mitigate the impacts of declining water availability.

By providing quantitative evidence of declining water resources, this study offers crucial insights for policymakers and water managers. The findings can inform the development of targeted interventions, such as water allocation policies, infrastructure investments, and early warning systems for drought. Additionally, the identification of factors influencing reservoir levels can support the optimization of water management practices and the exploration of alternative water sources.

Continued monitoring of reservoir levels and hydrological conditions, coupled with the expansion of data collection efforts, is essential for effective water resource planning and decision-making. This work not only contributes to the scientific understanding of hydrological changes but also offers practical tools for ongoing and future monitoring efforts, ensuring that crucial water resources are managed effectively in the face of changing environmental conditions.

Ultimately, this research contributes to building resilience to water scarcity and ensuring water security for future generations.

REFERENCES

Achite, M., Elshaboury, N., Jehanzaib, M., Vishwakarma, D., Pham, Q., Anh, D., Abdelkader, E., & Elbeltagi, A. (2023). Performance of Machine Learning Techniques for Meteorological Drought Forecasting in the Wadi Mina Basin, Algeria. *Water (Basel)*, 15(4), 765. https://doi.org/. DOI: 10.3390/w15040765

Aïtelghazi, A., Rhinane, H., Bensalmia, A., & Giuliani, G. (2019). Using the Landsat-7 data to study the correlation between the surface temperature and phytoplankton turbidity Case study: Al Massira Lake (Settat - Morocco). *Materials Today: Proceedings*, 13, 496–504. DOI: 10.1016/j.matpr.2019.04.006

Bachta, M. S., & Ben Nasr, J. (2020). Water Policy in Tunisia. In Zekri, S. (Ed.), *Water Policies in MENA Countries* (pp. 161–184). Springer International Publishing. DOI: 10.1007/978-3-030-29274-4_8

Berhail, S., & Katipoğlu, O. M. (2023). Spatiotemporal detection of abrupt change in trends of rainfall and dry and wet periods at different time scales: The case of the Medjerda basin in northeast Algeria. *Acta Geophysica*, 71(5), 2497–2516. DOI: 10.1007/s11600-023-01097-2

Bounif, M., Rahimi, A., & Bouassria, A. (2023) Characterization of the temporal variability of climate and climatic drought in the Doukkala irrigated perimeter (Morocco). https://doi.org/DOI: 10.5281/ZENODO.10184704

Cardoso De Matos, A. (2022). The Spread of Scientific Knowledge and Technology Transfer: André Coyne (1891–1960) and the Construction of Dams in 20th Century Portugal. In Borgato, M. T., & Phili, C. (Eds.), *In Foreign Lands: The Migration of Scientists for Political or Economic Reasons* (pp. 203–219). Springer International Publishing. DOI: 10.1007/978-3-030-80249-3_9

Deng, Y., Jiang, W., Tang, Z., Ling, Z., & Wu, Z. (2019). Long-Term Changes of Open-Surface Water Bodies in the Yangtze River Basin Based on the Google Earth Engine Cloud Platform. *Remote Sensing (Basel)*, 11(19), 2213. DOI: 10.3390/rs11192213

Derdous, O., Bouamrane, A., & Mrad, D. (2021). Spatiotemporal analysis of meteorological drought in a Mediterranean dry land: Case of the Cheliff basin–Algeria. *Modeling Earth Systems and Environment*, 7(1), 135–143. DOI: 10.1007/s40808-020-00951-2

El Qorchi, F., Yacoubi Khebiza, M., Omondi, O. A., Karmaoui, A., Pham, Q. B., & Acharki, S. (2023). Analyzing Temporal Patterns of Temperature, Precipitation, and Drought Incidents: A Comprehensive Study of Environmental Trends in the Upper Draa Basin, Morocco. *Water (Basel)*, 15(22), 3906. DOI: 10.3390/w15223906

Elair, C., Rkha Chaham, K., & Hadri, A. (2023). Assessment of drought variability in the Marrakech-Safi region (Morocco) at different time scales using GIS and remote sensing. *Water Science and Technology: Water Supply*, 23(11), 4592–4624. DOI: 10.2166/ws.2023.283

Elfadli, K. I., & Wahab, M. M. A. (2024) Impacts of climate change on drought in Libya. In: Hydroclimatic Extremes in the Middle East and North Africa. Elsevier.

European Commission. Joint Research Centre. (2020) Global warming and drought impacts in the EU: JRC PESETA IV project : Task 7. Publications Office, LU.

Gaitán, E., Monjo, R., Pórtoles, J., & Pino-Otín, M. R. (2020). Impact of climate change on drought in Aragon (NE Spain). *The Science of the Total Environment*, 740, 140094. DOI: 10.1016/j.scitotenv.2020.140094 PMID: 32562994

Gao, B. (1996). NDWI—A normalized difference water index for remote sensing of vegetation liquid water from space. *Remote Sensing of Environment*, 58(3), 257–266. DOI: 10.1016/S0034-4257(96)00067-3

Garnier, J., Cicerelli, R. E., De Almeida, T., Belo, J. C. R., Curto, J., Ramos, A. P. M., Valadão, L. V., Satge, F., & Bonnet, M.-P. (2024). Water Resources Monitoring in a Remote Region: Earth Observation-Based Study of Endorheic Lakes. *Remote Sensing (Basel)*, 16(15), 2790. DOI: 10.3390/rs16152790

Garrido-Perez, J. M., Vicente-Serrano, S. M., Barriopedro, D., García-Herrera, R., Trigo, R., & Beguería, S. (2024). Examining the outstanding Euro-Mediterranean drought of 2021–2022 and its historical context. *Journal of Hydrology (Amsterdam)*, 630, 130653. DOI: 10.1016/j.jhydrol.2024.130653

Hadri, A., Saidi, M. E. M., & Boudhar, A. (2021). Multiscale drought monitoring and comparison using remote sensing in a Mediterranean arid region: A case study from west-central Morocco. *Arabian Journal of Geosciences*, 14(2), 118. DOI: 10.1007/s12517-021-06493-w

Hakam, O., Baali, A., Azennoud, K., Lyazidi, A., & Bourchachen, M. (2023). Assessments of Drought Effects on Plant Production Using Satellite Remote Sensing Technology, GIS and Observed Climate Data in Northwest Morocco, Case of the Lower Sebou Basin. *International Journal of Plant Production*, 17(2), 267–282. DOI: 10.1007/s42106-023-00236-5

Kazemi Garajeh, M., Haji, F., Tohidfar, M., Sadeqi, A., Ahmadi, R., & Kariminejad, N. (2024). Spatiotemporal monitoring of climate change impacts on water resources using an integrated approach of remote sensing and Google Earth Engine. *Scientific Reports*, 14(1), 5469. DOI: 10.1038/s41598-024-56160-9 PMID: 38443699

Kesgin, E., Yaldız, S. G., & Güçlü, Y. S. (2024). Spatiotemporal variability and trends of droughts in the Mediterranean coastal region of Türkiye. *International Journal of Climatology*, 44(4), 1036–1057. DOI: 10.1002/joc.8370

Lemus-Canovas M, Insua-Costa D, Trigo RM, & Miralles DG (2024) Record-shattering 2023 Spring heatwave in western Mediterranean amplified by long-term drought. NPJ Clim Atmos Sci 7:25. DOI: 10.1038/s41612-024-00569-6

Mahajan, S., & Martinez, J. (2021). Water, water, but not everywhere: Analysis of shrinking water bodies using open access satellite data. *International Journal of Sustainable Development and World Ecology*, 28(4), 326–338. DOI: 10.1080/13504509.2020.1851803

Marcé, R., & Armengol, J. (2009). Water Quality in Reservoirs Under a Changing Climate. In Sabater, S., & Barceló, D. (Eds.), *Water Scarcity in the Mediterranean. Springer Berlin Heidelberg, Berlin* (pp. 73–94). DOI: 10.1007/698_2009_38

Meteorological Service of Catalonia. (2024) L'any 2023, el segon més càlid, i dels més secs, URL https://static-m.meteo.cat/wordpressweb/wp-content/uploads/2024/01/02141659/ABCA_2023_ABM_des.pdf

Moravec, V., Markonis, Y., Rakovec, O., Svoboda, M., Trnka, M., Kumar, R., & Hanel, M. (2021). Europe under multi-year droughts: How severe was the 2014–2018 drought period? *Environmental Research Letters*, 16(3), 034062. DOI: 10.1088/1748-9326/abe828

Moreno-de-las-Heras, M., Bochet, E., Vicente-Serrano, S. M., Espigares, T., Molina, M. J., Monleón, V., Nicolau, J. M., Tormo, J., & García-Fayos, P. (2023). Drought conditions, aridity and forest structure control the responses of Iberian holm oak woodlands to extreme droughts: A large-scale remote-sensing exploration in eastern Spain. *The Science of the Total Environment*, 901, 165887. DOI: 10.1016/j.scitotenv.2023.165887 PMID: 37517715

Moumane, A., Al Karkouri, J., Benmansour, A., El Ghazali, F. E., Fico, J., Karmaoui, A., & Batchi, M. (2022). Monitoring long-term land use, land cover change, and desertification in the Ternata oasis, Middle Draa Valley, Morocco. *Remote Sensing Applications: Society and Environment*, 26, 100745. DOI: 10.1016/j.rsase.2022.100745

Quirós, E., & Fragoso-Campón, L. (2024). Spatial analysis of remote sensing and meteorological indices in a drought event in southwestern Spain. *Theoretical and Applied Climatology*, 155(5), 3757–3770. DOI: 10.1007/s00704-024-04846-5

Ramat, G., Santi, E., Paloscia, S., Fontanelli, G., Pettinato, S., Santurri, L., Souissi, N., Da Ponte, E., Abdel Wahab, M. M., Khalil, A. A., Essa, Y. H., Ouessar, M., Dhaou, H., Sghaier, A., Hachani, A., Kassouk, Z., & Lili Chabaane, Z. (2023). Remote sensing techniques for water management and climate change monitoring in drought areas: Case studies in Egypt and Tunisia. *European Journal of Remote Sensing*, 56(1), 2157335. DOI: 10.1080/22797254.2022.2157335

Rossi, L., Naumann, G., Gabellani, S., & Cammalleri, C. (2023). A combined index to characterize agricultural drought in Italy at municipality scale. *Journal of Hydrology. Regional Studies*, 47, 101404. DOI: 10.1016/j.ejrh.2023.101404

Sawassi, A., Khadra, R., & Crookston, B. (2024). Water Banking as a Strategy for the Management and Conservation of a Critical Resource: A Case Study from Tunisia's Medjerda River Basin (MRB). *Sustainability (Basel)*, 16(9), 3875. DOI: 10.3390/su16093875

Sur, K., Verma, V. K., & Pateriya, B. (2021). Surface water estimation at regional scale using hybrid techniques in GEE environment-A case study on Punjab State of India. *Remote Sensing Applications: Society and Environment*, 24, 100625. DOI: 10.1016/j.rsase.2021.100625

Tamiminia, H., Salehi, B., Mahdianpari, M., Quackenbush, L., Adeli, S., & Brisco, B. (2020). Google Earth Engine for geo-big data applications: A meta-analysis and systematic review. *ISPRS Journal of Photogrammetry and Remote Sensing*, 164, 152–170. DOI: 10.1016/j.isprsjprs.2020.04.001

Tanarhte, M., De Vries, A. J., Zittis, G., & Chfadi, T. (2024). Severe droughts in North Africa: A review of drivers, impacts and management. *Earth-Science Reviews*, 250, 104701. DOI: 10.1016/j.earscirev.2024.104701

Tramblay Y, Llasat MC, Randin C, & Coppola E (2020) Climate change impacts on water resources in the Mediterranean. Reg Environ Change 20:83, s10113-020-01665-y. DOI: 10.1007/s10113-020-01665-y

Wang, Q., Song, K., Xiao, X., Jacinthe, P.-A., Wen, Z., Zhao, F., Tao, H., Li, S., Shang, Y., Wang, Y., & Liu, G. (2022). Mapping water clarity in North American lakes and reservoirs using Landsat images on the GEE platform with the RGRB model. *ISPRS Journal of Photogrammetry and Remote Sensing*, 194, 39–57. DOI: 10.1016/j.isprsjprs.2022.09.014

Wu, F., Yang, X., Cui, Z., Ren, L., Jiang, S., Liu, Y., & Yuan, S. (2024). The impact of human activities on blue-green water resources and quantification of water resource scarcity in the Yangtze River Basin. *The Science of the Total Environment*, 909, 168550. DOI: 10.1016/j.scitotenv.2023.168550 PMID: 37979857

Xu, H. (2006). Modification of normalised difference water index (NDWI) to enhance open water features in remotely sensed imagery. *International Journal of Remote Sensing*, 27(14), 3025–3033. DOI: 10.1080/01431160600589179

Chapter 5
Boosting Artificial Intelligence Performance in Sentinel-2 Imagery Analysis:
A DL Approach for Water Bodies Detection

Abdessamad Elmotawakkil
https://orcid.org/0009-0009-3395-1009
Department of Computer Science, University Ibn Tofail, Kenitra, Morocco

Nourddine Enneya
Department of Computer Science, Faculty of Sciences, University Ibn Tofail, Kenitra, Morocco

ABSTRACT

The combination of remote sensing with AI methods such as deep learning, along with tools like geographic information systems, has transformed both spatial analysis and how decisions are made in a variety of fields. In this chapter, the authors show how important these AI technologies and techniques can be for improving the accuracy of information produced by remote sensors when it comes to monitoring water resources managing them sustainably, assessing flooding risks and environmental monitoring, thus contributing to a more informed and sustainable future. Accurate identification of water bodies using satellite images is crucial for purposes such as managing water resources, the environment, urban planning, and reacting to disasters. This study evaluates the performance of three advanced deep learning models DeepLabV3+, U-Net, and FCN in the semantic segmentation when looking

DOI: 10.4018/979-8-3693-9651-3.ch005

at water bodies in pictures taken by Sentinel-2 satellites. The three models were assessed by metrics of accuracy, recall, precision, F1score, and Dice coefficient. The best performer was U-Net: it scored highest overall.

INTRODUCTION

The analysis of spatial data has been fundamentally transformed by the integration of artificial intelligence (AI), Geographic Information Systems (GIS), and remote sensing (RS). Janga et al., 2023. These technologies allow us to access geospatial information as never before from its acquisition right through interpretation leading to progress in areas like environmental monitoring, urban planning disaster response management for water resources and public health. Remote sensing provides a means of obtaining information about earths surface or atmosphere from a distance. Satellite imagery has been especially useful for monitoring Hu et al., 2023 an array of phenomena with great detail accuracy (land cover changes, vegetation patterns, water bodies, atmospheric properties).GIS enables storage manipulation analysis visualization spatial datasets it merges information from different sources so users can identify trends relationships patterns that help with decision-making processes Juhász, 2024. By using GIS alongside remote-sensing techniques new possibilities open up furthering our ability to solve problems. But when AI specifically deep learning (DL) techniques are included the full potential unleashed Pierdicca and Paolanti, 2022. DL subset machine learning methods has changed how we analyse information including geospatial details . These complex mathematical algorithms learn automatically from large amounts data such as patterns enabling prediction classification tasks be undertaken accurately. Combining all three models has opened doors for spatial analysis development fresh solutions complex real-world problems there is even greater capability than any single system by itself Fu et al., 2024.The Deep learning models such as Convolutional Neural Networks have shown impressive capabilities in extracting useful information from satellite and aerial images . This allows for more precise and efficient analysis of things like land use, bodies of water, urban areas and natural resources. In this study, researchers focus on analysing data that includes pictures of water taken by the Sentinel-2 satellite Drusch et al., 2012. There are also black and white images created using something called the Normalized Difference Water Index (NDWI). This method is often used to find out if there is plant life in a body of water McFeeters, 1996.The pictures are split into two groups: training images and testing images. 80% are for training, and the remaining 20% are for testing after researchers make sure the data is standardised so that they can compare different sets of results fairly. The article compare three types of cutting-edge deep learning model: DeepLabV3+, U-Net, and Fully Convolutional Network (FCN). All have been very successful at various tasks within computer vision. DeepLabV3+ etc have all been very successful at lots of jobs within computer vision Chen et al., 2018a; Long et al., 2015a tasks that involve making sense of images. It could be really helpful to use them with satellite data too such a lot, Getting satellite images to show where water ends and land begins is important for many projects. People might want to know the borders of rivers and lakes for activities including managing water supplies, working out which areas are at risk of flooding or carrying out environmental checks along coastlines before building there. By using advanced "deep learning" techniques, we can get computers to help us understand more from these pictures than ever before. Waterways are crucial for lots of different things: the health of our environment, town planning decisions plus it helps if we know exactly where the edges of things are when looking down from space onto earths surface Zhang et al., 2023.These kinds of calculations were already difficult and time consuming when experts had to do them all by hand. It will be shown how well each method works with this particular set of information by comparing results against maps that have already been drawn specially for this project. The researchers give numbers to prove which system is best but just showing examples of final images produced at each stage might also be useful for readers wanting an idea of how methods differ Taha and Hadi, 2019. The F1 score which blends recall and precision into a single number gives an overall gauge of model accuracy. Meantime, Dice coefficient indicates how much predicted segmentation masks resemble their true counterparts. Researchers combine GIS data with artificial intelligence techniques augmented by satellite data in order to provide readers with up-to-date information about environmental changes plus insights useful for managing freshwater supplies Cao et al., 2024. They hope not only to help specialists better understand those trends but also to arm policymakers with facts that may enable them to make decisions likely to affect lots of people; further work could include creating tools available via the internet so professionals around the world can take advantage of our findings for their own projects!

Related Works

The literature review highlights recent advancements in deep learning approaches for water body extraction and classification. Various models, including Visual Foundation Models (VFMs) such as SAM and CLIP, the Faster R-CNN algorithm, and specialized frameworks like WaterDeep and WaterSegDiff, demonstrate significant improvements in accuracy and efficiency across different environmental contexts, from highmountain lakes to urban water features. These methodologies showcase the increasing capability of deep learning models in precise and reliable water body mapping.

Gharbia, 2023 presents an automated method for detecting water bodies in satellite imagery utilizing the Faster R-CNN algorithm. This approach was evaluated on two datasets comprising water body images derived from Sentinel-2 and Landsat-8 (OLI) satellite imagery, encompassing a total of over 3500 images. The results demonstrate that the proposed method achieved accuracies of 98.7% and 96.1% for the two respective datasets, markedly outperforming the convolutional neural network (CNN) approach, which attained accuracies of 96% and 80% on the same datasets. These outcomes underscore the efficacy of the proposed method in precisely mapping water bodies from satellite images. Furthermore, the Sentinel-2 dataset outperformed the Landsat dataset in both the Faster R-CNN and CNN methods for water body detection.

In Ozdemir et al., 2024, a novel methodology for water body extraction and classification is presented, utilizing recent developments in deep learning, particularly Visual Foundation Models (VFMs) such as the Segment Anything Model (SAM) and Contrastive Language-Image Pre-training (CLIP). This innovative two-step approach first employs SAM's Automatic Mask Generator for image segmentation and then applies zero-shot classification with CLIP. The effectiveness of this methodology is evidenced by its remarkable performance in analyzing remote sensing and aerial imagery, achieving a mean Intersection over Union (IoU) of 94.41% and an F1 score of 96.97% for satellite images, and a mean IoU of 90.83% with an F1 score exceeding 94.56% for aerial images. These results highlight the method's superior capability in accurately identifying water bodies in complex scenarios.

Perez-Torres et al., 2024 explores the challenges of lake monitoring in high-mountain regions, focusing on the Ancash and Cuzco areas of the Peruvian Andes. It compares three deep learning models for lake segmentation: the established DeepWaterMapV2 and WatNet models, and the newly adapted WaterSegDiff model, which integrates diffusion and transformation mechanisms tailored for lake segmentation. Additionally, the Normalized Difference Water Index (NDWI) with Otsu thresholding is used as a benchmark. The study leverages a new dataset of Landsat-8 multispectral imagery from 2013 to 2023 and evaluates the models using metrics such as Mean Intersection over Union (MIoU), Pixel Accuracy (PA), and F1 Score. The findings reveal that DeepWaterMapV2 and WatNet perform equivalently well in the challenging high-mountain environment, with WatNet's lower computational complexity making it particularly suitable for implementation. Furthermore, a detailed temporal analysis of Lake Singrenacocha highlighted the robust performance of the WatNet model.

Sun et al., 2024 presents WaterDeep, a deep learning framework inspired by DeepLabV3+, designed to address the challenges of accurately delineating urban water features in high-resolution remote sensing imagery. By integrating multi-scale data through a fusion of highand low-level features, WaterDeep achieves superior performance metrics, including 99.284% accuracy and a 96.513% F1 score. This approach is particularly effective in edge delineation and shadow differentiation, making it highly valuable for water management in complex urban environments.

Despite the promising advancements in deep learning models for water body extraction and classification, several limitations persist. One key challenge is the dependency on high-quality, diverse training data. Models such as SAM, CLIP, and Faster R-CNN are highly sensitive to the data they are trained on, meaning that limited or biased datasets can result in reduced accuracy when applied to new or unrepresented environments. This data dependency can lead to generalization issues, where models trained on specific satellite platforms like Sentinel-2 or Landsat-8 may not perform as well on different datasets or across varying geographical regions. Additionally, the computational complexity of advanced models like WaterDeep and WaterSegDiff poses a barrier to their application in real-time scenarios or in settings with limited computational resources.

Furthermore, environmental challenges such as complex terrains, atmospheric conditions, and shadowing in high-mountain and urban areas present significant hurdles for these models. Even the best-performing models, like WatNet or DeepWaterMapV2, can struggle to maintain accuracy in these conditions. Another limitation is the interpretability of deep learning models. Often considered "black boxes," these models provide little insight into their decision-making processes, which can hinder their acceptance in critical applications such as environmental monitoring and water management. Addressing these limitations is essential for the continued advancement and practical application of deep learning in water body mapping.

MATERIALS AND METHODOLOGY

Overview

This section outlines the materials and methods utilized in our research. We detail our approach to ensuring both the precision and dependability of deep learning models for semantic segmentation of water bodies using Sentinel-2 satellite imagery. The discussion starts with an overview of data acquisition and preprocessing techniques, then proceeds to a thorough description of the deep learning models and evaluation metrics applied in the study.

Materials

In this research, we gathered aerial images from Sentinel-2 that show bodies of water in Europe. Each picture has an accompanying black-and-white mask: areas that aren't water appear black while water is shown as white. To create these masks, we used something called the Normalized Water Difference Index (NWDI). This is an index commonly employed to figure out where plants are in satellite pictures but we applied it in a slightly different way so we could spot bodies of water with particular clarity.We needed lots of pictures and their masks to train our DL model. Happily, there were publicly available data sets that did just that Escobar, 2019. For each of the 2841 Sentinel-2 images covered by these sets there was a corresponding mask so thats how many pairs we started out with. These satellite snaps were all taken by the same spacecraft and offer resolutions ranging from 10m to 60m on the ground Caballero et al., 2019; Jiang et al., n.d. All the information is grouped into two main folders: one for images and another one dedicated solely to their masks. Pixels within these masks are binary; when put through NWDI they come out as either "water" (black) or "not water" (white).

It should be noted that when the Normalized Difference Water Index (NDWI) Cvik, 2011 is mentioned, it may actually be referring to various water-related indices that have been calculated from satellite imagery. One of these was created by Gao in 1996 for monitoring how amounts of water in leaves change over time; his method uses nearinfrared and shortwave-infrared light Gao, 1995. We used NDWI but with a difference: The formula we applied to produce study area masks was developed specifically for this purpose and will be referred to as Normalized Water Difference Index (NWDI) throughout this paper. Mathematically, it can be expressed as:

$$NDWI = \frac{W_{NR} - W_{\mathrm{SWIR}}}{W_{NR} + W_{\mathrm{SWIR}}}$$

where W_{NIR} and W_{SWIR} are the reflectances of the near-infrared band and the shortwave infrared band, respectively.

$$NDWI = \frac{W_{G} - W_{\mathrm{NIR}}}{W_{G} + W_{\mathrm{NIR}}}$$

where W_{G} is the reflectance in the green band.

Both equations (1) and (2) are used to monitor water level changes. This NDWI produces a numerical and visual interpolation of the output image, where values from (-1) to 0 represent a shiny surface without any traces of water content and values near +1 represent water content Surekha and Gupta, 2023.

Methodology

Figure 1. AI Model Development and Evaluation Pipeline

The deep learning workflow outlined in Figure 1 starts with essential steps of dataset preprocessing and normalization, which are fundamental for preparing the data for efficient model training. Following these steps, the dataset is divided into training, validation, and testing subsets. The training subset is utilized to train various deep learning models, while the validation subset is used to monitor model performance throughout the training phase. To ensure the models generalizability, final evaluations are carried out using the testing subset.

This study involves training three deep learning models DeepLabV3+, U-Net, and Fully Convolutional Networks (FCNs) and assessing their performance through various metrics such as precision, accuracy, recall, F1-score, and Dice coefficient. Among these, the U-Net model demonstrated superior overall performance and was subsequently finetuned through hyperparameter optimization with grid search cross-validation.

The U-Net model, which achieved the best results, was tested on an independent dataset to determine its final performance metrics. The evaluation revealed that U-Net surpassed the other models in precision, recall, F1-score, and Dice coefficient, underscoring its proficiency in the semantic segmentation of water bodies from satellite imagery.

DeepLabV3+

Iis an advanced semantic segmentation model proposed by Chen et al., 2018b. It combines an encoder-decoder architecture with atrous separable convolution to accurately classify pixels in images. This approach allows DeepLabV3+ to capture multiscale contextual information, resulting in superior performance in image understanding tasks 2.

Figure 2. Convolutional Neural Network (CNN) Architecture for Water Body Detection from Satellite Imagery

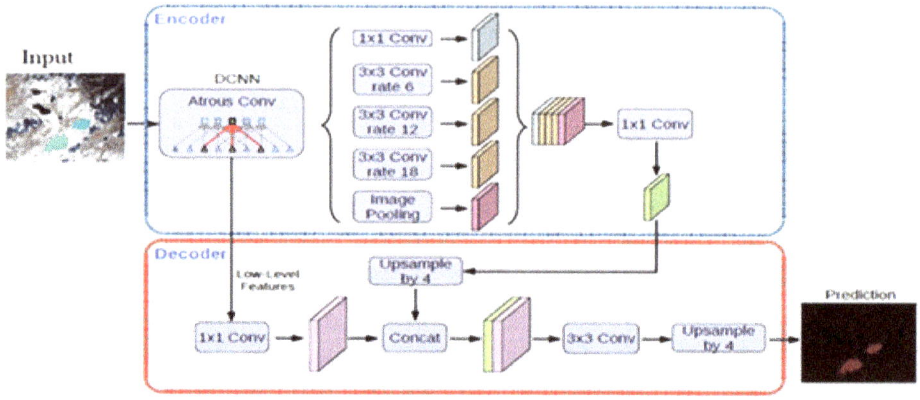

The mathematical equation for the Atrous Spatial Pyramid Pooling (ASPP) module in DeepLabV3+ can be representešd as follows:

$$ASPP(x) = concat\left(Conv_{\{1\times1\}(x)}, Conv_{\{3\times3\}(x)}, Conv_{\{3\times3\}(x,\ \backslash text\{atrous\ rate\}=6)},\right)$$

$$Conv\_\{3\times3\}(x, \backslash text\{atrous\ rate\} = 12), Conv\_\{3\times3\}(x, \backslash text\{atrous\ rate\} = 18))$$

Here, x represents the input feature map, and $\text{Conv}_{k\times k}$ denotes a convolutional layer with a kernel size of k k. The ASPP module performs convolutions with different atrous rates (holes in the filter) to capture contextual information at multiple scales. Finally, the results from these convolutions are concatenated to form the output feature map.

Fully Convolutional Networks (FCNs)

Are a distinct category of deep learning models specifically designed for semantic segmentation tasks. Unlike conventional convolutional neural networks (CNNs), which focus on image classification, FCNs can process input images of various sizes and generate segmentation maps with pixel-level accuracy Long et al., 2015b. The typical FCN architecture includes an encoder-decoder structure, where the encoder is responsible for feature extraction from the input image, and the decoder upsamples the extracted feature maps to generate the final segmentation output.

The mathematical formulation of FCNs involves convolutional layers for feature extraction and upsampling layers for the creation of segmentation maps. A simplified depiction of the FCN architecture illustrates this process:

Figure 3. Convolutional Neural Network (FCN) Architecture for Water Body Detection from Satellite Imagery

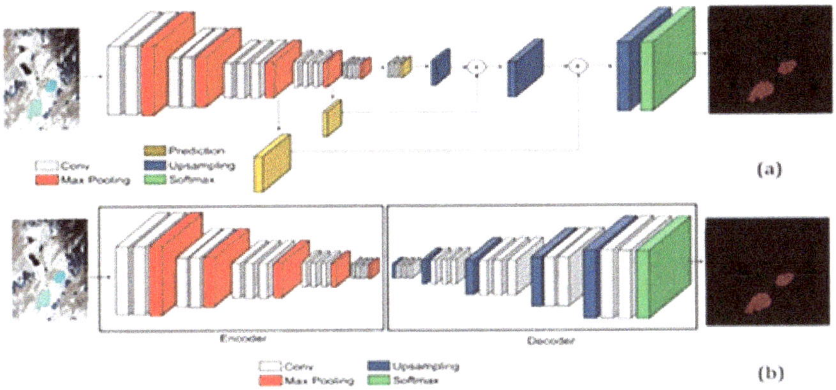

Let x denote the input image and $f(x)$ represent the output segmentation map. The FCN can be represented mathematically as:

$$f(x) = \text{Upsample}(\text{Conv}_1(x)) \quad (3)$$

Here, Conv_1 denotes the convolutional layers responsible for feature extraction, and Upsample denotes the upsampling operation applied to the feature maps to generate the segmentation map.

This equation encapsulates the basic idea of FCNs: processing the input image through convolutional layers to extract features and then upsampling the resulting feature maps to produce the segmentation output.

U-Net

The U-Net architecture, proposed by Ronneberger et al., 2015, is a convolutional neural network (CNN) model that was created specifically for problems involving semantic segmentation; it is especially well-suited for segmenting satellite images. It has skip connections and an encoder-decoder design.

Figure 4. U-Net Architecture Overview (Example for 32x32 Pixels)

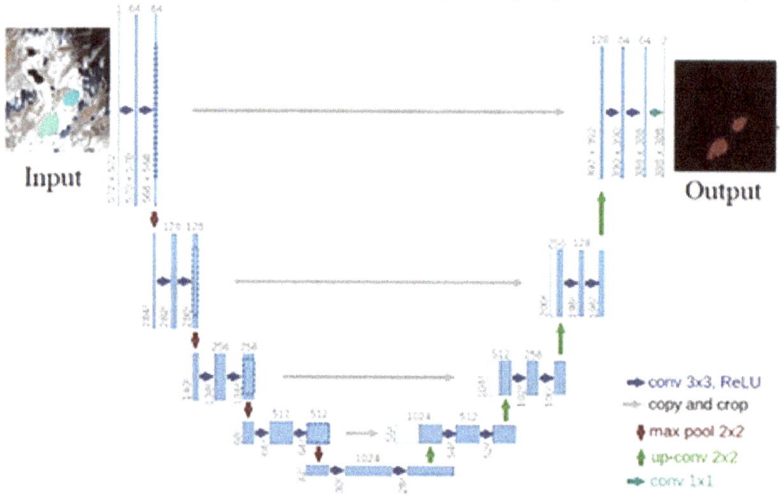

Convolutional neural network (CNN) model U-Net architecture is specifically tailored for semantic segmentation tasks, making it an excellent choice for biomedical picture segmentation applications. The architecture consists of skip connections in an encoder-decoder design. An overview of each element is provided below:

Encoder (Contracting Path)

The role of the encoder is to extract features from the input image while simultaneously decreasing its spatial dimensions.

Typically, it is comprised of sequential convolutional layers, which are succeeded by max-pooling layers.

These convolutional layers apply a series of trainable filters to the input, facilitating the extraction of increasingly abstract features.

Decoder (Expanding Path)

The decoder's function is to restore the spatial dimensions of the feature maps to match the size of the original input.

This process involves a sequence of upsampling layers, frequently implemented as transposed convolutions, followed by subsequent convolutional layers.

The upsampling layers are responsible for enhancing the spatial resolution of the feature maps.

Skip Connections

Skip connections facilitate direct linking between encoder layers and their corresponding layers in the decoder.

These connections are crucial for maintaining spatial information and mitigating issues related to vanishing gradients during training.

By implementing skip connections, the decoder can leverage both the high-level semantic data from the encoder and the detailed spatial information.

Final Layer

The concluding layer in the U-Net architecture generally includes a 1x1 convolutional layer paired with a softmax activation function.

This final layer is responsible for producing the segmentation mask by assigning class probabilities to every pixel in the input image.

Evaluation Metric

valuation metrics are essential tools in various fields to assess the performance of models, systems, or processes. These metrics provide a quantifiable measure to determine how well an entity meets certain criteria or objectives Sokolova and Lapalme, 2009. Here's a brief overview of some common evaluation metrics used in this chapter:

Accuracy

Is a metric utilized to assess the performance of a classification model. It indicates the proportion of correct predictions, encompassing both true positives and true negatives, out of the total number of predictions made. The formula for calculating accuracy is:

$$\text{Accuracy} = \frac{\text{True positives} + \text{True Negatives}}{\text{Total Number of Instances}}$$

True Positives (TP): The number of instances that were correctly predicted as positive. *True Negatives (TN)*: The number of instances that were correctly predicted as negative. *Total Number of Instances*: The sum of all the occurrences in the dataset which include false positives, negatives, true positives, and true negativesa˘.

Precision

Precision measures a classification model's effectiveness in accurately identifying positive instances. It is calculated as the ratio of true positives to the total of true positives and false positives Powers, 2008. The formula for precision is:

$$Precision = \frac{True\ Positives}{True\ Positives + False\ Positives}$$

True Positives (TP): The count of instances that are accurately predicted as positive.

False Positives (FP): The count of instances that are mistakenly predicted as positive.

F1-Score

Is a statistic that balances precision and recall by taking the harmonic mean of the two. It is helpful for evaluating models when the class distribution is unbalanced since it combines the two measurements into a single value. Sokolova and Lapalme, 2009.

The formula is:

$$F1 - score = * \frac{Precision \times Recall}{Precision + Recall}$$

Recall:

Recall, alternatively termed as sensitivity or the true positive rate, quantifies the proportion of actual positive instances that the model successfully identifies. It is computed by dividing the number of true positives by the combined total of true positives and false negatives Powers, 2008. The formula is:

$$Recall = \frac{True\ Positives}{True\ Positives + False\ Negatives}$$

Dice Coefficient:

The Dice Coefficient is frequently applied in image segmentation tasks to assess the similarity between two sets. It is calculated by taking twice the intersection of the two sets and dividing it by the sum of their sizes Powers, 2008. The formula is:

$$Dice\ Coefficient = \frac{2*|A \cap B|}{|A| + |B|}$$

A and *B* represent the sets being compared. In the context of image segmentation, *A* might represent the pixels in the ground truth segmentation mask, and *B* might represent the pixels in the predicted segmentation mask . $|A \cap B|$ denotes the size of the intersection of sets *A* and *B*, which represents the number of pixels that are correctly segmented by both the ground truth and the predicted segmentation. $|A|$ and $|B|$ represent the sizes of sets *A* and *B* respectively, which represent the total number of pixels in each segmentation mask.

A higher Dice coefficient value, ranging from 0 to 1, indicates better agreement between the predicted and ground truth segmentation masks Powers, 2008.

RESULTS

the results achieved by three deep learning models: DeepLabV3+, U-Net, and Fully Convolutional Network. They were all used to perform semantic segmentation of water bodies from satellite images captured during Sentinel-2 mission. For example, these models will be compared in terms of accuracy, precision, recall, F1-score and Dice coefficient.

Table 1. Performance Metrics of Different Models

Models	*Accuracy*	*Precision*	*Recall*	*F1-score*	*Dice Coefficient*
DeepLabV3+	0.6535	0.8654	0.7563	0.8208	0.7608
U-Net	0.7554	0.9037	0.8403	0.8700	0.8600
FCN	0.6789	0.8453	0.6363	0.7206	0.7206

The table 1 show the U-Net model performed best on all counts: accuracy at 0. 75; precision at 0. 90; recall at 0. 84 ; F1-score at 0. 87; and Dice coefficient at 0. 86. This suggests that U-Nets architecture which includes skip connections and a distinctive U shape was especially good at understanding and making use of context in the images. In other words, by looking at neighbouring pixels (or picture elements), the network was able to correctly outline where bodies of water were in the satellite pictures.

The DeepLabV3+ and FCN models showed good results as well; while DeepLabV3+ had slightly better precision than FCN, which in turn had a better recall rate. Nevertheless it was U-Net that topped them all when looking at overall F1-score plus Dice coefficient: both being popular ways to measure how well things like images can be split into parts with similar features (semantic segmentation tasks).

Figure 5. The U-net model Performance Evaluation During Training

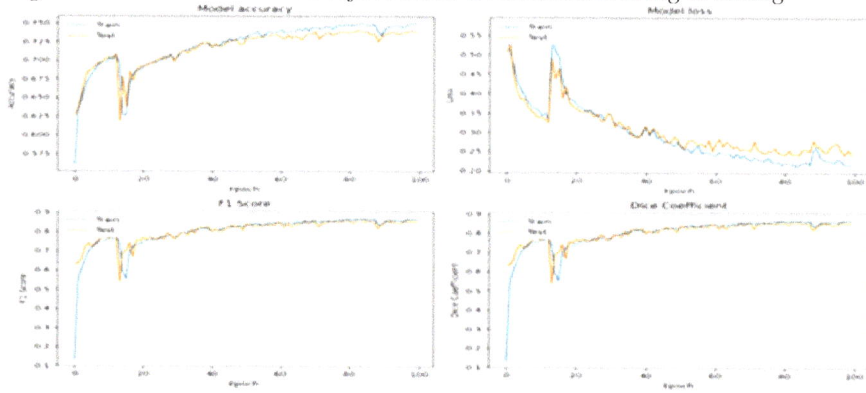

Figure 6. The predicted mask shown in the image appears to be generated by the U-Net model

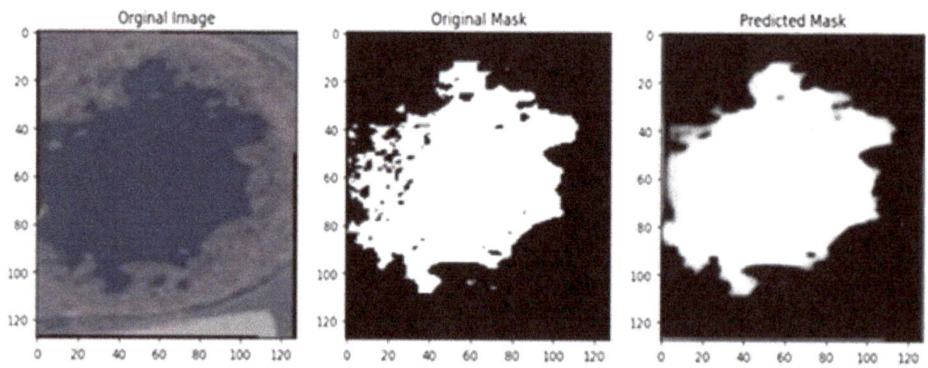

The figure 5 shows the summery of accuracy, loss, F1 score and Dice coefficients performance evaluation of deep learning model on training dataset as well as on testing dataset with respect to the training epochs. The curves suggests that the performance of the model is increasing with training, on the fact that the training metrics are slightly greater than the testing metrics so it makes clear the future overfitting. The two curves of accuracy and F1 score also have similar patterns which means the ability of the model to classify instances are also same. The loss curves are decreasing, this shows that the model is learning to minimize the loss function. Figure 11: Dice coefficient (i.e., overlap between predicted and ground truth segmentation masks) throughout the training (note the difference in scale as compared to the previous graph, and note the comparison to the results of the segmentation approach) Trend similar to F1 score, but more fluctuations.

In the figure 6, we can observe that the predicted mask closely matches the original mask, accurately capturing the intricate details and boundaries of the water body. This remarkable performance of the U-Net model can be attributed to its ability to combine low-level features from the encoder path with high-level features from the decoder path, facilitated by the skip connections. This multi-scale feature fusion allows the model to capture both fine-grained details and broader contextual information, resulting in precise segmentation masks.

DISCUSSION

The results of this study highlight the substantial capabilities of deep learning models, particularly the U-Net architecture, in precisely identifying water bodies from satellite images. U-Nets distinctive design, which integrates the advantages of both traditional convolutional neural networks (CNNs) and fully convolutional networks (FCNs), proves highly effective for tasks involving semantic segmentation. The incorporation of skip connections within U-Net enables it to capture features at multiple scales, ensuring the preservation of spatial information and enhancing the accuracy of water body detection across different environments.

Real-World Applications

In real-world scenarios, the application of these models can have profound impacts on environmental management and policy-making. For instance, in Morocco, accurate water body detection is crucial for managing scarce water resources, particularly in regions like the Atlas Mountains and the Sahara Desert, where water is a vital but limited resource. The U-Net model could be employed to monitor the water levels in critical reservoirs, such as the Al Wahda Dam, the second largest in Africa. Precise monitoring of water levels in this dam is essential for optimizing irrigation, drinking water supply, and hydroelectric power generation. By providing accurate, real-time data on water body changes, U-Net could help authorities make informed decisions that enhance water resource management and mitigate the impacts of droughts, a recurring challenge in the region.

Moreover, the model's application could extend to coastal management in Morocco, where monitoring the encroachment of seawater into freshwater bodies due to climate change is critical. By accurately mapping these changes, the model could inform policies aimed at protecting freshwater resources and managing the impacts of rising sea levels. Additionally, in urban areas such as Casablanca, U-Net could be used to plan and manage stormwater infrastructure, ensuring that new developments are resilient to flooding and other water-related challenges.

Limitations and Future Work

While the results of this study are promising, it is important to acknowledge the limitations to provide a comprehensive understanding of the research:

Dataset Bias: One major limitation is the potential bias in the dataset used for training the models. Since the study primarily relies on Sentinel-2 imagery, the models may not generalize well to other types of satellite imagery or different geographical regions with varying environmental conditions. For example, the specific atmospheric conditions and seasonal variations in Morocco might not be fully captured by the dataset, which could affect the model's performance in these contexts.

Computational Complexity: The U-Net model, while efficient in many respects, still requires significant computational resources, which could limit its applicability in resource-constrained settings. In rural areas of Morocco, where access to highperformance computing infrastructure may be limited, deploying such models could be challenging.

Generalizability: The model's performance across different environmental conditions, such as varying levels of vegetation cover, snow, or urbanization, needs further evaluation. The current study's scope may not fully capture these variations, which could affect the model's robustness in different scenarios.

Future Research Directions

To address these limitations and enhance the applicability of the research, future work could focus on several key areas:

Diverse Satellite Imagery: Exploring the use of different types of satellite imagery, including data from sources like Landsat, MODIS, or commercial high-resolution satellites, could improve the model's generalizability across various environmental conditions and regions.

Advanced Techniques: Implementing advanced techniques such as ensemble learning, which combines the strengths of multiple models, could further enhance the accuracy and robustness of water body detection. This approach could be particularly beneficial in complex environments, where single models may struggle to capture all relevant features.

Model Optimization: Developing more efficient algorithms that reduce the computational demands of these models is crucial for broader adoption, especially in regions with limited access to high-end computing resources. Techniques such as model pruning, quantization, or the use of lightweight architectures could make these models more accessible and practical for real-world applications.

CONCLUSION

This study underscores the significant efficacy of deep learning models, particularly the U-Net architecture, in the precise delineation of water bodies from satellite imagery. The U-Net model's architecture, characterized by its integration of skip connections and multi-scale feature extraction, has demonstrated superior performance in semantic segmentation tasks, establishing it as a critical tool for applications in water resource management, flood risk assessment, and urban planning.

The implications of these findings are particularly relevant in regions like Morocco, where effective water management is essential. However, this research also acknowledges certain limitations, including potential biases inherent in the dataset and the computational intensity of the models, which may constrain their generalizability and applicability across different contexts.

Future research endeavors should aim to mitigate these limitations by incorporating diverse satellite imagery sources, employing advanced methodologies such as ensemble learning, and optimizing models for enhanced computational efficiency. Addressing these areas will further extend the impact of this research, contributing to more informed and sustainable decision-making in the management of water resources and related domains.

Competing Interests

The authors declare that they have no conflict of interest.

REFERENCES

Caballero, I., Ruiz, J., & Navarro, G. (2019). Sentinel-2 satellites provide near-real time evaluation of catastrophic floods in the west mediterranean. *Water (Basel)*, 11(12), 2499. DOI: 10.3390/w11122499

Cao, H., Tian, Y., Liu, Y., & Wang, R. (2024). Water body extraction from high spatial resolution remote sensing images based on enhanced u-net and multi-scale information fusion. *Scientific Reports*, 14(1), 16132. DOI: 10.1038/s41598-024-67113-7 PMID: 38997473

Chen, L.-C., Zhu, Y., Papandreou, G., Schroff, F., & Adam, H. (2018a). Encoder-decoder with atrous separable convolution for semantic image segmentation. *Proceedings of the European conference on computer vision (ECCV)*, 801–818.

Chen, L.-C., Zhu, Y., Papandreou, G., Schroff, F., & Adam, H. (2018b). Encoder-decoder with atrous separable convolution for semantic image segmentation.

Cvik, J. E. (2011). Normalized difference water index. *Remote Sensing Letters*, 2(4), 225–236.

Drusch, M., Del Bello, U., Carlier, S., Colin, O., Fernandez, V., Gascon, F., Hoersch, B., Isola, C., Laberinti, P., Martimort, P., Meygret, A., Spoto, F., Sy, O., Marchese, F., & Bargellini, P. (2012). Sentinel-2: Esa's optical high-resolution mission for gmes operational services. *Remote Sensing of Environment*, 120, 25–36. DOI: 10.1016/j.rse.2011.11.026

Escobar, F. (2019). Images and masks of satellite images of water bodies for image segmentation.

Fu, S., Wang, L., Khalil, U., Cheema, A. H., Ullah, I., Aslam, B., Tariq, A., Aslam, M., & Alarifi, S. S. (2024). Prediction of surface urban heat island based on predicted consequences of urban sprawl using deep learning: A way forward for a sustainable environment. *Physics and Chemistry of the Earth Parts A/B/C*, 135, 103682. DOI: 10.1016/j.pce.2024.103682

Gao, B.-C. (1995). Normalized difference water index for remote sensing of vegetation liquid water from space. *Imaging spectrometry*, 225–236.

Gharbia, R. (2023). Deep learning for automatic extraction of water bodies using satellite imagery. *Photonirvachak (Dehra Dun)*, 51(7), 1511–1521. DOI: 10.1007/s12524-023-01705-0

Hu, L., Zhang, C., Zhang, M., Shi, Y., Lu, J., & Fang, Z. (2023). Enhancing fair data services in agricultural disaster: A review. *Remote Sensing (Basel)*, 15(8), 2024. Advance online publication. DOI: 10.3390/rs15082024

Janga, B., Asamani, G. P., Sun, Z., & Cristea, N. (2023). A review of practical ai for remote sensing in earth sciences. *Remote Sensing (Basel)*, 15(16), 4112. Advance online publication. DOI: 10.3390/rs15164112

Juhász, L. (2024). Assessing publication trends in selected giscience journals. *International Journal of Geographical Information Science*, 38(8), 1443–1467. DOI: 10.1080/13658816.2024.2347306

Long, J., Shelhamer, E., & Darrell, T. (2015). Fully convolutional networks for semantic segmentation. *Proceedings of the IEEE conference on computer vision and pattern recognition*, 3431–3440.

McFeeters, S. (1996). The use of the normalized difference water index (ndwi) in the delineation of open water features. *International Journal of Remote Sensing*, 17(7), 1425–1432. DOI: 10.1080/01431169608948714

Ozdemir, S., Akbulut, Z., Karsli, F., & Kavzoglu, T. (2024). Extraction of water bodies from high-resolution aerial and satellite images using visual foundation models. *Sustainability (Basel)*, 16(7), 2995. DOI: 10.3390/su16072995

Perez-Torres, W. I., Uman-Flores, D. A., Quispe-Quispe, A. B., Palomino-Quispe, F., Bezerra, E., Leher, Q., Paixão, T., & Alvarez, A. B. (2024). Exploratory analysis using deep learning for water-body segmentation of perus high-mountain remote sensing images. *Sensors (Basel)*, 24(16), 5177. Advance online publication. DOI: 10.3390/s24165177 PMID: 39204871

Pierdicca, R., & Paolanti, M. (2022). Geoai: A review of artificial intelligence approaches for the interpretation of complex geomatics data. *Geoscientific Instrumentation, Methods and Data Systems*, 11(1), 195–218. DOI: 10.5194/gi-11-195-2022

Powers, D. (2008). Evaluation: From precision, recall and f-factor to roc, informedness, markedness & correlation. *Mach. Learn. Technol., 2.*

Ronneberger, O., Fischer, P., & Brox, T. (2015). U-net: Convolutional networks for biomedical image segmentation. *Lecture Notes in Computer Science*, 9351, 234–241. DOI: 10.1007/978-3-319-24574-4_28

Sokolova, M., & Lapalme, G. (2009). A systematic analysis of performance measures for classification tasks. *Information Processing & Management*, 45(4), 427–437. DOI: 10.1016/j.ipm.2009.03.002

Sun, D., Gao, G., Huang, L., Liu, Y., & Liu, D. (2024). Extraction of water bodies from high-resolution remote sensing imagery based on a deep semantic segmentation network. *Scientific Reports*, 14(1), 14604. DOI: 10.1038/s41598-024-65430-5 PMID: 38918493

Surekha, J. N., & Gupta, N. (2023). Ndr-unet: Segmentation of water bodies in remote sensing using nested dense. DOI: 10.21203/rs.3.rs-3159538/v1

Taha, Z. A., & Hadi, A. (2019). Performance evaluation metrics for semantic image segmentation: A comprehensive survey. *Information (Basel)*, 10(10), 319.

Zhang, X., Ge, Y., Chen, J., Ling, F., Wang, Q., Du, D., & Xiang, R. (2023). Highquality super-resolution mapping using spatial deep learning. *iScience*, 26(6), 106875. DOI: 10.1016/j.isci.2023.106875 PMID: 37288344

Chapter 6

Integration of Remote Sensing and GIS for Optimal Site Selection of Dams in Arid to Semi-Arid Environments:

A Case Study of the Rheraya Basin and Enhancement of the Empirical Model With the Curve Number Model

Fatima Ezzahra El Ghazali
https://orcid.org/0000-0002-1980-3028
Cadi Ayyad University, Marrakech, Morocco

Abdellah Khouz
Higher School of Technology Essaouira, Cadi Ayyad University, Marrakech, Morocco

ABSTRACT

This chapter is dedicated to presenting the optimization method employed, which primarily utilizes the functionalities provided by GIS (geographic information systems) and remote sensing for data structuring, cross-referencing information layers, and spatial analysis across various themes. It incorporates a multicriteria hierarchical analysis approach, enabling the integration of multiple decision criteria into a unified model. This approach involves conducting comparative evaluations of each pair of

DOI: 10.4018/979-8-3693-9651-3.ch006

criteria and calculating their weights for the comparative assessment of each pair of options concerning each sub-criterion. Both methods employed showcase a robust multicriteria decision support tool. It is noteworthy that ensuring the validity of the SCS-CN procedure necessitates the incorporation of other supply models to enhance the accuracy of determining suitable locations for water accumulation structures.

I. INTRODUCTION

Surface runoff is influenced by several factors such as the duration and intensity of precipitation, soil type, land use, vegetation cover, slope, and the density of the hydrographic network (Faregh and Benkhaled 2016). Generally, significant runoff originates in a natural watershed. This watershed, which is an extremely heterogeneous and complex physical system, converts precipitation into flow in rivers and channels towards outlets, often urban areas (Brooks et al. 2003).

In studies on urban flooding caused by intense precipitation, the production function often simplifies to a runoff coefficient because the volume of water runoff affected by initial losses is negligible compared to the total volume (Brooks et al. 2003).

Global water demand, especially in developing countries like Morocco, is increasing due to population growth, rapid urbanization, climate change, and agricultural techniques and crop patterns that consume a lot of water(Abdelmajid et al. 2021). In this case, the production function is more complex due to environmental heterogeneity. The volume of water from the natural watershed is greater than that produced within the city, and estimating it is one of the challenges. Water scarcity tends to be a universal problem, with many regions suffering from severe water shortages (Malik et al. 2020).

The Haouz plain faces high demand for managing and supplying water resources. Water accessibility in this region is typically tied to heavy intensities and short periods of precipitation. (Ouassanouan et al. 2022) Generally, arid regions are characterized by irregular distribution of precipitation in terms of location and time, high temperatures, evaporation, and a lack of groundwater and surface water resources (Musy and Higy 2004). Water scarcity in the Haouz plain is the primary concern limiting water management and planning (Bzioui 2004).

II. METHODOLOGICAL APPROACH

Remote Sensing (RS) data and GIS data were combined to create thematic layers, which were then assigned appropriate weights using the Multi-Criteria Analysis (MCA) technique. This process was used to optimize the empirical model developed in the previous study. (El Ghazali et al. 2021a), this chapter will focus on presenting the optimization method used. This method primarily leverages the capabilities of GIS for structuring data, overlaying information layers, and conducting spatial analysis across various themes. It incorporates a multicriteria hierarchical analysis approach, which enables the integration of multiple decision criteria into a unified model. This approach involves comparing each pair of criteria, calculating their weights, and assessing each pair of options relative to each sub-criterion.(El Ghazali et al. 2021b)

To develop a synthesis map of land use suitable for rainwater harvesting in an arid to semi-arid area, the method used for estimating runoff in arid to semi-arid zones is the SCS method (USDA 1986) widely used due to its simplicity and limited number of required parameters (Mack 1995); (Banasik and Woodward 2010); (Bo et al. 2011) ; (Mishra, Kumar Kansal, and Aggarwal 2012); (Jeon, Lim, and Engel 2014); (Giridhar and Viswanadh 2014)

The SCS method is based on two parameters to estimate runoff in the form of depth from rainfall. These two parameters are the maximum retention potential S and the initial abstraction Ia. They are expressed in terms of Curve Number or CN (Elhakeem and Papanicolaou 2009). The CN term depends on soil type, land use, vegetation cover, and antecedent moisture conditions (Meißl et al. 2023) (USDA 1986). The SCS-CN method has been used in ungauged rural watersheds and has evolved from its initial objective to be adopted for estimating runoff in urban and forested watersheds (Walega et al. 2020) (Hawkins et al. 2008).

1. Study area:

The Haouz-Tensift basin, located in central Morocco (Fig. 1), experiences low and irregular rainfall, with an average ranging from 240 mm/year in the plains to over 461 mm/year in the mountains.

The southern part, comprising the High Atlas mountains, is the water production zone characterized by very high altitudes, significant precipitation, and highly active surface hydrology (Boudhar 2009). This area is where the main tributaries of the Tensift River originate: N'Fis, Ourika, Rheraya, Zat, and R'dat. (Lahcen, Nidal, and Mohamed 2020)

Figure 1. Localization of the Rheraya watershed and the Haouz plain (Hajhouji 2018)

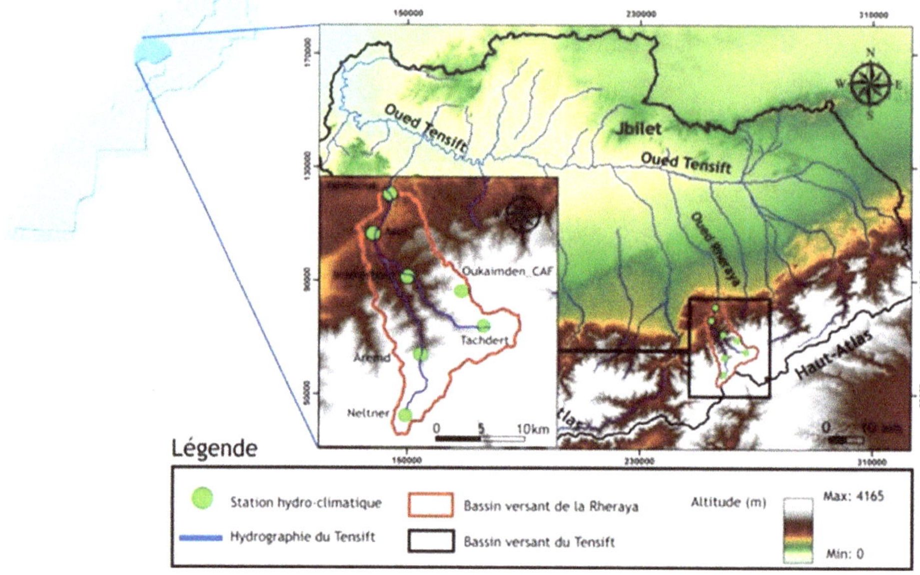

The southern part of the basin consists of a succession of nine sub-basins oriented from south to north. The Rheraya basin is one of the most active sub-basins of the Tensift River.(Lahcen et al. 2020) Its productive upstream water area is located in the High Atlas Mountains and culminates at Jbel Toubkal (4165 m). The Rheraya watershed covers an area of 225 km^2. The elevations of the basin range from 1030 m up to the highest point of the High Atlas at 4165 m (Toubkal). Its outlet is defined by the hydrometric station of Tahanaout at 1030 m (31.3°N, 7.9°E).

The geological delimitation of the Haouz plain (Fig. 2) is mainly based on the extension of Plio-Quaternary alluvial deposits, although it includes a few rare pre-Pliocene outcrops. (Khouz et al. 2023) The Haouz plain is flanked by the Paleozoic massif of Jbilet to the north and Paleozoic outcrops of the High Atlas to the south, which correspond to residual reliefs of the Hercynian basement cleared of the Neogene cover (Soulaimani and Ouanaimi 2011). These terrains are mainly composed of schists, pelites, sandstones, and quartzites. (Rhoujjati et al. 2021)

Initially, the domain of the Haouz plain was a tectonic depression filled with silicoclastic deposits resulting from the erosion of the High Atlas chain during its uplift in the Neogene and Quaternary periods (Hajhouji 2018). These deposits are generally alluvial cones and fluvial formations, transported and deposited by the network of Atlasian wadis. The plain is overlooked by the mountain range of the

High Atlas, subdivided into three zones ranging from the chain's peaks towards the plain (Moreau 1976).

The Rheraya watershed is in direct contact with the Haouz plain in its northern part, and the Atlasian watersheds of N'Fis to the west and Ourika to the east. From a hydrogeological perspective, 60% of the basin's surface is occupied by poorly permeable formations (metamorphic and eruptive rocks).

Figure 2. Geological map of the Tensift watershed (Hajhouji 2018)

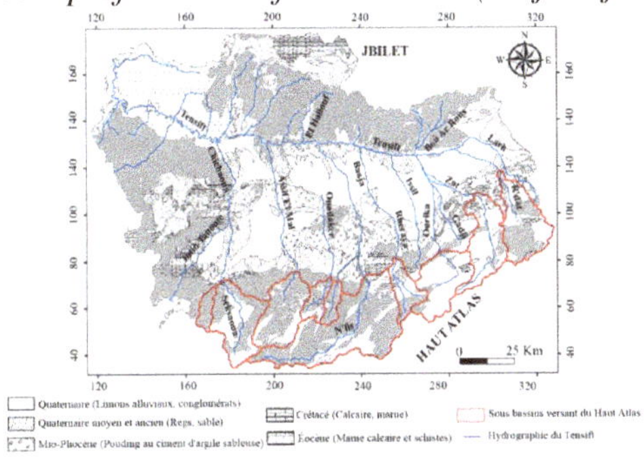

On the geological map of the Rheraya basin (Fig. 2), the following features can be distinguished:

The sub-Atlas zone, which covers the northern third of the basin, is composed of a rigid Precambrian and Paleozoic basement made of grès-schist and limestone. This basement is exposed in the gorges of Moulay Brahim in the north of the basin where the outlet is located. (Salma et al. 2023) Overlying this basement is a significant Mesozoic cover with a red detrital facies consisting of Triassic conglomerates, sandstones, and siltstones. This soft rock forms the current Asni basin. To the north and south of the Asni basin, the Kik and Tihallatine plateaus are characterized by harder carbonate and marl facies. This sub-Atlas zone forms the promontory of the northern slope of the High Atlas, serving as the initial barrier overlooking the Haouz plain. This formation is prevalent throughout most of the Atlas basins. (JADOUD et al. 2023)

Further south, the Rheraya horst consists of primary and Permo-Triassic rock materials. It is an elevated block compared to the sub-Atlas zone, sloping towards the west. Permo-Triassic materials are visible in the two branches that create a Y-shape within this basin: the Imlil and Imennane valleys.

These valleys consist of deposits (doleritic basalts) resting on a fossilized pre-Permo-Triassic surface (conglomerates and sandstones).(Elghouat et al. 2024)

Figure 3. Geological map of the Rheraya watershed

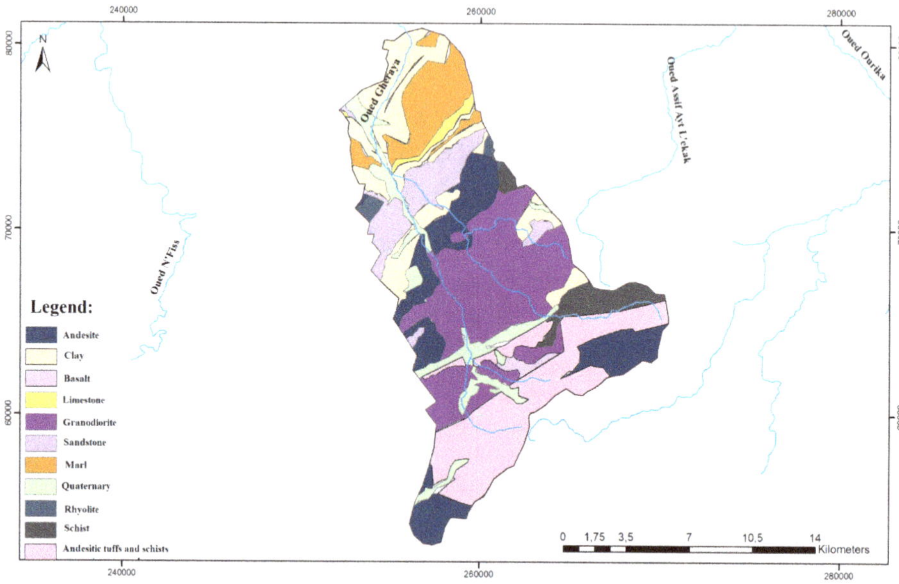

The permeability of the substratum largely contributes to the understanding of water circuits and to the analysis of the differentiated feeding of flows in the wadis. On the surface, the rocks of the substratum can promote either runoff (impermeable rocks) or infiltration (permeable rocks). At depth, this permeability also defines the possibilities of underground reserves that can feed the wadis (springs, aquifer drainage, etc.). The geological map for hydraulic services, regarding the existence of the pedological permeability map. The results are at 1/500,000 scale and some reports on the nature of the rocks, as well as the analysis of the geological diagram in Figure 3, have made it possible to classify the outcrops of the Rheraya basin according to their degree of permeability.

Table 1. Rock permeability table based on Figure 4

	Rheraya
Permeable	15.6 km^2
	4.2%

continued on following page

Table 1. Continued

	Rheraya
Semi-permeable	121.3 km^2
	36.1%
Impermeable	197 km^2
	59.9%

The basin is characterized by sedimentary formations of sandstone-clay and limestone, which are either permeable or impermeable (Fig. 3). The vegetation cover is very sparse on the slopes, with irrigated crop areas localized near the thalwegs. The slopes are steeper in the upstream part. The average annual precipitation at the Tahanaout outlet station (1064 m altitude) is 356 mm, and it is 445 mm at the Oukaimeden CAF station at an altitude of 2612 m. The average flow recorded at the Tahanaout hydrometric station is around 1.15 m^3/s. (Bahija, Abdelmoumen, and Souad 2023)

2. METHODOLOGY

The general approach involves using a method to estimate surface runoff by analyzing the physical characteristics of the watershed. The first step is the creation of spatial databases of land use and soil types using a GIS, based on direct and indirect investigation methods. This is followed by the processing of raw data to prepare it for simulation. The last stage involves obtaining and interpreting the results. To streamline this process, operators were combined to rank alternative zones based on their alignment with the evaluation criteria. A synthesis map was created through visual interpretation of satellite imagery, highlighting various permanent structures, watershed boundaries, drainage networks, and more. The numerical analysis was conducted using GIS and image processing software. (ESRI 2005) (ArcGIS Ver. 10.3).

Figure 5 illustrates the steps involved in designing the runoff coefficient map in the form of a flowchart.

Figure 4. Flowchart for the synthesis map design

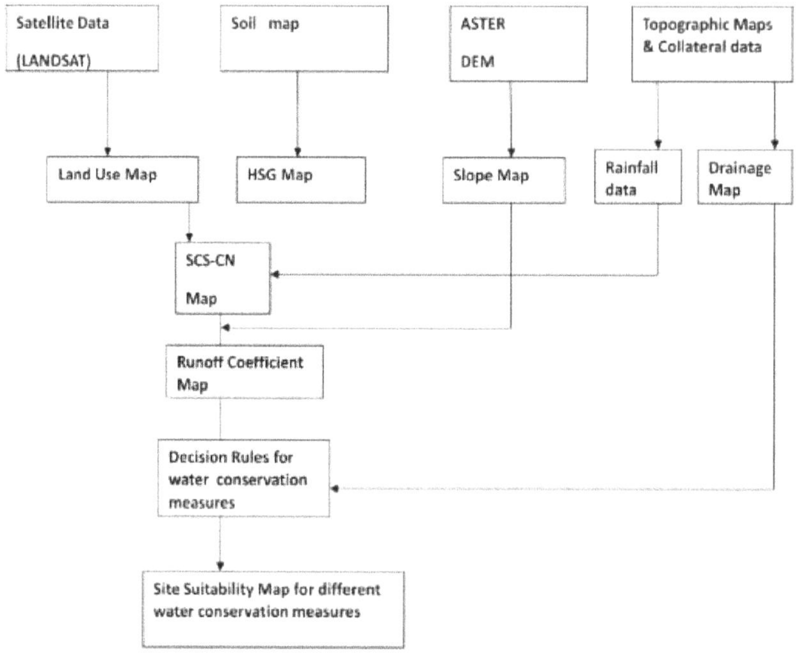

The methodology used for site selection in this study is the SCS Curve Number method. (Forootan 2023) The SCS (Soil Conservation Service) method is the most commonly used technique for estimating surface runoff from small watersheds for a given rainfall event. This method examines the relationship between land cover (including cover type, soil treatment, and hydrological condition) and the hydrological soil group, which together determine the curve number. (Gabriel and Diabé 2021).

The SCS-CN method (USDA 1986) (USDA-SCS, 1972), As described in Appendix B, this method allows for the determination of surface runoff resulting from a rainfall event, taking into account the characteristics of the slope and watershed where it occurs. It is an effective technique for estimating the approximate volume of runoff during a rainfall event across a specific area. (Forootan 2023) Like all empirical models, the following Table 2 summarizes the advantages and limitations of our model.(Forootan 2023)

Table 2. Advantages and Limitations of Our Model

The advantages	The limitations
• The adopted meteorological module has overcome the issue of lacking pluviograph data. • The SCS-CN method is straightforward, reliable, and does not require large amounts of data. • The SCS-CN method is representative when using NRCS tables directly (our case).	• It is less suitable for analyzing small precipitation events, and it is recognized that it becomes less accurate for rainfall amounts less than 10 mm. • Using hydrological soil groups directly without adaptation to the structure, texture, and land use/land cover (LULC) of the study area can potentially introduce errors. The values in these tables are derived from the analysis of small watersheds in the United States and may not be fully suitable for other contexts.

The Curve Number is influenced by several factors, including the soil's hydrological group, land use, hydrological conditions, and initial moisture levels. Soils are divided into four groups based on their properties and infiltration capacities. The NRCS (Natural Resources Conservation Service) classifies soils into hydrological groups A, B, C, and D, with permeability decreasing from A to D. Group D soils, which have the lowest permeability, correspond to the highest Curve Numbers. (Lasco and Moglen 2023) The Curve Number itself serves as an index that reflects how a watershed responds to rainfall, indicating the portion of rainfall that contributes to surface runoff. Curve Numbers range from 0 to 100, where higher values denote a higher percentage of runoff. In this study, the concepts underlying the SCS method's Curve Numbers were utilized to map runoff potential.

Surface runoff in arid and semi-arid regions is a vital water source for rivers, lakes, and streams. It also plays a key role in determining criteria for rainwater harvesting. (Hashim and Sayl 2022).

The Curve Number map is developed by combining soil hydrological data with land use maps. The runoff is then determined using this formula: (Jahan, Pradhanang, and Bhuiyan 2021) (Diskin and Nazimov 1995):

$$Q = \frac{(P - Ia)^2}{(P - Ia + S)}$$

Where:

Q is the runoff volume [L],

P is the precipitation volume [L],

S is the maximum potential retention [L],

Ia is the initial abstraction (the volume of water in the soil before runoff, including infiltration, rainfall intercepted by vegetation, or evapotranspiration) [L] and generally Ia = 0.2S. The CN is expressed by the following equation (El Garouani, Aharik, and El Garouani 2020) (Dorel 1993):

$$S = \frac{1000}{CN} - 10$$

The Curve Number (CN) ranges from 30 to 100, with lower values indicating soils with low runoff potential and higher values indicating soils with high runoff potential. The SCS method relies on soil texture data to classify soils into hydrologic groups (A, B, C, and D). Curve Numbers are determined for each hydrologic soil group by considering different types of land cover. A Curve Number map will be produced by combining the hydrologic soil groups with the grid surfaces representing land cover.

Selecting appropriate sites for rainwater harvesting requires evaluating several factors and objectives. To facilitate this, remote sensing data is used to generate a thematic input database in ArcGIS. This study follows four main steps, outlined as:

1. Selection of rainwater harvesting criteria:
2. Assessment levels of the suitability of the selected criteria.
3. Assignment of weights to these criteria.
4. Generating maps using GIS tools.

Based on the six factors outlined by the Food and Agriculture Organization (FAO)—hydrology, climate, topography, soils, socio-economics, and agronomy—as cited by (Sayl, Mohammed, and Ahmed 2020), six criteria were selected for this study, guided by literature review, the specific characteristics of the study area, and expert judgment. These criteria, which include soil texture, slope, runoff, vegetation cover, and drainage density, were employed to identify optimal locations for rainwater harvesting using ArcGIS software.

The Curve Number (CN) was estimated on a per-pixel basis for the study area, utilizing a spatial guide and a soil map reclassified into hydrologic soil groups and conditions (Figure 5). A higher CN value signals higher runoff potential in the area, as noted by (Krois and Schulte 2014). The runoff map for the study area, depicted in Figure 6, is divided into four classes, and was generated using ArcGIS. According to the USDA-SCS (1972), the four major hydrologic soil groups are described as follows:

Group A: Soils in this group exhibit low runoff potential when wet, allowing water to easily percolate through. Typically, these soils contain less than 10% clay and over 90% sand and gravel, characterized by sandy and gravelly textures. Loamy sand, sandy loam, or silt loam soils may also be included if they are well-structured with low density. The hydraulic conductivity of all soil layers exceeds 14.4 cm/h, with permeable soil depths greater than 50 cm and a water table depth exceeding 60 cm. Soils with over 100 cm of permeable depth or a water table deeper than 100 cm are classified in this group if the hydraulic conductivity of the top 100 cm is greater than 3.6 cm/h.

Group B: Soils in this group have a relatively low runoff potential when wet. Typically, they contain 10% to 20% clay and 50% to 90% sand, featuring loamy sand and sandy loam textures. Soils with loam, silt loam, silt, or sandy clay loam textures can also be included if well-structured and low-density. The hydraulic conductivity in the top 50 cm ranges from 3.6 to 14.4 cm/h, with a permeable soil depth exceeding 50 cm and a water table depth greater than 60 cm. Soils with more than 100 cm of permeable depth or a water table deeper than 100 cm belong in this group if the hydraulic conductivity of the top 100 cm is between 1.45 and 3.6 cm/h.

Group C: Soils in this group have a relatively high runoff potential when wet. They generally contain 20% to 40% clay and less than 50% sand, with textures such as loam, silt loam, silt, sandy clay loam, clay loam, and silty clay loam. Well-structured, low-density soils with clay, silty clay, or sandy clay textures may also be included. The hydraulic conductivity in the top 50 cm falls between 0.35 and 3.6 cm/h, with permeable soil depths exceeding 50 cm and a water table depth greater than 60 cm. Soils with more than 100 cm of permeable depth or a water table deeper than 100 cm are placed in this group if the hydraulic conductivity of the top 100 cm is between 0.15 and 1.45 cm/h.

Group D: Soils in this group have high runoff potential when wet, typically containing more than 40% clay and less than 50% sand, with clayey textures. These soils may also have significant swelling and shrinking potential in certain regions. Soils with less than 50 cm of permeable depth or a water table within 60 cm of the soil surface are classified in this group. The hydraulic conductivity of horizons between 50 and 100 cm deep is less than 0.35 cm/h. Soils with more than 100 cm of permeable depth or a water table deeper than 100 cm belong in this group if the hydraulic conductivity of the top 100 cm is less than 0.15 cm/h.

Table 3 summarizes the characteristics described above:

Table 3. Characterization of hydrological soil groups

Hydrological Group	Runoff Potential	Texture
A	Minimal	Sand
B	Low	Sandy soil Silty sand Sandy clay Sandy clay loam

continued on following page

Table 3. Continued

Hydrological Group	Runoff Potential	Texture
C	Medium	Silty clay Silty clayey sand Clayey silt Silty sandy clay Silty sandy loam Silty sand Sandy clay loam
D	High	Clay

Figure 5. Hydrological soil map of the study area

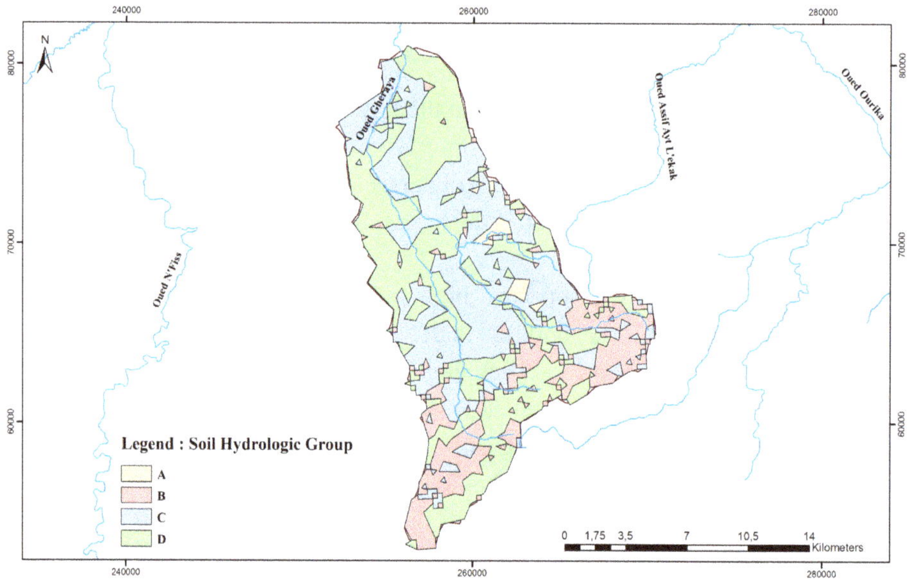

Figure 6. Runoff map of the study area

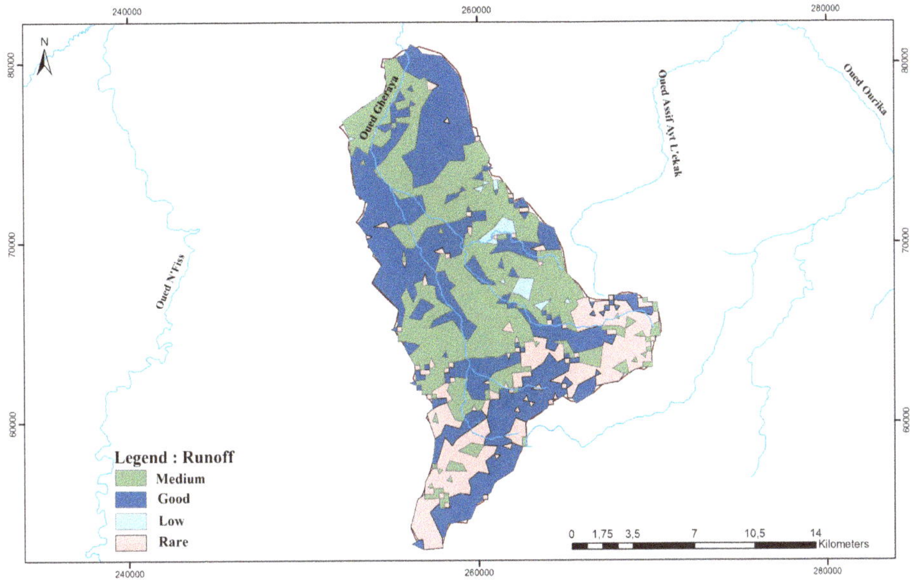

a) The Slope

The traditional method for delineating watersheds from topographic maps is imprecise and time-consuming. In recent years, this method has been replaced by the automatic extraction of information from a DEM (Chehbouni et al. 2008) . It is recommended that rainwater harvesting areas have slopes of less than 5% as they require extensive earthworks and are likely to undergo extraordinary erosion due to the uneven distribution of runoff water. DEM data was used to obtain topographic characteristics such as slope, altitudes, and the watershed boundary (Figure 7). The DEM used is ASTER-GDEM with an 8m resolution. The DEM allowed for the extraction of all physical characteristics of the watershed under ArcGIS using the Arc Hydro Tool. The slope map is categorized into four distinct classes, as illustrated in Figure 7.

Figure 7. Slope map of the Oued Rheraya watershed.

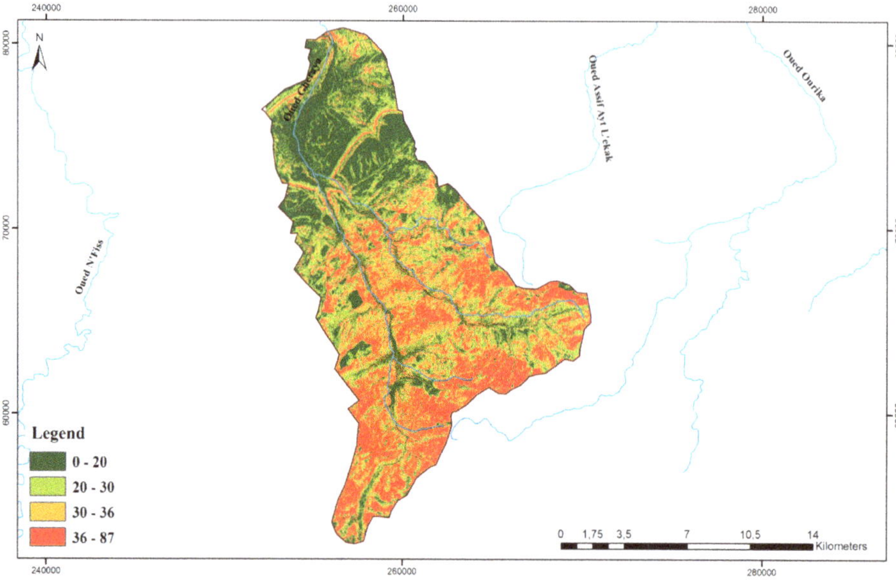

b) Soil Texture:

The characteristics of watershed soils are crucial for determining suitable sites for rainwater harvesting, as they influence both surface runoff and infiltration rates. Soils with medium to fine textures are generally more favorable for rainwater harvesting due to their improved ability to retain and infiltrate water. (Mbilinyi et al. 2007). Clayey soil, with its high-water retention capacity and low permeability, is particularly suitable for water conservation. Sites with clayey soil are optimal for rainwater harvesting as they effectively hold and manage collected water. (Adham et al. 2018). Therefore, soil texture is a crucial criterion for selecting sites for rainwater harvesting. The soil texture map, classified according to the FAO World Map, reveals that the basin contains four distinct soil types, as illustrated in Figure 8

Figure 8. Soil texture map of the Rheraya watershed.

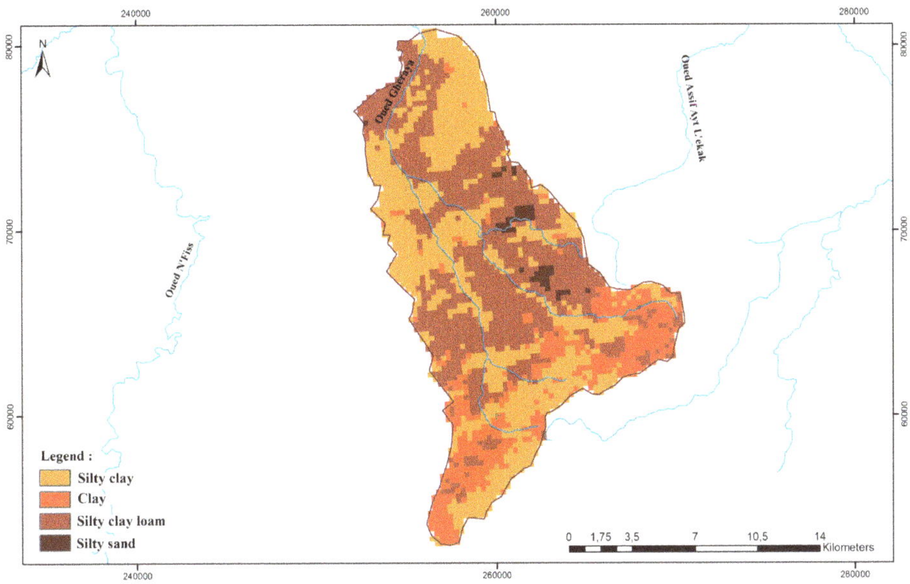

c) Vegetation Index

The Normalized Difference Vegetation Index (NDVI) is a key measure used to assess vegetation. It is characterized by a distinct spectral signature, with a prominent peak in the near-infrared spectrum and lower reflectance in the red spectrum. NDVI emphasizes the contrast between the visible red band and the near-infrared band.(De La Iglesia Martinez and Labib 2023)

$$\mathrm{NDVI} = \frac{(\mathbf{PIR} - \mathbf{R})}{(\mathbf{PIR} + \mathbf{R})}$$

The Normalized Difference Vegetation Index (NDVI) is highly useful for identifying vegetation presence, evaluating biomass, and measuring photosynthetic activity. NDVI values range from -1 to +1, with higher values indicating denser vegetation. Negative values are associated with non-vegetated surfaces like snow, water, or clouds, where reflectance in the red band exceeds that in the near-infrared band. Various vegetation indices can be derived from satellite data, with NDVI being one of them, calculated from reflectances in the red and near-infrared channels.

According to NDVI maps, vegetation cover in the Rheraya watershed is relatively sparse, indicating degraded quality in both coverage and regeneration. The land cover in the Rheraya watershed is characterized by three types:

- Bare soil: NDVI < 0.15
- Dense cover (forest): NDVI > 0.25
- Moderately dense cover (grassland): NDVI between 0.15 and 0.25

A rise in positive NDVI values signifies an increase in green vegetation, while negative or near-zero NDVI values indicate non-vegetated areas, which could include rocks, bare soil, or water bodies. Figure 9 depicts the spatial distribution of vegetation cover throughout the study area during the spring season.

Figure 9. Vegetation index (NDVI) map.

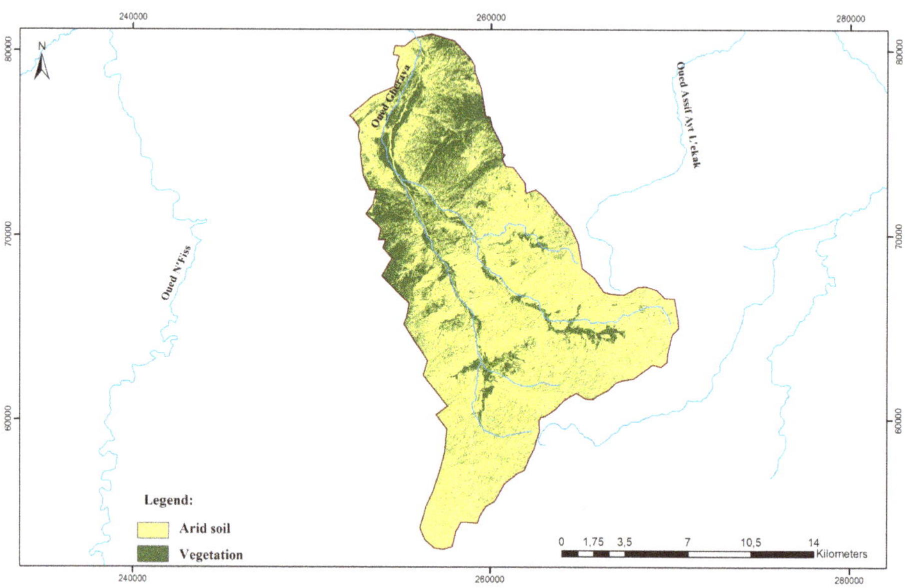

d) Land Use:

The effect of land use on hydrology has been a significant research area over the past decade, particularly on runoff generation Liu et al. 2014. Therefore, it was essential to create a land use database called "Land Use." To determine the land use of the area through remote sensing, the LANDSAT 8 OLI satellite image with a 4-3-2 (red-green-blue) band combination was used to provide natural colors. The supervised classification was established using the Maximum Likelihood Classification (MLC) method (Krois and Schulte 2014). Supervised classification was used as a satellite image processing technique to extract land use data in our study area. This technique has been widely used by many authors and in various fields (Maimouni et al., 2012). As its name suggests, supervised classification requires operator supervision, who must have prior knowledge of the area to set the number of different classes that essentially correspond to the main land use types. On the image, the operator must identify samples of the various known classes in training areas (also called regions of interest) and perform a statistical analysis of their spectral signature (Benchelha et al. n.d.). Supervised classification is based on the principle that the spectral signatures of a given class are identical and cannot exist in another class. The result of the supervised classification on the Landsat-8 satellite image is shown in Figure 10.

Figure 10. Land use map

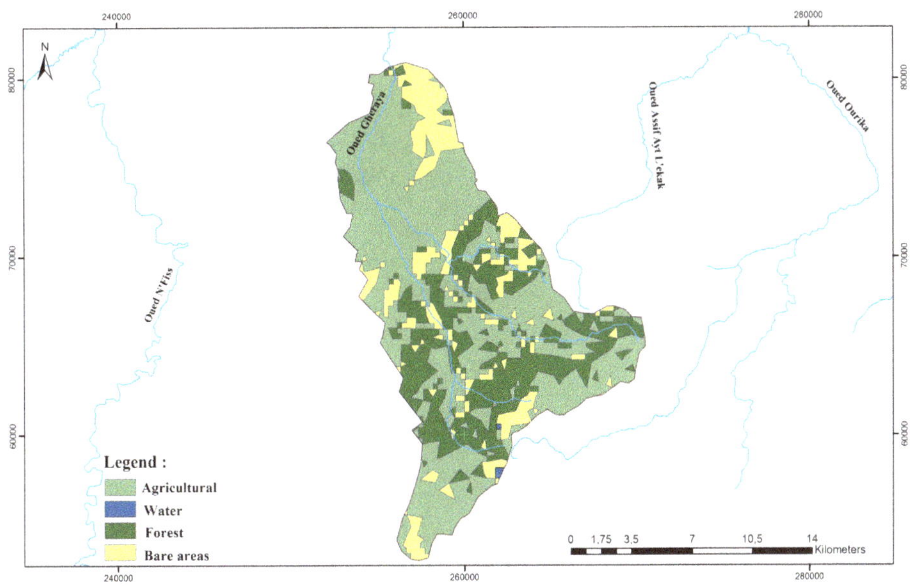

e) Drainage Density

Drainage density allows for the assessment of the importance of the hydrographic network and the possibilities for runoff evacuation in the basin. (Sab u et al. 2023) Considering the area, the drainage density formula is written as follows:

$$Dd\ (km-1) = \frac{\text{Total length of the hydrographic network (km)}}{\text{Area of the basin (km}^2\text{)}}$$

A complex and well-developed hydrographic network leads to a very high accumulation, which is an inverse factor. (Ganie et al. 2023) The distribution varies due to the water flow density in each square, as shown in Figure 11.

Regarding the Rheraya basin, a simple examination of its figures shows that it has a relatively low drainage density compared to the surrounding basins. The relatively small area and the presence of several high peaks, such as Toubkal (4167 m), can favor the formation of a large reservoir of water in either solid or liquid form. In other words, the high altitudes and the modest surface area can create a slope that promotes rapid runoff of streams, concentrating into the main drain. In fact, the Rheraya is relatively less drained despite the conditions favoring significant drainage, which it encompasses, compared to other basins (the high altitudes where precipitation accumulates in solid form or lakes). Given the various scales and measurements used in this study, each criterion has been categorized based on the different suggested criteria. Table 1 outlines the factors and categorizes pixel values according to the scales recommended in the literature. (Al-Adamat et al. 2012) (Baban and Wan-Yusof 2003). Each map created for the factors contains pixel values, which are classified according to their suitability.

Figure 11. Drainage density map.

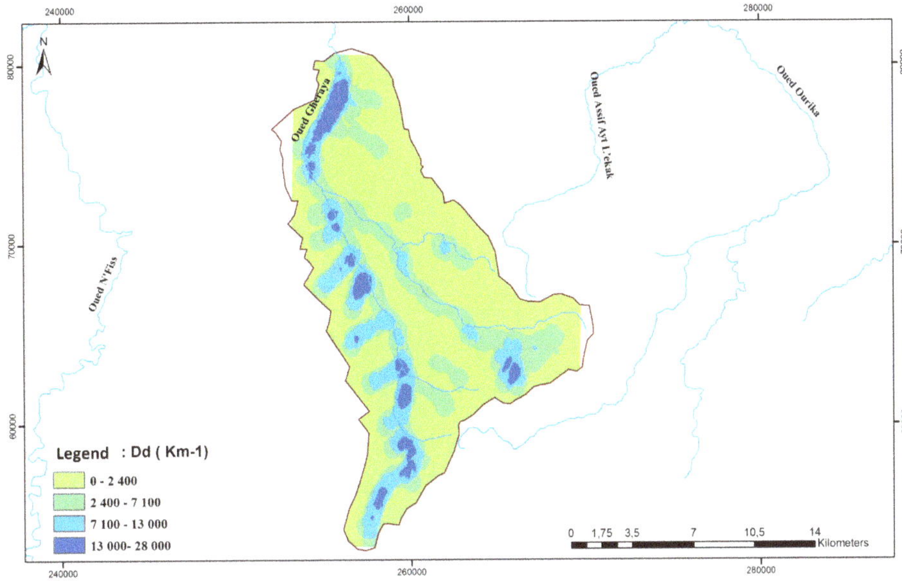

f) Generating Maps Using GIS

The point allocation method assigns weights to criteria based on their priority by having the decision-maker allocate a specific number of points to each criterion. Criteria that receive more points are considered more important relative to others. (Baban and Wan-Yusof 2003). The process of calculating the relative importance of each criterion is known as criteria normalization (Saaty 1984). Thus, to integrate multiple factors (both qualitative and quantitative) into the model, they need to be made comparable, i.e., to formulate the suitability of the different factors on a common scale. Once the evaluation criteria are identified concerning the search for suitable rainwater harvesting areas, it is necessary to determine the layer of each factor that indicates the suitability degree of each spatial unit. For each identified criterion, the procedure calculates a criterion map that measures the level of suitability or unsuitability of the spatial unit for rainwater harvesting for the specified criterion. These suitability classifications must then be ranked based on the relative importance of the contribution made by this particular criterion to achieve the final goal. This will result in a series of maps that will be aggregated by weighted linear combination.

After creating the criteria maps in GIS, each criterion is ranked according to its significance in the water collection process. During reclassification, the criterion with the greatest impact on water collection is assigned the highest value. The criteria's weighting reflects their importance in rainwater collection, as outlined in Table 4. This study adopts the weights and ratings established by the literature. (Al-Adamat 2008) (Baban and Wan-Yusof 2003), which have similar environmental and physical conditions to those in the current study.

Table 4. Classification and Weighting for Each Criterion

Criteria	Classification	Weight
Runoff (mm)	4	6
High	3	
Moderately high	2	
Moderately low	1	
Low		
HSG	4	5
D	3	
C	2	
B	1	
A		
La pente (%)	4	4
0 - 20	3	
20 - 30	2	
30 - 36	1	
36 - 87		
La texture de sol	4	3
Clay loam	3	
Loam	2	
Sandy clay loam	1	
Sandy loam		
NDVI	2	2
Vegetation	1	
Bare soil		
la densité de drainage	3	1
Much order	2	
Average order	1	
Less order		

III. RESULTS AND DISCUSSION

The initial results of this study provide foundational data on soil types and land use. The combination of land use and soil type is the most influential factor in run-off generation. Soil type, defined by grain size composition, plays a critical role. The Rheraya watershed predominantly features two soil types: clayey and fine silty clay. These soil types, classified as hydrological soil group "D," are characterized by very low infiltration rates and high runoff potential.

Using ArcGIS and the SCS-CN model, each pixel is assigned a score across the six main layers. As shown in Figure 12, the model outputs a map categorized into three rainwater harvesting zones: suitable, moderate, and unsuitable. The map indicates that 17% of the total research area is suitable for rainwater harvesting structures, while 55% is moderately suitable, and 28% is unsuitable for rainwater harvesting implementation.

The slopes of the primary suitable areas range from 1% to 3%, and the predominant soil textures are silty clay to sandy silty clay. These characteristics are crucial in runoff production due to their low permeability. Analysis of the topographic map and high-resolution satellite images reveals that most of the rocky lands are mountainous with steep slopes and lack vegetation cover. The findings of this study align with those of the study by (Mbilinyi et al. 2007), which indicated that areas with soils such as silty clay combined with gentle slopes are ideal for implementing rainwater harvesting structures.

The final step involves integrating various maps—such as hydrologic soil groups (HSG), drainage, slope, soil texture, NDVI, and runoff calculations—to determine the most suitable sites for runoff water harvesting using a weighted linear combination approach. Criteria were assigned weights based on their importance, employing a point allocation method. This method assigns a specific number of points to each criterion, with higher points reflecting greater relative importance.

Figure 12. Map of Suitable Zones for Rainwater Harvesting

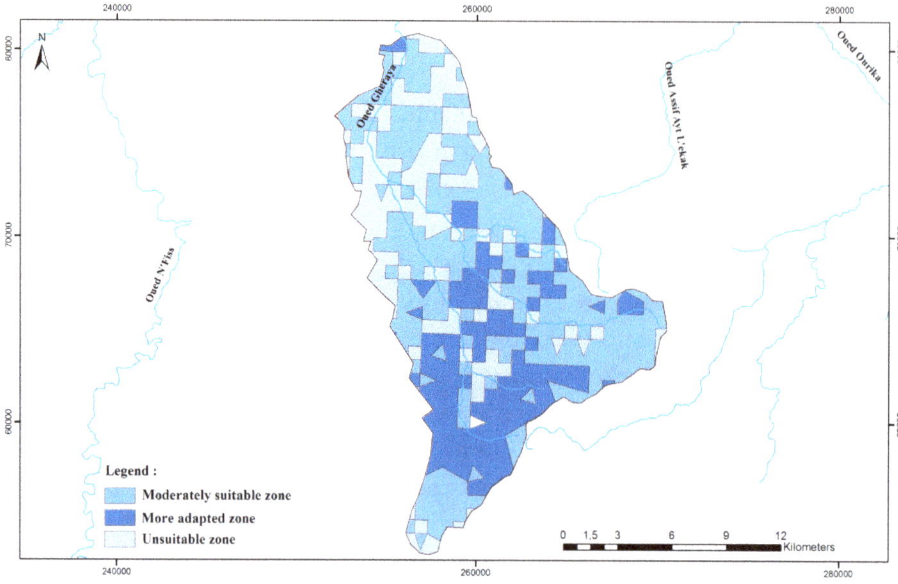

The comparative results of the multi-criteria analysis and the SCS-CN model in the Rheraya watershed (fig. 13) have demonstrated that: The multi-criteria analysis was validated by the SCS-CN model; the most suitable areas are shown in dark blue on both maps. Moreover, a hillside dam is currently under construction in this watershed, with the boundary of the moderately suitable zone located precisely in the commune of Asni, on the Imlil River. Our study reveals other locations for the implementation of new hillside dams, whether to the north or south of the watershed.

Figure 13. Comparative Results of Multi-Criteria Analysis and SCS-CN Model

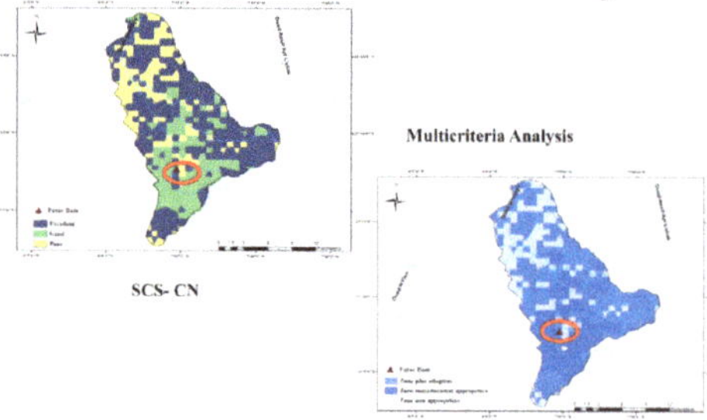

At the end of this work, two models were used to determine water accumulation rates using multi-criteria analysis (El Ghazali et al. 2021a) and SCS-CN, which included several major and minor factors. The long-term objective is to use these models as decision support tools in the implementation of future dams.

The multi-criteria analysis was validated by the SCS-CN model based on the compilation of various parameters governing water collection, using remote sensing and GIS, which enabled decision-making for better management and planning of water resources over large areas in arid and semi-arid climates, while reducing necessary investments.

The driest regions suffer from water scarcity. These models are therefore very valuable decision-making tools for the effective governance of this natural resource, as they can cover thousands of square kilometers with a very limited budget. The cost-effectiveness of this methodology is highly satisfactory.

IV. CONCLUSION

The design of a map of suitable zones for rainwater harvesting in the Rheraya watershed revealed three recharge classes, namely, an area with very low potential below 5%, an area with low potential between 5% and 10%, and an area with medium potential ranging from 10% to 20%. The summary map serves as a decision-making aid for planning future artificial accumulation projects by identifying high-potential dam sites. The tool allows for the calculation of the actual annual accumulated rainfall volume.

The model output produces a map with three categories of rainwater harvesting zones: suitable, moderate, and unsuitable. The resulting map shows that the downstream area of the watershed is suitable for the implementation of rainwater harvesting structures. The suitable area represented 17% of the total research area, while 55% represent moderately suitable areas for rainwater harvesting, and 28% represent areas that are not suitable for the implementation of rainwater harvesting structures.

The empirical model was validated by the SCS-CN model based on the compilation of various parameters governing water collection using remote sensing and GIS, which enabled decision-making for better management and planning of water resources over large areas in arid and semi-arid climates, while reducing necessary investments. As a data layer, the result was presented in percentages.

REFERENCES

Abdelmajid, S., Mukhtar, A., Baig, M. B., & Reed, M. R. (2021). Climate change, agricultural policy and food security in Morocco. Emerging Challenges to Food Production and Security in Asia, Middle East, and Africa: Climate Risks and Resource Scarcity, 171–196. DOI: 10.1007/978-3-030-72987-5_7

Adham, A., Sayl, K. N., Abed, R., Abdeladhim, M. A., Wesseling, J. G., Riksen, M., Fleskens, L., Karim, U., & Ritsema, C. J. (2018). A GIS-based approach for identifying potential sites for harvesting rainwater in the Western Desert of Iraq. *International Soil and Water Conservation Research*, 6(4), 297–304. DOI: 10.1016/j.iswcr.2018.07.003

Al-Adamat, R. (2008). GIS as a decision support system for siting water harvesting ponds in the Basalt Aquifer/NE Jordan. *Journal of Environmental Assessment Policy and Management*, 10(02), 189–206. DOI: 10.1142/S1464333208003020

Al-Adamat, R., AlAyyash, S., Al-Amoush, H., Al-Meshan, O., Rawajfih, Z., Shdeifat, A., Al-Harahsheh, A., & Al-Farajat, M. (2012). The combination of indigenous knowledge and geo-informatics for water harvesting siting in the Jordanian Badia. DOI: 10.4236/jgis.2012.44042

Baban, S. M. J., & Wan-Yusof, K. (2003). Modelling optimum sites for locating reservoirs in tropical environments. *Water Resources Management*, 17(1), 1–17. DOI: 10.1023/A:1023066705226

Bahija, Abdelmoumen, & Souad. (2023). Analysis of the Hydrological Functioning of the Oued Reghaya Watershed in the Context of Climate Change (Tensift, Morocco). African Journal of Advanced Pure and Applied Sciences, 160–65.

Banasik, K., & Woodward, D. (2010). Empirical Determination of Runoff Curve Number for a Small Agricultural Watershed in Poland. 2nd Joint Federal Interagency Conference, Las Vegas, NV. DOI: 10.3390/w6051118

Bo, X. I. A. O., Qing-Hai, W. A. N. G., Jun, F. A. N., Feng-Peng, H. A. N., & Quan-Hou, D. A. I. (2011). Application of the SCS-CN Model to Runoff Estimation in a Small Watershed with High Spatial Heterogeneity. *Pedosphere*, 21(6), 738–749. DOI: 10.1016/S1002-0160(11)60177-X

Boudhar, A. (2009). *Télédétection Du Manteau Neigeux et Modélisation de La Contribution Des Eaux de Fonte Des Neiges Aux Débits Des Oueds Du Haut Atlas de Marrakech*. Université Cadi Ayyad, Faculté Des Sciences et Techniques. DOI: 10.1016/j.jag.2009.09.008

Brooks, K. N., Ffolliott, P. F., Gregersen, H. M., & DeBano, L. F. (2003). *Hydrology and the Management of Watersheds*. Iowa State University Press. DOI: 10.5923/j.re.20140402.04

Bzioui, M. (2004). Rapport National 2004 Sur Les Ressources En Eau Au Maroc. *UN Water-Africa*, 94. Advance online publication. DOI: 10.4314/aga.v19i3.1723

Chehbouni, A., Escadafal, R., Duchemin, B., Boulet, G., Simonneaux, V., Dedieu, G., Mougenot, B., Khabba, S., Kharrou, H., Maisongrande, P., Merlin, O., Chaponnière, A., Ezzahar, J., Er-Raki, S., Hoedjes, J., Hadria, R., Abourida, A., Cheggour, A., Raibi, F., & Sobrino, J. A. (2008). An Integrated Modelling and Remote Sensing Approach for Hydrological Study in Arid and Semi-arid Regions: The SUDMED Programme. *International Journal of Remote Sensing*, 29(17–18), 5161–5181. DOI: 10.1080/01431160802036417

Diskin, M. H., & Nazimov, N. (1995). Linear Reservoir with Feedback Regulated Inlet as a Model for the Infiltration Process. *Journal of Hydrology (Amsterdam)*, 172(1–4), 313–330. DOI: 10.1016/0022-1694(95)02709-X

Dorel, M. (1993). Travail Du Sol En Bananeraies: Cas Des Andosols. Fruits, 48(2), 77–82. https://doi.org/ RTC.2409DOI: 10.62498.A

El Garouani, A., Aharik, K., & El Garouani, S. (2020). Water Balance Assessment Using Remote Sensing, Wet-Spass Model, CN-SCS, and GIS for Water Resources Management in Saïss Plain (Morocco). *Arabian Journal of Geosciences*, 13(15), 738. DOI: 10.1007/s12517-020-05730-y

El Ghazali, F. E., Laftouhi, N.-E., Fekri, A., Randazzo, G., & Benkirane, M. (2021a). Enhancing the Success of New Dams Implantation under Semi-Arid Climate, Based on a Multicriteria Analysis Approach: Case of Marrakech Region (Central Morocco). *Open Geosciences*, 13(1), 1494–1508. DOI: 10.1515/geo-2020-0313

Elghouat, A., Algouti, A., Algouti, A., Baid, S., Ezzahzi, S., Kabili, S., & Agli, S. (2024). Integrated Approaches for Flash Flood Susceptibility Mapping: Spatial Modeling and Comparative Analysis of Statistical and Machine Learning Models. A Case Study of the Rheraya Watershed, Morocco. *Journal of Water and Climate Change*, 15(8), jwc2024726. Advance online publication. DOI: 10.2166/wcc.2024.726

Elhakeem, M., & Papanicolaou, A. N. (2009). Estimation of the Runoff Curve Number via Direct Rainfall Simulator Measurements in the State of Iowa, USA. *Water Resources Management*, 23(12), 2455–2473. DOI: 10.1007/s11269-008-9390-1

Forootan, E. (2023). GIS-Based Slope-Adjusted Curve Number Methods for Runoff Estimation. *Environmental Monitoring and Assessment*, 195(4), 489. DOI: 10.1007/s10661-023-11039-6 PMID: 36939939

Gabriel, N., & Diabé, E. (2021). Évaluation de La Susceptibilité Des Sols Au Ruisseau Dans Le Bassin Versant Du Knam (Ouest Cameroun), En Appliquant La Méthode Services de Conservation Des Sols Numéro de Courbe. *Revista de Geomorfologie*, 23(1), 39–53. DOI: 10.21094/rg.2021.136

Ganie, P. A., Posti, R., Kunal, K., Kunal, G., Bharti, V. S., Sehgal, V. K., Sarma, D., & Pandey, P. K. (2023). Modelling of the Himalayan Mountain River Basin through Hydro-Morphological and Compound Factor-Based Approaches Using Geoinformatics Tools. *Modeling Earth Systems and Environment*, 9(3), 3053–3084. DOI: 10.1007/s40808-023-01691-9

Giridhar, M. V. S. S., & Viswanadh, G. K. (2014). Runoff Estimation in an Ungauged Watershed Using RS and GIS. *Journal of Indian Association Water Works*, 3, 2009. DOI: 10.1016/j.jhydrol.2014.02.028

Hajhouji, Y. (2018). Modélisation Hydrologique Du Bassin Versant de l'oued Rheraya et Sa Contribution à La Recharge de La Nappe Du Haouz (Bassin Du Tensift, Maroc). https://ird.hal.science/ird-04351673

Hashim, H. Q., & Sayl, K. N. (2022). Incorporating GIS Technique and SCS-CN Approach for Runoff Estimation in the Ungauged Watershed: A Case Study West Desert of Iraq. *Iraqi Journal of Civil Engineering*, 14(2), 1–6. DOI: 10.37650/ijce.2020.172862

Hawkins, R. H., Ward, T. J., Woodward, D. E., & Van Mullem, J. A. (2008). *Curve Number Hydrology: State of the Practice*. American Society of Civil Engineers. DOI: 10.1061/9780784410042

Jahan, K., Pradhanang, S. M., & Md, A. E. B. (2021). Surface Runoff Responses to Suburban Growth: An Integration of Remote Sensing, GIS, and Curve Number. *Land (Basel)*, 10(5), 452. DOI: 10.3390/land10050452

Jeon, J.-H., Lim, K. J., & Engel, B. A. (2014). Regional Calibration of SCS-CN L-THIA Model: Application for Ungauged Basins. *Water (Basel)*, 6(5), 1339–1359. DOI: 10.3390/w6051339

Khouz, A., Jaffal, M., Trindade, J., Bougadir, B., El Bchari, F., Kchikach, A., El Ghorfi, M., Ibouh, H., Jadoud, M., Kadiri, O., & Manar, A. (2023). Understanding the Deep Structure of the Essaouira Basin Using Gravity Data: Hydrogeological Inferences for a Semiarid Region in Central-Western Morocco. *Geosciences*, 13(11), 345. DOI: 10.3390/geosciences13110345

Krois, J., & Schulte, A. (2014). GIS-Based Multi-Criteria Evaluation to Identify Potential Sites for Soil and Water Conservation Techniques in the Ronquillo Watershed, Northern Peru. *Applied Geography (Sevenoaks, England)*, 51, 131–142. DOI: 10.1016/j.apgeog.2014.04.006

Lahcen, M., Nidal, F., & Mohamed, G. (2020). *Floristic Study and Assessment of the Environmental Factors Governing the Distribution of Riparian Plants in the Zat Sub-Basin*. Tensift Watershed. DOI: 10.35759/JAnmPlSci.v45-2.3

Lasco & Moglen. (2023). Hydrologic Conservatism as a Rationale for Selecting NRCS Initial Abstraction Ratio. Journal of Hydrologic Engineering, 28(12), 04023041. DOI: 10.1061/JHYEFF.HEENG-6016

Liu, Z., Yao, Z., Huang, H., Wu, S., & Liu, G. (2014). Land Use and Climate Changes and Their Impacts on Runoff in the Yarlung Zangbo River Basin, China. *Land Degradation & Development*, 25(3), 203–215. DOI: 10.1002/ldr.1159

Mack, M. J. (1995). HER—Hhydrologic Evaluation of Runoff; The Soil Conservation Service Curve Number Technique as an Interactive Computer Model. *Computers & Geosciences*, 21(8), 929–935. DOI: 10.1016/0098-3004(95)00029-8

Maimouni, S., Bannari, A., El-Harti, A., & El-Ghmari, A. (2012). "Indices spectraux et" spectral angle mapper" pour la cartographie des risques d'érosion hydrique à partir des données ali eo-1. Teledetection, 10(4), 213–23. https://hal.science/hal-01948890

Malik, S., Pal, S. C., Chowdhuri, I., Chakrabortty, R., Roy, P., & Das, B. (2020). Prediction of Highly Flood Prone Areas by GIS Based Heuristic and Statistical Model in a Monsoon Dominated Region of Bengal Basin. *Remote Sensing Applications: Society and Environment*, 19, 100343. DOI: 10.1016/j.rsase.2020.100343

Martinez, A. I., & Labib, S. M. (2023). Demystifying Normalized Difference Vegetation Index (NDVI) for Greenness Exposure Assessments and Policy Interventions in Urban Greening. *Environmental Research*, 220, 115155. DOI: 10.1016/j.envres.2022.115155 PMID: 36584843

Mbilinyi, B. P., Tumbo, S. D., Fm Mahoo, H., & Mkiramwinyi, F. O. (2007). GIS-Based Decision Support System for Identifying Potential Sites for Rainwater Harvesting. *Physics and Chemistry of the Earth Parts A/B/C*, 32(15–18), 1074–1081. DOI: 10.1016/j.pce.2007.07.014

Meißl, G., Klebinder, K., Zieher, T., Lechner, V., Kohl, B., & Markart, G. (2023). Influence of Antecedent Soil Moisture Content and Land Use on the Surface Runoff Response to Heavy Rainfall Simulation Experiments Investigated in Alpine Catchments. *Heliyon*, 9(8), e18597. Advance online publication. DOI: 10.1016/j.heliyon.2023.e18597 PMID: 37560701

Mishra, S. K., Kansal, A. K., & Aggarwal, N. (2012). Assessment of Design Runoff Curve Number for a Watershed. *Water Practice & Technology*, 7(4), wpt2012066. Advance online publication. DOI: 10.2166/wpt.2012.066

Musy, A., & Higy, C. (2004). Hydrologie: Une Science de La Nature. Vol. 21. PPUR presses polytechniques. DOI: 10.4000/geocarrefour.1794

Ouassanouan, Y., Fakir, Y., Simonneaux, V., Kharrou, M. H., Bouimouass, H., Najar, I., Benrhanem, M., Sguir, F., & Chehbouni, A. (2022). Multi-Decadal Analysis of Water Resources and Agricultural Change in a Mediterranean Semiarid Irrigated Piedmont under Water Scarcity and Human Interaction. *The Science of the Total Environment*, 834, 155328. DOI: 10.1016/j.scitotenv.2022.155328 PMID: 35452720

Rhoujjati, N., Hanich, L., Bouchaou, L., Patris, N., N'da, A. B., & Chehbouni, A. (2021). Isotopic Tracers to Assess the Snowmelt Contribution to the Groundwater Recharge: A Case from the Moroccan High and Middle Atlas Mountains. *Arabian Journal of Geosciences*, 14(23), 2611. DOI: 10.1007/s12517-021-08737-1

Saaty. (1984). The Analytic Hierarchy Process: Decision Making in Complex Environments. In Quantitative assessment in arms control. Springer. DOI: 10.3390/conservation3010003

Sab u, D. A., & Gh Şerban, P. (2023). Combining Radar Quantitative Precipitation Estimates (QPEs) with Distributed Hydrological Model for Controlling Transit of Flash-Flood Upstream of Crowded Human Habitats in Romania. *Natural Hazards*, 116(1), 1209–1238. DOI: 10.1007/s11069-022-05718-9

Salma, K., Ahmed, A., Abdellah, A., & Akram, E. G. (2023). Quantification of Water Erosion Using Empirical Models RUSLE and EPM in the Rheraya Basin in the High Atlas of Marrakech. *Disaster Advances*, 16(5), 19–28. DOI: 10.25303/1605da019028

Sayl, K. N., Mohammed, A. S., & Ahmed, A. D. (2020). GIS-Based Approach for Rainwater Harvesting Site Selection. B IOP Conference Series: Materials Science and Engineering. DOI: 10.1088/1757-899X/737/1/012246

Soulaimani, A., & Ouanaimi, H. (2011). Anti-Atlas et Haut Atlas, Circuit Occidental. Nouveaux Guides Géologiques et Miniers Du Maroc. *Notes et Mémoires Du Service Géologique Du Maroc*, 3, 9–72.

Walega, A., Amatya, D. M., Caldwell, P., Marion, D., & Panda, S. (2020). Assessment of Storm Direct Runoff and Peak Flow Rates Using Improved SCS-CN Models for Selected Forested Watersheds in the Southeastern United States. *Journal of Hydrology. Regional Studies*, 27, 100645. DOI: 10.1016/j.ejrh.2019.100645

Chapter 7
Monitoring Agricultural Dynamics and Its Impact on Water Resources Using GIS and Remote Sensing Data:
A Case Study in the Ben Mansour–Mnasra Region, Kenitra, Morocco

Tarik Bahouq
Département de Géographie, Faculté des Sciences Humaines et Sociales, Université Ibn Tofaïl, Kénita, Morocco

Amina Moumane
https://orcid.org/0009-0001-9188-6964
Département de Géographie, Faculté des Sciences Humaines et Sociales, Université Ibn Tofaïl, Kénita, Morocco

Nadia Touhami
Département de Géographie, Faculté des Sciences Humaines et Sociales, Université Ibn Tofaïl, Kénita, Morocco

ABSTRACT

The Ben Mansour region in the Kénitra province of northwestern Morocco, a vital agricultural hub, is facing significant environmental and hydrological challenges due to its intensive agricultural practices and climatic conditions. This study examines land use and land cover changes over two decades (2000, 2010, and 2020)

DOI: 10.4018/979-8-3693-9651-3.ch007

and assesses groundwater quality and levels in the region. Using Landsat satellite imagery, the authors analyzed vegetation dynamics through the NDVI, revealing substantial shifts in land use. Groundwater assessments, including depth and electrical conductivity measurements, reveal critical issues of overexploitation and salinization. Groundwater levels vary widely, with notable depletion and increased salinity due to intensive extraction and seawater intrusion. These findings underscore the urgent need for sustainable water management practices and land use strategies to address the environmental impacts and ensure the long-term viability of agricultural activities in the region.

INTRODUCTION

In recent decades, the intensification of agriculture and increased water extraction have emerged as critical challenges with profound implications for water resources and agricultural sustainability, particularly in arid and semi-arid regions (Molle and Tanouti 2017). Morocco, located in North Africa with a predominantly arid to semi-arid climate, faces significant water stress driven primarily by human activities rather than solely by natural factors (Benabderrazik et al. 2021; Ouassanouan et al. 2022).

The Mediterranean region, characterized by its distinctive climate with hot, dry summers and mild, wet winters (Tramblay et al. 2020), is experiencing escalating pressures due to intensified agricultural practices (Devkota and Yigezu 2020). In Morocco, these pressures are manifested in the overextraction of groundwater to support high-yield farming and greenhouse cultivation (El Mountassir and Bahir 2023). This has led to unsustainable extraction rates, especially in areas where groundwater resources are already limited (Moumane et al. 2021). The situation is further exacerbated by climate change-induced droughts, which intensify the competition for limited water resources and strain existing water management systems (El Assaoui et al. 2021).

Within this context, the Ben Mansour- Mnasra region in the Kénitra province of northwestern Morocco represents a microcosm of the broader challenges facing the Mediterranean region (Hssaisoune et al. 2020). This area, integral to Morocco's agricultural sector, is characterized by its fertile sandy-loam soils and favorable climatic conditions. However, these advantages are overshadowed by significant water-related challenges. The region's reliance on groundwater for irrigation, driven by the expansion of private wells and intensified agricultural practices, has led to critical concerns about the sustainability of water resources (Aguedai et al. 2022).

Groundwater resources in coastal zones like the Ben Mansour- Mnasra are particularly vulnerable to overexploitation. The sandy sediments that characterize the region's aquifers offer high recharge potential but are also prone to salinization when subjected to excessive extraction. The intrusion of seawater into the aquifer further exacerbates the problem (Aguedai et al. 2020), compromising water quality and posing a threat to agricultural productivity. The delicate balance between groundwater recharge and extraction is disrupted, leading to adverse impacts on both the quantity and quality of water available for irrigation. (Mohammed et al. 2022).

This study aims to address these challenges by analyzing land cover changes, assessing vegetation dynamics, and evaluating groundwater levels and salinity in the Ben Mansour- Mnasra region. Using Landsat satellite imagery and geostatistical modeling, the research provides insights into the spatial and temporal variations in land use and vegetation cover, offering a comprehensive understanding of the region's evolving landscape. The findings will inform strategies for sustainable land and water resource management, crucial for maintaining the productivity and ecological health of the Ben Mansour- Mnasra region.

2. MATERIALS AND METHODS

2.1 Study Area

The Ben Mansour- Mnasra region, located within the Kénitra region along the northwestern coast of Morocco (Figure 1), plays a crucial role in the country's agricultural sector. Situated in the Rharb basin along the Atlantic coast., with a focus on crops like avocado, bananas and peanuts, as well as various other crops. The expansion of agriculture in the region has led to a significant increase in the development of private wells, essential for maintaining a reliable water supply for irrigation and sustaining agricultural activities (Figure 3).

The region experiences a Mediterranean climate with a strong oceanic influence, receiving an average annual precipitation of approximately 551 mm.

The geology of the study area during the Plio-Quaternary period witnessed the accumulation of a diverse range of sediments over the underlying Mio-Pliocene marls. These deposits are characterized by a mix of clays, sands, and gravels of continental origin, along with sands, sandstones, and limestones formed in coastal marine settings. Stratigraphically, these deposits are divided into four units: During the Early Quaternary, coastal sedimentation was marked by the formation of littoral biocalcarenites in small gulfs, with dune sands and sandstones along the coast. In contrast, the interior of the basin predominantly saw the accumulation of continental gravelly deposits. The stratigraphic and facies variations of these units are illustrated in Figure 2 (Benseddik et al. 2017; Kili et al. 2006).

Figure 1. Location of the study area

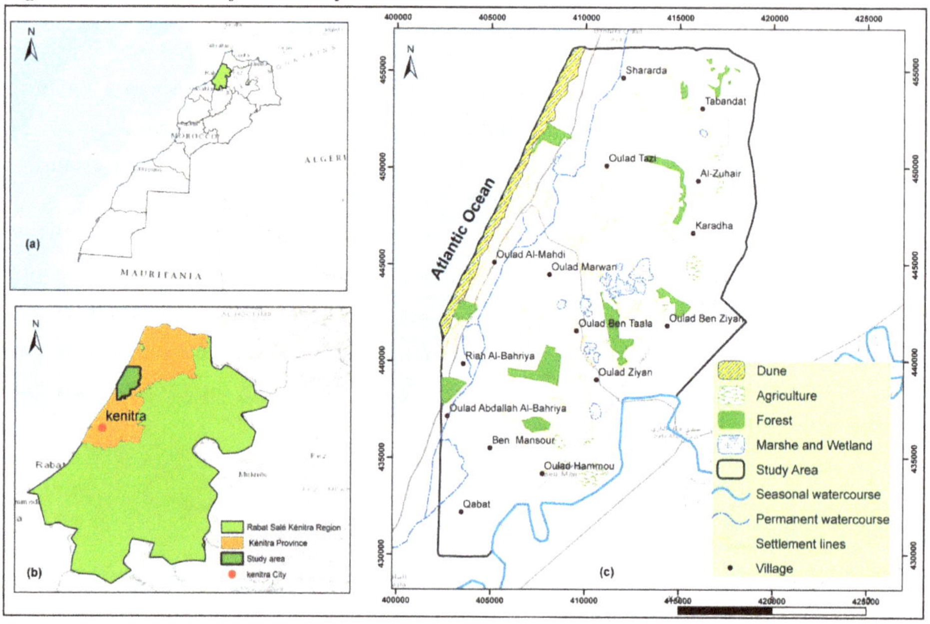

Sandy soils cover a significant portion of the coastal zone of the Gharb Plain (Qanza et al. 2019). The Ben Mansour- Mnasra region is primarily composed of sandy-clay and silty-clay soils (Sanad et al. 2024). These soils play a crucial role in groundwater recharge and influence the overall hydrological behavior of the region. However, the Ben Mansour- Mnasra region faces critical challenges related to groundwater quality. Intensive farming practices have severely impacted the groundwater (Marouane et al. 2014; Omrania et al. 2019; Saâdi and Maslouhi 2003), leading to overexploitation, disrupted hydrodynamic equilibrium, and declining water quality. Of particular concern is the issue of seawater intrusion, which poses a significant threat to the health of the aquifer.

Figure 2. Geological cross-section NW-SE of the coastal zone of Gharb (Mnasra) according to Batchi et al. (2017)

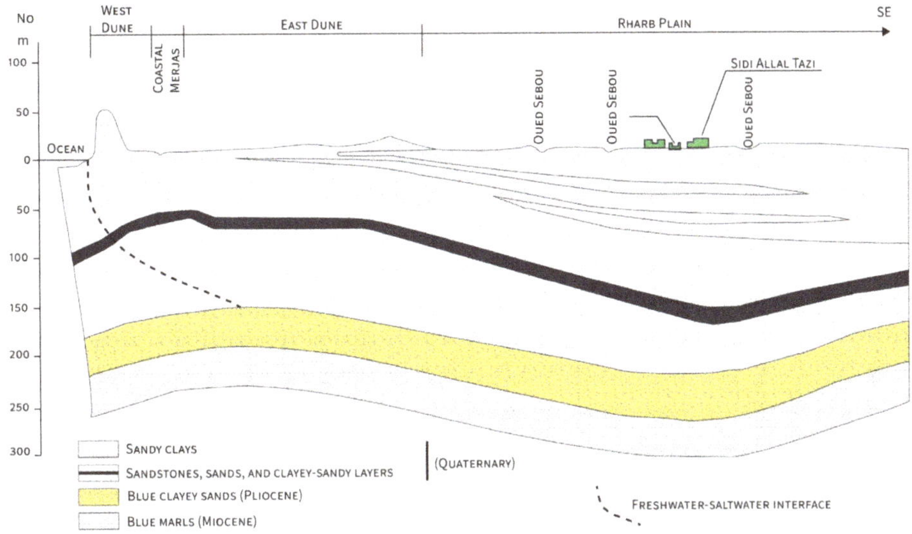

Figure 3. Photos depicting the study area: (a) Field Measurements, (b) Irrigation Basin Filled, (c) Peanuts and Avocado Farm, and (d) Large Tomato Farm

2.1.3 Satellite Image Selection

To monitor land cover change in the study area, three different Landsat satellite images were selected, corresponding to April 24, 2000, April 27, 2010, and May 1, 2020. The images were chosen based on their temporal relevance to the study period and their cloud-free nature, ensuring high-quality data for analysis. The satellite images were obtained from the Earth Explorer website (http://earthexplorer.usgs .gov/), with a spatial resolution of 30 meters (Table 1).

Table 1. Satellite Images Used for Land Cover Change Analysis

Satellite	Sensor	Path/Row	Acquisition Date	Resolution
Landsat 5	TM	201/036	April 24, 2000	30 m
Landsat 5	TM	202/036	April 27, 2010	30 m
Landsat 8	OLI/TIRS	201/036	May 1, 2020	30 m

This table summarizes the key details of the satellite images used in the study, including the satellite name, sensor type, path/row location, acquisition date, spatial resolution. These images were then analyzed to detect changes in land cover over the study period, providing insights into the dynamics of the landscape in the area of interest.

2.2 Maximum Likelihood Classification (MLC)

During this study, the Maximum Likelihood Classification (MLC) method was employed using ArcGIS to analyze the remotely sensed imagery of the study area (Shivakumar and Rajashekararadhya 2018). MLC was chosen due to its robust ability to classify pixels based on the statistical probability of class membership, taking into account the variance and covariance of the spectral signatures of different land cover types. This approach was particularly effective in distinguishing the diverse land cover types present in the study area, including agricultural fields, water bodies, and urban areas. By using well-defined training samples, the classification process was able to generate accurate land cover maps (Heinl et al. 2009), which were essential for further spatial analysis and environmental assessment.

2.3 Vegetation Land Cover Monitoring Using NDVI

To assess long-term changes in vegetation, cover and land use, various methodologies have been developed using satellite imagery. Among these, the Normalized Difference Vegetation Index (NDVI) is a widely used metric for monitoring vegetation dynamics, given its effectiveness in tracking changes in land cover. NDVI is calculated as the ratio between the red band (Red) and the near-infrared band (NIR) (Robinson et al. 2017). The formula 1 for NDVI is as follows:

$$\mathrm{NDVI} = \frac{(\mathrm{NIR} - \mathrm{Red})}{(\mathrm{NIR} + \mathrm{Red})} \tag{1}$$

For Landsat 5 imagery, the NIR band (0.77–0.90 µm) corresponds to band 4, and the red band (0.63–0.69 µm) to band 3. In contrast, for Landsat 8, the NIR band (0.85–0.88 µm) is band 5, and the red band (0.64–0.67 µm) is band 4, as specified by USGS (USGS.gov).

The NDVI values span from -1 to +1, with a value of 0 signifying surfaces devoid of chlorophyll, while positive values reflect increasing levels of chlorophyll content. As NDVI measures photosynthetic activity, it is responsive to environmental fluctuations, resulting in possible yearly variations in vegetation indices. To achieve precise land cover classification, ground truth data was obtained during field visits conducted alongside groundwater surveys. A total of 120 points were randomly selected and verified on-site using a Garmin GPS device to confirm their location and land cover type, leading to an overall accuracy assessment of 89%, which is deemed satisfactory.

2.4 Groundwater Level Measurement

The groundwater level was measured relative to the planting level, which was established as the reference point ($Z = 0$). Measurements were recorded as meters below ground level (m bgl). A manual water level meter was utilized to accurately determine the GL. This device operates by signaling visually and audibly when its electrode contacts the water surface in a well, allowing for precise depth readings. For this analysis, 200 wells were systematically selected throughout the study area during field surveys conducted in April 2020. Data collected for each well included GPS coordinates, elevation (Z), date and time of the measurement, the intended use of the well (agricultural or potable), and any notable field observations.

2.5 Electrical Conductivity Measurement

In parallel with the groundwater level measurements, the electrical conductivity (EC) of the groundwater was also measured in situ. A conductivity meter was employed to assess the salinity levels of the groundwater directly at each well. This method provided immediate insights into the water quality, enabling an understanding of the groundwater's suitability for agricultural or drinking purposes. The integration of GL and EC data offers a comprehensive view of the groundwater characteristics in the study area, highlighting the spatial variability and potential challenges related to water resource management.

2.6 Geostatistical Modeling

This study utilized geostatistical modeling to generate continuous surface maps for Groundwater Level (GL) and Electrical Conductivity (EC) through the Inverse Distance Weighting (IDW) interpolation technique, chosen for its precision in depicting spatial variations based on the proximity of measured points (Bajjali 2023).

3. RESULTS AND DISCUSSION

3.1 Land Use and Land Cover (LULC) Changes in the Area (2000, 2010, 2020)

Between 2000 and 2020, the Land Use and Land Cover (LULC) in the area underwent significant transformations, reflecting broader socio-economic and environmental shifts. The most striking change was the substantial expansion of greenhouse farming, which grew from a mere 2% of the total land area in 2000 to 10% by 2010, and further to 13% by 2020 (Table 2 and figure 4). This increase suggests a strong move towards intensive agriculture and controlled-environment farming, driven by the need to maximize crop yields and adapt to varying climatic conditions. Greenhouses offer a way to protect crops from adverse weather while ensuring consistent production, which is likely why this method has gained popularity over the past two decades.

Table 2. Land Use and Land Cover (LULC) Changes in the Area (2000, 2010, 2020)

LULC Type	2000 (%)	2010 (%)	2020 (%)	Change (2000-2010)	Change (2010-2020)	Overall Change (2000-2020)
Greenhouse	2	10	13	+8	+3	+11
Dune	4	4	4	0	0	0
Rainfed Agriculture	19	10	22	-9	+12	+3
Build-up	9	8	11	-1	+3	+2
Forest	17	14	3	-3	-11	-14
Marshes	10	9	7	-1	-2	-3
Irrigated Land	39	45	40	+6	-5	+1

In contrast, the area covered by dunes remained stable at 4% throughout the 20-year period. This stability might be attributed to successful conservation efforts or natural stabilization processes that prevented further encroachment or erosion of these sandy landscapes. The persistence of dune areas could also indicate that they are not seen as prime targets for agricultural or urban development, thus maintaining their size over Rainfed agriculture, which depends on natural rainfall rather than irrigation, experienced significant fluctuations. In 2000, rainfed agriculture accounted for 19% of the land area. However, by 2010, this had decreased sharply to 10%. This decline could be linked to changes in land use policies, a shift towards more reliable irrigated farming, or possibly a series of dry years that made rainfed farming less viable. Interestingly, by 2020, rainfed agriculture saw a resurgence, increasing to 22%. This rebound suggests a renewed interest in or necessity for rainfed farming, potentially due to the revival of traditional practices, shifts in market demand, or improved rainfall patterns during the latter period.

Urbanization has also been a noticeable trend, albeit progressing at a slower pace. The proportion of built-up areas, which includes residential, commercial, and industrial developments, slightly decreased from 9% in 2000 to 8% in 2010. This minor reduction might be due to a temporary halt in urban expansion or reclassification of land use categories. However, from 2010 to 2020, urban areas expanded to cover 11% of the land. This increase is consistent with the gradual urban sprawl often seen in growing economies, where the demand for housing and infrastructure pushes the boundaries of cities into previously undeveloped areas. A particularly concerning trend is the drastic decline in forest cover. Forests covered 17% of the land in 2000, but this figure dropped to 14% by 2010 and plummeted further to just 3% by 2020. This dramatic reduction points to significant deforestation, likely driven by agricultural expansion, logging, or urban development. The loss of forested areas could have severe environmental impacts, including biodiversity loss, soil erosion, and changes to local climate patterns, highlighting the urgent need for reforestation efforts and stronger conservation policies. Marshes and wetlands also experienced a gradual decline, decreasing from 10% of the land area in 2000 to 9% in 2010, and then to 7% in 2020. These areas are often drained for agricultural use, and their reduction could be a result of such practices. Additionally, climate change impacts, such as altered precipitation patterns and rising temperatures, might have contributed to the shrinkage of these vital ecosystems. The loss of marshes and wetlands can have significant ecological consequences, including reduced water quality, loss of habitat for wildlife, and decreased flood protection.

Figure 4. Land Use and Land Cover (LULC) Changes in the Area (2000, 2010, 2020)

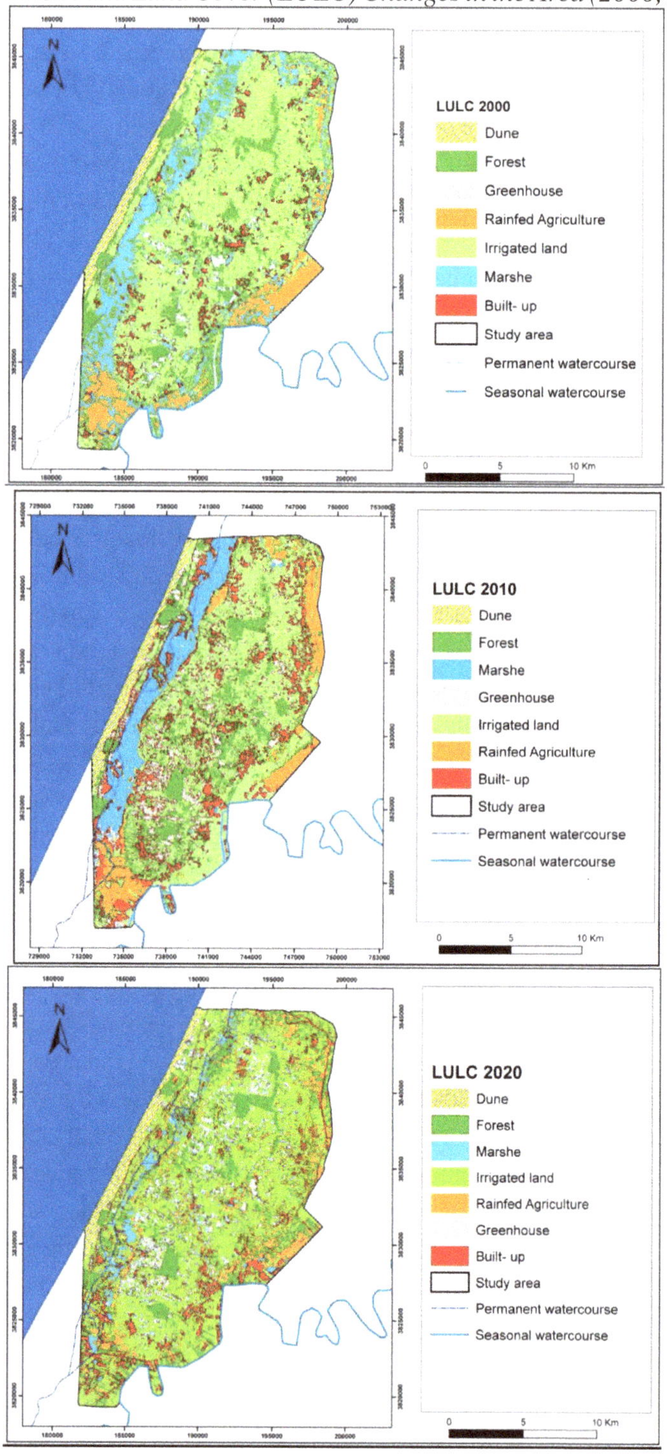

Figure 5. Land Use and Land Cover (LULC) Changes in the Area (2000, 2010, 2020)

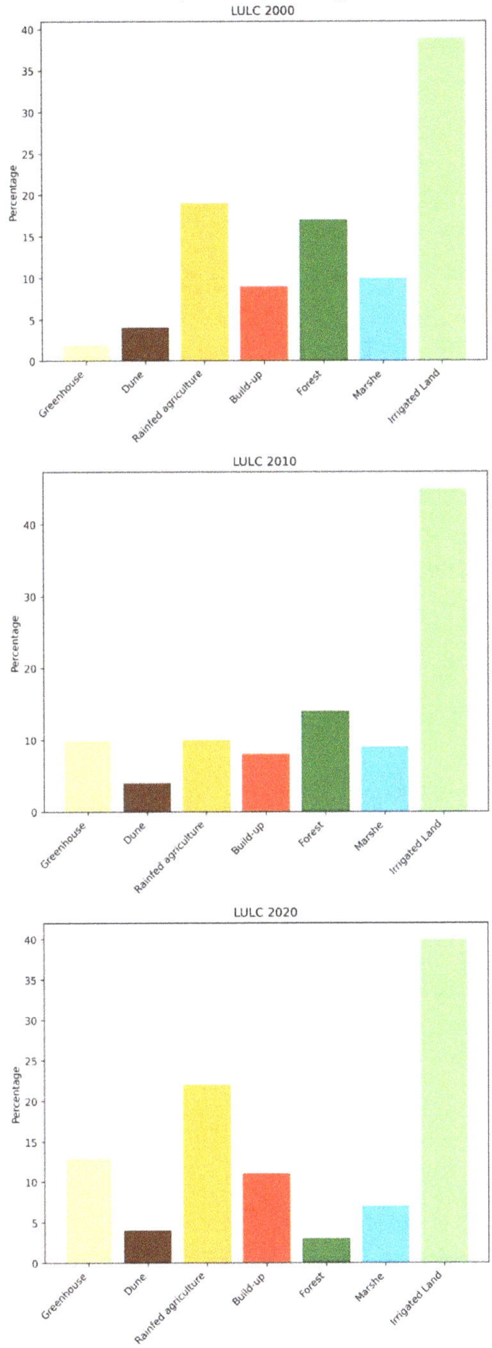

Irrigated land, which is crucial for high-yield agriculture in arid and semi-arid regions, showed a complex pattern of change. In 2000, irrigated land constituted 39% of the area, which increased to 45% by 2010. This increase reflects a push towards more intensive, water-dependent agriculture, possibly in response to growing food demands or shifts in agricultural policy. However, by 2020, the extent of irrigated land slightly decreased to 40%. This reduction could indicate emerging challenges such as water scarcity, the economic cost of irrigation, or a strategic shift towards other forms of land use.

Overall, the LULC changes between 2000 and 2020 indicate a dynamic landscape (Figure 5) shaped by various factors, including economic development, environmental pressures, and agricultural practices. The significant increase in greenhouse farming and the dramatic decline in forest cover are particularly notable, suggesting a shift towards more intensive land use practices that could have long-term implications for the environment (Loulad et al. 2023). The stability of dune areas, fluctuations in rainfed agriculture, gradual urbanization, and the reduction of marshes and wetlands further illustrate the complex interplay of natural and human factors influencing land use in the region. These changes underscore the importance of sustainable land management strategies to mitigate negative environmental impacts and ensure the continued productivity and health of the landscape.

3.2 Detection of Vegetation Changes using NDVI

We classified the vegetation cover using NDVI (Normalized Difference Vegetation Index) values based on the following table 3:

Table 3. NDVI Classification

NDVI Index	Land Use Types
Between 0 and -1	Water – Snow – Clouds
Between 0 and 0.2	Rock – Barren Land – Buildings
Between 0.2 and 1	Vegetation Cover

Figure 6. NDVI Changes in the Area (2000, 2010, 2020)

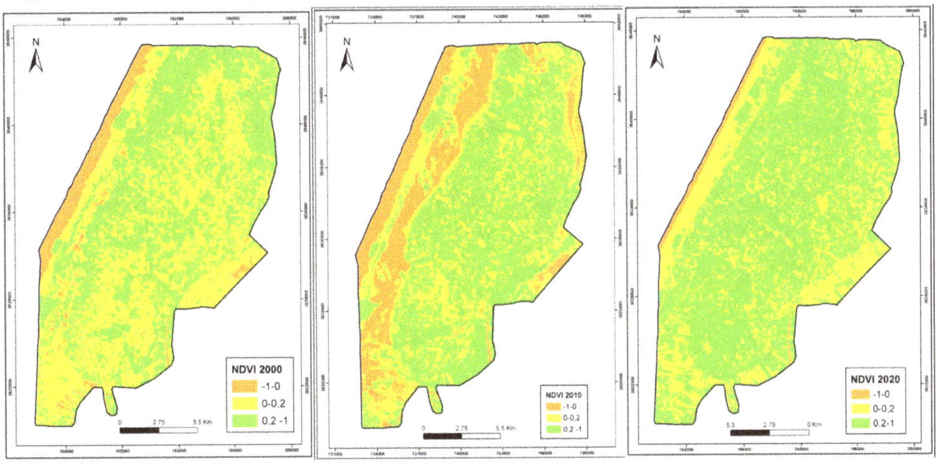

This classification allows for the systematic assessment of vegetation changes and the quantification of different land use types over time, facilitating a detailed understanding of how vegetation cover has evolved in the area

Figure 6 illustrates the variation in vegetation coverage and density in the study area. The NDVI values for the satellite images captured in 2000 range from -1 to 1. Higher NDVI values indicate greater vegetation density and intensity, representing areas with crops and some small forests (depicted in green). As the NDVI values decrease, the colour shifts towards yellow, reflecting lower vegetation density and intensity, signifying the presence of barren land and residential areas, which were prevalent during this period. The lowest NDVI values indicate areas with minimal vegetation coverage, primarily representing surface water bodies such as marshes and wadis (depicted in white), highlighting the scarcity of green spaces.

In 2010, significant changes occurred in land use and vegetation cover within the Ben Mansour community. The analysis reveals that the area covered by surface water bodies and barren lands expanded notably. NDVI values during this period ranged between -1 and 0, especially on the western edge of the region, where marshes were fully inundated due to heavy rainfall and flooding. The area of these marshes reached approximately 3426 hectares. In contrast, the area of barren lands and forests decreased. The total area of barren lands and forests was reduced to 5995 hectares, down from about 8960 hectares in 2000. Concurrently, agricultural intensification led to an increase in cultivated land, with the area of crops expanding to 10728 hectares. This reflects a shift in land use patterns, with a notable increase in agricultural activities and a reduction in the extent of forested and barren areas.

In 2020, there was a near-complete disappearance of areas previously covered by marshes and wadis due to successive years of drought experienced over the past five years. This environmental shift prompted local farmers and external investors to utilize these lands for agriculture, particularly for seasonal crops such as maize, sunflower, and wheat. Consequently, the area under cultivation increased to 13,554 hectares by 2020.

Additionally, there was a notable expansion in non-traditional land uses, including greenhouse agriculture. These new uses fall within NDVI values between 0 and 0.2, which are associated with built-up areas and barren lands. This reflects a diversification of agricultural practices and a shift towards controlled environment agriculture as a response to changing climatic conditions and land availability.

Figure 7. Land Use Area Over Time using NDVI

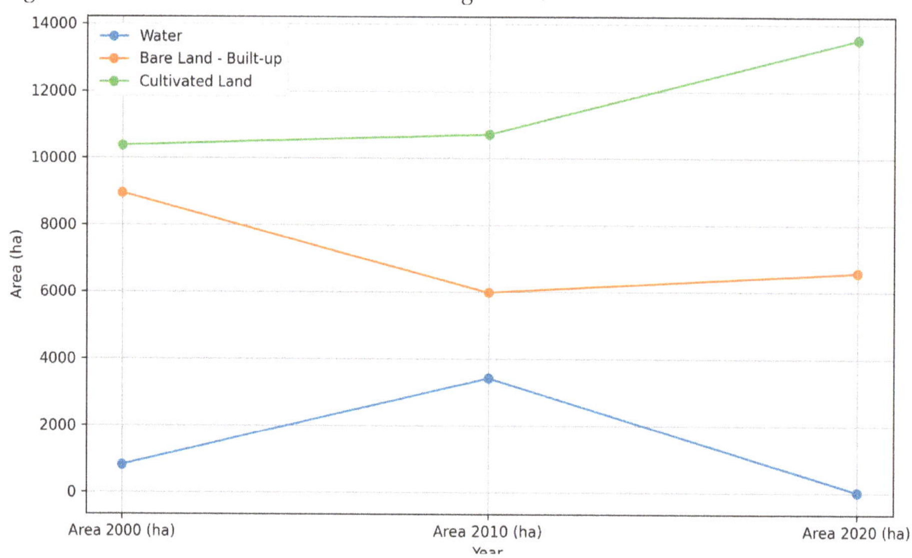

The results from comparing vegetation cover areas based on land use types (Figure 7) indicate a significant increase in cultivated land, which expanded from 10,390 hectares in 2000 to 13,554 hectares in 2020. Conversely, the area of surface water bodies decreased substantially, from 819 hectares to just 23 hectares, due to the drying up of major wadis and marshes in the region. These areas now exhibit sparse vegetation and some new crops. Additionally, most of the previously vacant or pasture lands have been converted into agricultural fields due to the recent intensification of agricultural activities in the Ben Mansour community.

3.3 Groundwater Depth

Wells are the principal source of water in the Ben Mansour community, complementing rainfall and serving multiple purposes such as irrigation, drinking, and other uses. As the population and land holdings in the region have grown, the number of wells has increased correspondingly. This rise, coupled with advancements in irrigation techniques and pumping technology, has led to the intensive and unsustainable use of these wells. Consequently, the groundwater table in the area is experiencing significant issues, including depletion, declining water levels, and increased salinity. To assess the extent of these problems, we conducted measurements of water depth in 158 wells. The depths varied from a minimum of 1.1 meters to a maximum of 18.5 meters, with an average depth of 5.5 meters. This depth distribution reveals two main patterns of water movement: in the southwestern part of the region, water flows from west to east, while in the northern and central areas, it flows from east to west. The median well depth, calculated as 5.3 meters, is close to the average, indicating that the depth distribution across the studied area is relatively uniform (Figure 8). This data provides a comprehensive view of the current state of groundwater resources and their sustainability in Ben Mansour.

Figure 8. Map of Groundwater Depth at Springs for Spring 2020

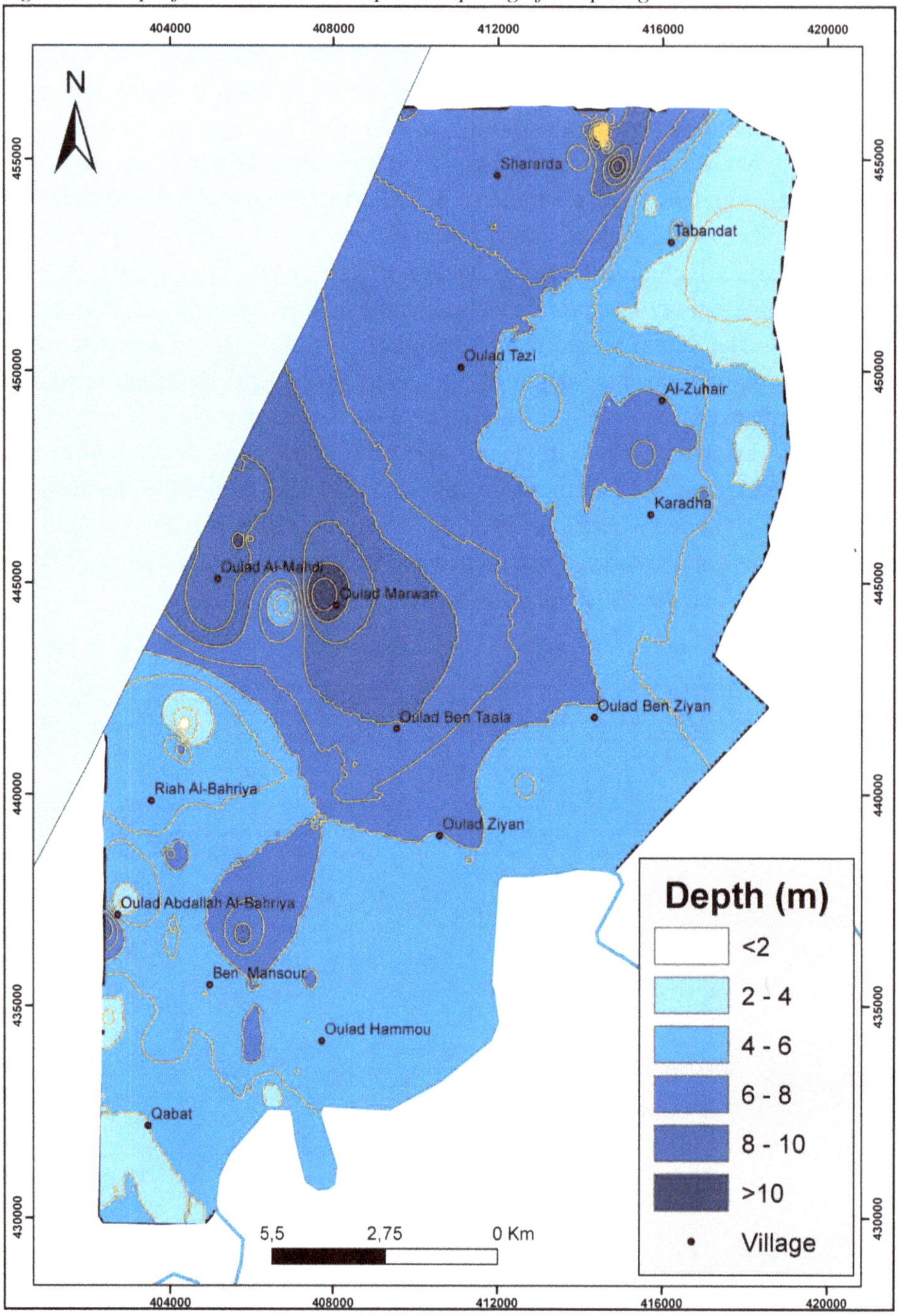

3.4 Electrical Conductivity of Groundwater

Our analysis of the electrical conductivity map of groundwater in the studied area reveals that the coastal regions—such as Qabat, Riah Al-Bahriya, and Shararda—are the most affected by high salinity levels, with electrical conductivity exceeding 3000 μS/cm (Figure 9). The impact of salinity in these areas is evident in the agricultural lands, which have developed a white crust, and crops irrigated with this saline water often exhibit symptoms like yellowing leaves, root burning, and plant death. Moving away from the coast, the electrical conductivity of groundwater gradually decreases to between 1500 and 3000 μS/cm in areas such as Oulad Ben Taala, Oulad Ziyan, and Karadha. Here, the salinity impact is moderate, affecting various types of crops with symptoms like leaf yellowing and the death of salt-sensitive plants. In contrast, some inland regions of the studied area, such as Oulad Hammou and Oulad Tazi, show improved groundwater salinity levels, ranging from 1000 to 1500 μS/cm. This reduced salinity has only a minor effect on agricultural productivity in these areas. Based on the measurement of salinity in over 180 wells, the average salinity in the studied area is 2100 μS/cm.

Figure 9. Map of Groundwater Salinity at Springs for Spring 2020

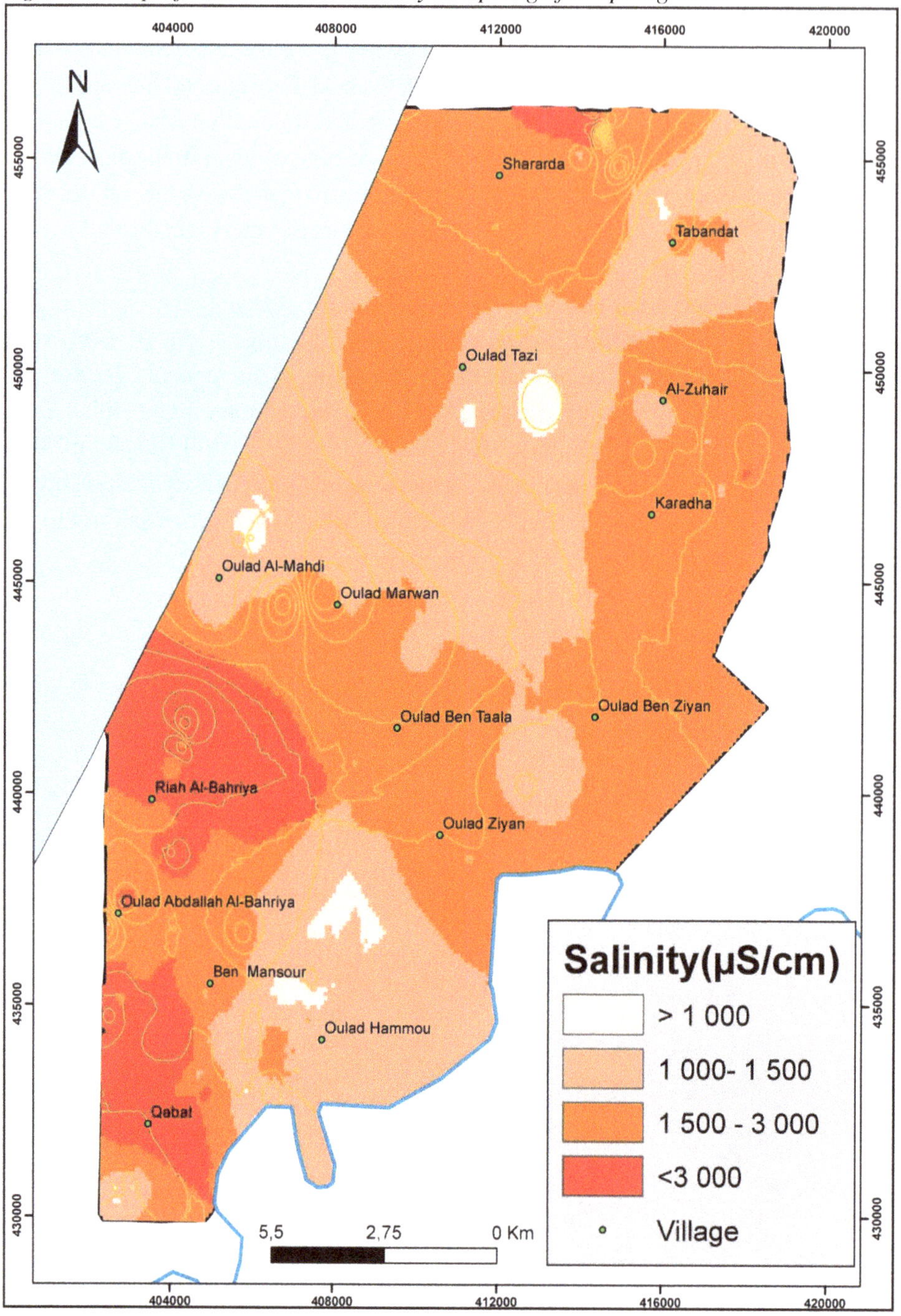

3.5 Impact of Salinity on Crops by Soil Type

Results from Figure 9 indicate that more than 50% of the area is affected by increased salinity, which has become a severe problem, leading to soil degradation, fertility loss, and reduced productivity. This poses a significant threat to the sustainability of agriculture in the region.

The degree of electrical conductivity of groundwater in the studied area and its impact on agriculture vary according to soil type and permeability (Sanad et al. 2024). Salinity levels are higher in areas with Vertisols (Tirs Soils) due to its low permeability. Consequently, when these soils are irrigated with saline water, they retain moisture for extended periods. This allows plants to absorb the necessary water and mineral salts, while excess water evaporates, leaving salt deposits in the soil. Over time, this process leads to the accumulation of salts, resulting in soil salinization and reduced agricultural productivity. Vertisols, characterized by their heavy and thick structure, have a high water retention capacity and are suitable for crops with long growth cycles, such as various types of wheat, and spring crops like maize. However, if there is excessive rainfall during the agricultural season, adequate drainage is required to prevent waterlogging. Given the cohesive nature of crusty soils, their tillage requires significant mechanical effort, necessitating the use of machinery for plowing.

In contrast, sandy soils with high permeability enable plants to absorb the necessary water while excess moisture drains away, minimizing salt accumulation. As a result, salinity levels are lower in sandy soils, promoting agricultural productivity and intensification. These soils are commonly used for high-value crops like vegetables and fruit trees (e.g., bananas and avocados). Local farmers report that avocado trees cannot tolerate irrigation water with a salinity exceeding 1000 uS/cm, when well water becomes too salty, farmers often resort to chemical treatments containing calcium, potassium oxide, and carboxylic acid to reduce salinity levels.

3.6 Groundwater Depletion Causes

Fieldwork has identified several factors contributing to the decline in groundwater levels in the region. These include intensive consumption of surface water aquifers due to extensive irrigation practices (Loulad et al. 2023), particularly for crops such as bananas, fruit trees, and vegetables during summer, and increased agricultural activity with multiple cropping cycles per season on the same land. The proliferation of wells and boreholes, driven by land fragmentation and limited alternative water sources, further depletes groundwater reserves. Additionally, the transition from diesel to electric or natural gas-powered mechanical pumps has influenced water extraction methods, with natural gas-powered pumps being predominant and showing varied consumption patterns. The evolution of drilling techniques has also played a role, shifting from labor-intensive manual drilling to more efficient mechanized methods.

Other contributing factors include reduced and irregular precipitation in recent years, recurrent droughts (Acharki et al. 2024; Gumus et al. 2024), and the lack of cooperation among farmers in sharing wells for irrigation. Competition among farmers for water from wells using high-capacity pumps, the perception of well water as individual property with unrestricted use, and a lack of awareness about water value exacerbate the situation. Urbanization, including street paving and the creation of drainage systems directing water away from the aquifers to rivers and the sea, further limits groundwater recharge. Additionally, the region's farmers have not benefited from support programs under the Green Morocco Plan. These issues have resulted in significant groundwater depletion, with some wells drying up, and the need for long-distance travel to access potable water.

3.7 Deterioration of Water Quality and Increased Salinity

The deterioration of water quality and increased salinity in the region stem from several interrelated factors. Climate change and rising sea levels have led to seawater intrusion, particularly affecting coastal aquifers in the Gharb Plain and areas like Ben Mansour (El Hamidi et al. 2021). This intrusion increases groundwater salinity, exacerbated by the over-extraction of aquifers, which allows saline seawater to encroach upon freshwater reserves, negatively impacting coastal agriculture (El Mahmouhi et al. 2018). Additionally, the irrational use of pesticides and fertilizers contributes significantly to water quality issues (Aziane et al. 2020). The excessive application of these substances, necessary for controlling pests and enriching soil, can lead to contamination of groundwater and further degradation of water quality. Other factors include the drainage of excess rainwater into Oued Sebou and the sea, reducing groundwater recharge, and the discharge of contaminated irrigation water from rice fields. The proximity of sewage channels to wells and pollutants from agricultural waste further exacerbate contamination. The lack of agricultural advisors, random seed selection, and high salinity levels due to intensive irrigation practices also contribute to poor plant health and soil degradation, compounding the overall impact on water quality.

The salinity problem is acute and severe, with a widespread impact on various types of crops. This growing issue threatens the future of agriculture in the area. Therefore, it is crucial to implement interventions and take necessary measures to prevent further exacerbation of the salinity problem in the future.

Intensive extraction of fresh groundwater from wells, particularly those near the coast, has led to a decrease in water table levels and the drying up of some wells. This situation has facilitated the intrusion of saline water from the sea. Saline water, being denser and heavier due to its higher concentration of salts and minerals compared to fresh groundwater, exerts significant pressure. As a result, this saline water can move inland, displacing fresh groundwater. When this saline water is used for irrigation, it poses a substantial risk to agricultural production, leading to a decline in crop yields and overall agricultural productivity.

3.8 Proposed Solutions and Strategies

To address the challenges posed by both human and natural factors that contribute to the depletion, deterioration, and increased salinity of groundwater resources—especially in the context of climate change impacts that undermine sustainable agriculture in the region—a range of solutions has been proposed. These solutions include:

- Efficient Irrigation Techniques: Implementing drip irrigation or small sprinklers to provide adequate water to plants and ensure the sustainability of water resources.
- Farmer Training: Providing training to farmers in modern irrigation techniques.
- Promoting Collaborative Water Management: Encouraging cooperative irrigation management among different farmers in the region.
- Introducing Low-Water-Use Crops: Growing economically important crops that require less water.
- Modern Irrigation Methods: Mandating the use of modern irrigation techniques to conserve water.
- Water Awareness: Raising awareness about the critical importance of water.
- Legal Compliance: Ensuring adherence to water management laws as a means of protecting groundwater resources.
- Focus on Major Consumers and Governance: Targeting large water consumers and improving water resource governance.
- Reducing Chemical Inputs: Minimizing the use of chemical pesticides and fertilizers.
- Distance from Sewage: Keeping wells or boreholes at least 100 meters away from sewage channels.
- Soil Improvement: Bringing in fertile, high-permeability soil from forested areas to cover saline-affected soils, using it for one to two years before it also becomes saline.

- Saline-Tolerant Crops: Choosing crops that can withstand salinity, such as grains, sunflowers, and corn.
- Encouraging Scientific Research: Promoting research in these areas to develop effective solutions.

4, CONCLUSION

This study provides a comprehensive analysis of land use and land cover (LULC) changes, vegetation dynamics, and groundwater conditions in the Ben Mansour-Mnasra region of Kénitra, Morocco. Through the examination of Landsat satellite imagery and geostatistical modeling, significant transformations in the landscape over the past two decades were identified, with critical implications for environmental sustainability and agricultural practices.

The LULC analysis reveals a substantial shift in the region's landscape. The dramatic expansion of greenhouse agriculture, coupled with the significant reduction in forest cover and marshes, underscores a trend towards intensified agricultural practices and urbanization. The increase in irrigated land, contrasted with a decline in rainfed agriculture, highlights a growing reliance on irrigation to support high-yield farming. This shift reflects broader socio-economic changes and increasing pressure on land resources, necessitating a reevaluation of land management strategies to mitigate environmental impacts and ensure the sustainability of agricultural productivity.

The NDVI analysis indicates notable changes in vegetation cover, with a substantial increase in cultivated land and a decrease in natural vegetation and water bodies. The expansion of greenhouse agriculture and the conversion of previously barren or pasture lands into agricultural fields reflect adaptive responses to changing climatic conditions and land availability. However, the near-complete disappearance of marshes and wadis due to prolonged droughts emphasizes the need for enhanced water management practices to address the challenges posed by reduced vegetation cover and changing land use patterns.

The assessment of groundwater depth and electrical conductivity highlights critical issues of overexploitation and salinization. The significant variation in groundwater levels, coupled with high salinity levels in coastal areas, reveals the adverse effects of intensive groundwater extraction and seawater intrusion. The decline in groundwater quality, evidenced by increased salinity and the impact on agricultural productivity, underscores the urgent need for sustainable water management practices. Geostatistical models provide valuable insights into the spatial distribution of groundwater conditions, emphasizing the importance of targeted interventions to address salinity issues and ensure the long-term viability of water resources.

Overall, the findings from this study underscore the complex interplay between land use changes, vegetation dynamics, and groundwater conditions in the Ben Mansour- Mnasra region. Addressing these challenges requires a holistic approach to land and water resource management, integrating sustainable practices to mitigate environmental impacts and support the continued productivity of the region's agricultural sector. Future research should focus on developing and implementing strategies to balance agricultural demands with environmental conservation, ensuring the sustainable management of natural resources in the face of ongoing climatic and socio-economic pressures.

REFERENCES

Acharki S, Arjdal Y, El Mansouri B (2024) Meteorological drought forecasting via soft computing models in Gharb perimeter (Northwest Morocco). E3S Web Conf 489:04015. DOI: 10.1051/e3sconf/202448904015

Aguedai, H., Jelbi, M., Lahlou, F., & Abdelaziz, M. (2022). Hydrochemical and geophysical characterization of the Mnasra coastal aquifers (Rharb basin NW Morocco). *Arabian Journal of Geosciences*, 15(18), 1480. DOI: 10.1007/s12517-022-10738-7

Aguedai, H., Lahlou, F., & El Mansouri, B. (2020). Intrusion Zones Identification in the Mnasra Aquifer (Morocco) Using the Seawater Intrusion Models and the Geophysics Data. In Jarar Oulidi, H., Fadil, A., & Semane, N. E. (Eds.), *Geospatial Technology* (pp. 51–68). Springer International Publishing. DOI: 10.1007/978-3-030-24974-8_4

Aziane, N., Larif, M., & Khaddari, A. (2020) State of nitric pollution of the Mnasra aquifer, coastal zone of the Gharb plain (Morocco). Moroccan Journal of Chemistry 8. https://doi.org/DOI: 10.48317/IMIST.PRSM/MORJCHEM-V8I4.17149

Bajjali, W. (2023). Spatial Interpolation. In *ArcGIS Pro and ArcGIS Online* (pp. 223–242). Springer International Publishing. DOI: 10.1007/978-3-031-42227-0_11

Batchi M, Karkouri JA, Fenjiro I, & Maaqili ME (2017) Étude comparative de deux modèles (DRASTIC et SI) pour l'évaluation de la sensibilité de la nappe phréatique de Mnasra (Maroc nord-occidental) à la pollution d'origine agricole. physio-geo 43–64. DOI: 10.4000/physio-geo.5213

Benabderrazik, K., Kopainsky, B., Tazi, L., Joerin, J., & Six, J. (2021). Agricultural intensification can no longer ignore water conservation – A systemic modelling approach to the case of tomato producers in Morocco. *Agricultural Water Management*, 256, 107082. DOI: 10.1016/j.agwat.2021.107082

Benseddik, B., El Mrabet, E., El Mansouri, B., Chao, J., & Kili, M. (2017). Delineation of artificial recharge zones in Mnasra Aquifer (NW, Morocco). *Modeling Earth Systems and Environment*, 3(1), 10. DOI: 10.1007/s40808-017-0267-2

Devkota, M., & Yigezu, Y. A. (2020). Explaining yield and gross margin gaps for sustainable intensification of the wheat-based systems in a Mediterranean climate. *Agricultural Systems*, 185, 102946. DOI: 10.1016/j.agsy.2020.102946

El Assaoui, N., Sadok, A., & Merimi, I. (2021). Impacts of climate change on Moroccan's groundwater resources: State of art and development prospects. *Materials Today: Proceedings*, 45, 7690–7696. DOI: 10.1016/j.matpr.2021.03.220

El Hamidi MJ, Larabi A, Faouzi M (2021) Modeling and Mapping of coastal aquifer vulnerability to seawater intrusion using SEAWAT code and GALDIT index technique: the case of the Rmel aquifer – Larache, Morocco. E3S Web Conf 298:05002. DOI: 10.1051/e3sconf/202129805002

El Mahmouhi, N., El Wartiti, M., & Azelmad, R. (2018). Spatial Imaging of the Water Resource and Natural Environment Coastal Gharb Basin: Downstream Part of the Hydraulic Basin of the Sebou River (Morocco). In Singh, D. N., & Galaa, A. (Eds.), *Contemporary Issues in Geoenvironmental Engineering* (pp. 169–175). Springer International Publishing. DOI: 10.1007/978-3-319-61612-4_13

El Mountassir, O., & Bahir, M. (2023). The Assessment of the Groundwater Quality in the Coastal Aquifers of the Essaouira Basin, Southwestern Morocco, Using Hydrogeochemistry and Isotopic Signatures. *Water (Basel)*, 15(9), 1769. DOI: 10.3390/w15091769

Gumus, V., El Moçayd, N., Seker, M., & Seaid, M. (2024). Future projection of droughts in Morocco and potential impact on agriculture. *Journal of Environmental Management*, 367, 122019. DOI: 10.1016/j.jenvman.2024.122019 PMID: 39106801

Heinl, M., Walde, J., Tappeiner, G., & Tappeiner, U. (2009). Classifiers vs. input variables—The drivers in image classification for land cover mapping. *International Journal of Applied Earth Observation and Geoinformation*, 11(6), 423–430. DOI: 10.1016/j.jag.2009.08.002

Hssaisoune, M., Bouchaou, L., Sifeddine, A., Bouimetarhan, I., & Chehbouni, A. (2020). Moroccan Groundwater Resources and Evolution with Global Climate Changes. *Geosciences*, 10(2), 81. DOI: 10.3390/geosciences10020081

Kili, M., El Mansouri, B., Chao, J., & Fora, A. A. (2006). De nouveaux éléments structuraux du complexe aquifère profond du bassin du Rharb (Maroc) : Implications hydrogéologiques. *Comptes Rendus Geoscience*, 338(16), 1194–1202. DOI: 10.1016/j.crte.2006.09.008

Loulad, S., Nguyen, T. T., Simou, M. R., Rhinane, H., & Buerkert, A. (2023). Monitoring rural-urban transformation in the coastal region of Rabat-Sale-Kenitra, Morocco. *PLoS One*, 18(8), e0290829. DOI: 10.1371/journal.pone.0290829 PMID: 37651348

Marouane, M., Dahchour, M., Dousset, S., & El Hajjaji, S. (2014) Monitoring the Risk of Nitrate Pollution in Mnasra Groundwater Under Field Condition–Morocco. Moroccan Journal of Chemistry 2. https://doi.org/DOI: 10.48317/IMIST.PRSM/MORJCHEM-V2I4.2137

Mohammed, J., Abdelaziz, M., Al Mazini, I., Magrane, B., El Mansouri, B., & Mohamed, E. B. (2022). Hydrogeophysical and hydrochemical assessment of groundwater quality of the Rharb shallow aquifer (NW Morocco). *Arabian Journal of Geosciences*, 15(7), 660. DOI: 10.1007/s12517-022-09956-w

Molle, F., & Tanouti, O. (2017). Squaring the circle: Agricultural intensification vs. water conservation in Morocco. *Agricultural Water Management*, 192, 170–179. DOI: 10.1016/j.agwat.2017.07.009

Moumane, A., El Ghazali, F. E., Al Karkouri, J., Delorme, J., Batchi, M., Chafiki, D., & Karmaoui, A. (2021). Monitoring spatiotemporal variation of groundwater level and salinity under land use change using integrated field measurements, GIS, geostatistical, and remote-sensing approach: Case study of the Feija aquifer, Middle Draa watershed, Moroccan Sahara. *Environmental Monitoring and Assessment*, 193(12), 769. DOI: 10.1007/s10661-021-09581-2 PMID: 34735624

Omrania, S., El Khodrani, N., Zouahri, A., Douaik, A., Iaaich, H., Lahmar, M., & El Hajjaji, S. (2019). Evaluation of the Groundwater Quality of the M'nasra, Kenitra (Morocco). *Materials Today: Proceedings*, 13, 1092–1101. DOI: 10.1016/j.matpr.2019.04.076

Ouassanouan, Y., Fakir, Y., Simonneaux, V., Kharrou, M. H., Bouimouass, H., Najar, I., Benrhanem, M., Sguir, F., & Chehbouni, A. (2022). Multi-decadal analysis of water resources and agricultural change in a Mediterranean semiarid irrigated piedmont under water scarcity and human interaction. *The Science of the Total Environment*, 834, 155328. DOI: 10.1016/j.scitotenv.2022.155328 PMID: 35452720

Qanza, H., Maslouhi, A., Abboudi, S., Mustapha, H., & Hmimou, A. (2019). Unsaturated Soil Hydraulic Properties Identification using Numerical Inversion and In-Situ Experiments from Mnasra Area, Morocco. *KSCE Journal of Civil Engineering*, 23(11), 4949–4959. DOI: 10.1007/s12205-019-0420-8

Robinson, N. P., Allred, B. W., Jones, M. O., Moreno, A., Kimball, J., Naugle, D., Erickson, T., & Richardson, A. (2017). A Dynamic Landsat Derived Normalized Difference Vegetation Index (NDVI) Product for the Conterminous United States. *Remote Sensing (Basel)*, 9(8), 863. DOI: 10.3390/rs9080863

Saâdi, Z., & Maslouhi, A. (2003). Modeling nitrogen dynamics in unsaturated soils for evaluating nitrate contamination of the Mnasra groundwater. *Advances in Environmental Research*, 7(4), 803–823. DOI: 10.1016/S1093-0191(02)00055-2

Sanad, H., Mouhir, L., Zouahri, A., Moussadek, R., El Azhari, H., Yachou, H., Ghanimi, A., Oueld Lhaj, M., & Dakak, H. (2024a). Assessment of Groundwater Quality Using the Pollution Index of Groundwater (PIG), Nitrate Pollution Index (NPI), Water Quality Index (WQI), Multivariate Statistical Analysis (MSA), and GIS Approaches: A Case Study of the Mnasra Region, Gharb Plain, Morocco. *Water (Basel)*, 16(9), 1263. DOI: 10.3390/w16091263

Sanad, H., Moussadek, R., Mouhir, L., Oueld Lhaj, M., Dakak, H., El Azhari, H., Yachou, H., Ghanimi, A., & Zouahri, A. (2024b). Assessment of Soil Spatial Variability in Agricultural Ecosystems Using Multivariate Analysis, Soil Quality Index (SQI), and Geostatistical Approach: A Case Study of the Mnasra Region, Gharb Plain, Morocco. *Agronomy (Basel)*, 14(6), 1112. DOI: 10.3390/agronomy14061112

Shivakumar, B. R., & Rajashekararadhya, S. V. (2018). Investigation on Land Cover Mapping Capability of Maximum Likelihood Classifier: A Case Study on North Canara, India. *Procedia Computer Science*, 143, 579–586. DOI: 10.1016/j.procs.2018.10.434

Tramblay, Y., Koutroulis, A., Samaniego, L., Vicente-Serrano, S. M., Volaire, F., Boone, A., Le Page, M., Llasat, M. C., Albergel, C., Burak, S., Cailleret, M., Kalin, K. C., Davi, H., Dupuy, J.-L., Greve, P., Grillakis, M., Hanich, L., Jarlan, L., Martin-StPaul, N., & Polcher, J. (2020). Challenges for drought assessment in the Mediterranean region under future climate scenarios. *Earth-Science Reviews*, 210, 103348. DOI: 10.1016/j.earscirev.2020.103348

Chapter 8

Monitoring Surface Water Variations at the Ain Kwachia Dam in Morocco's Northwest Central Plateau With the Use of Remote Sensing and GIS

Mohamed Gramz
https://orcid.org/0000-0003-4614-2576
Ibn Tofail University, Morocco

Lahcen Ouiaboub
https://orcid.org/0000-0002-9506-0297
Ibn Tofail University, Morocco

Mehdi Mettouchi
Hassan 2 University, Morocco

Moulay Hicham Azagane
Ibn Tofail University, Morocco

Hanane Meziane
Mohammed V University, Morocco

Lamiae El Bezzari
Ibn Tofail University, Morocco

ABSTRACT

Morocco's water situation is unique, with arid climatic zones dominating most of the country. This means that water resources need to be managed equitably, and decisions need to be taken to ensure the sustainability of this natural potential, which is essential for life in general. This is the case for the management of water

DOI: 10.4018/979-8-3693-9651-3.ch008

scarcity, especially rainwater, by hydraulic works in the central plateau in north-west Morocco, with a view to combating water stress, particularly since climate change is having an impact on rainfall, with a succession of drought years in our study area. The aim of this research is to highlight the role of the Ain Kwachia dam located on the Khellata river in terms of the storage of surface water from rainfall, by quantifying the surface area occupied by its water using satellite images obtained from remote sensing, and mapping the evolution of the dam using a GIS geographic information system over the period from 2013 to 2023. The results show that this structure has made it possible to store rainwater, with an increase of 65.22% in 2022 and 43.80% in 2023.

1. INTRODUCTION

The situation of water resources is under discussion, with a serious state of decline in terms of quantities at surface level as well as underground. This reality of water resources is due to several factors, namely climatic in terms of global warming (Wang et al. 2024), (Jiang et al. 2024), (Wu et al. 2024) as well as anthropogenic with the excessive use of water resources in economic activities during the production chain (Elfoul et Ghachi 2018). Faced with this particular hydrological situation in terms of dwindling freshwater resources, as well as water scarcity, short-term mitigation and medium- to long-term adaptation (Ongoma et al. 2024) measures remain necessary to ensure the sustainability of water resources (Hekmatnia et al. 2023), (Sherif et al. 2023), and of course environmental stability, as well as the biodiversity of ecosystems, and to combat water vulnerability. The water resources situation in countries with climates tending towards aridity, especially those located in the 30-degree part of the northern hemisphere and in the southern hemisphere, i.e., north and south of the equatorial line, respectively, as is the case on the Asian continent (Chen et al. 2024), also that the African continent, especially the Mediterranean area (Villani et al. 2024), as well as the South American continent (Garreaud et al. 2009).

In the Moroccan context, the water situation is in deficit after a succession of drought years (Thi et al. 2023), following climate change and its impact on the whole country, in addition to the dominance of arid climatic stages (Boutallaka et al. 2024). This is also the case for the Ykem watershed located to the north-west of Morocco's central plateau (Gramz et al. 2024), an area characterized by a homogeneous topography in terms of heights varying between 200 m and 400 m, over a considerable length between the Atlantic coastal dunes to the north and the intermediate part of the central plateau to the south. Economic activities in this area revolve around the irrigated agricultural sector, the case of farms also raising poultry, in addition to that of industry in terms of factories manufacturing nutrients as livestock feed. Previous studies have looked at the use of water resources in the large Bouregreg basin in general (Ouharba, Mabrouki, et Triqui 2024), but on a smaller scale in the watersheds located in the north of the central plateau, in the case of Ykem watershed. Given the function of this hydraulic structure, constructed downstream of the Khellata River, in capturing rainwater to deal with extreme situations like water stress, it makes sense to study the spatiotemporal variations of the Ain Kwachiua dam reservoir. This is accomplished by the competent services in charge of water resource management in this area, in this case the Bouregreg and Chaouia water basin agency ABHBC, with the goal of irrigating a 500 ha area of cultivated land. How effective can equitable surface water management be through hydraulic development in the case of the Ain Kwachia dam in Morocco's northwest central plateau, given the above information regarding the water situation in the study area? This question makes us think about potential adaptation strategies that could be used to protect surface-level water resources, to increase resilience in the face of severe rainfall shortages, and, of course, to consider the impact of climate change on water resources in terms of water stress.

In order to help decision-makers take the required actions for water sustainability, it is important to have a forward-looking vision of water potential that gives them visibility over surface water in the study area.The objective of this study is to quantify the surface area of the dam's impoundment over the period of 2013–2023, taking into account inter-annual variations, in order to highlight the economic contribution that the dam has brought about. To achieve this, the adoption of a methodological approach that aims to meet the expectations of this research, is based essentially on multiple data from the processing of information returned from satellite images, via the use of geospatialization tools (Kana et al. 2022), (Kumar et Jha 2022), such as remote sensing RS to acquire satellite images, because the different sensors on board the satellites allow images to be taken over different periods, making it easy to obtain data, especially for older dates, such as tens of years and more, meaning that official access to data remains a constraint to scientific research.and the geographic information system GIS (Mabrouki et al. 2023), (Rafik et al. 2022), (Ait El Mekki et Laftouhi 2016), which remains a powerful tool for processing, analyzing and representing data at several scales, for processing and mapping the results, was employed.

In addition, to ensure high quality analysis and accurate results, Landsat satellite images (Gómez-Fernández et al. 2024), due to their open access availability, and Google Earth Pro (Teixeira et al. 2024), at a resolution high enough and clearly enough, were used to facilitate the mapping of the Ain Kwachia dam reservoir variations in time and space over the research period. The resolution of Google Earth Pro images is higher in terms of pixels than that of free satellite images, which remain blurred in terms of their components, with a resolution of 30 m per pixel, giving clearer captures of the Ain Kwachia dam area.

2. GEOGRAPHIC SITUATION AND CLIMATE

1-2 Study Area

The Ain Kwachia dam is located on the Khellata stream in the Ykem watershed on the lower plateau of the central Moroccan plateau, exactly to the northwest. The normal reservoir volume is 11Mm3, with a normal reservoir surface area of 180 ha. The Ykem watershed, with a surface area of 162 km^2 and average interannual inflows of 7Mm^3, is bounded to the north by the Atlantic Ocean's coastal dunes, to the south by the intermediate part of the Cantral Plateau, to the east by the Bouregrag basin (Laouina et al. s. d.) (El Aoula et al. 2021) and to the west by the Lower Chaouia (Figure 1).

Figure 1. Map of localization study area using geospatialization tools

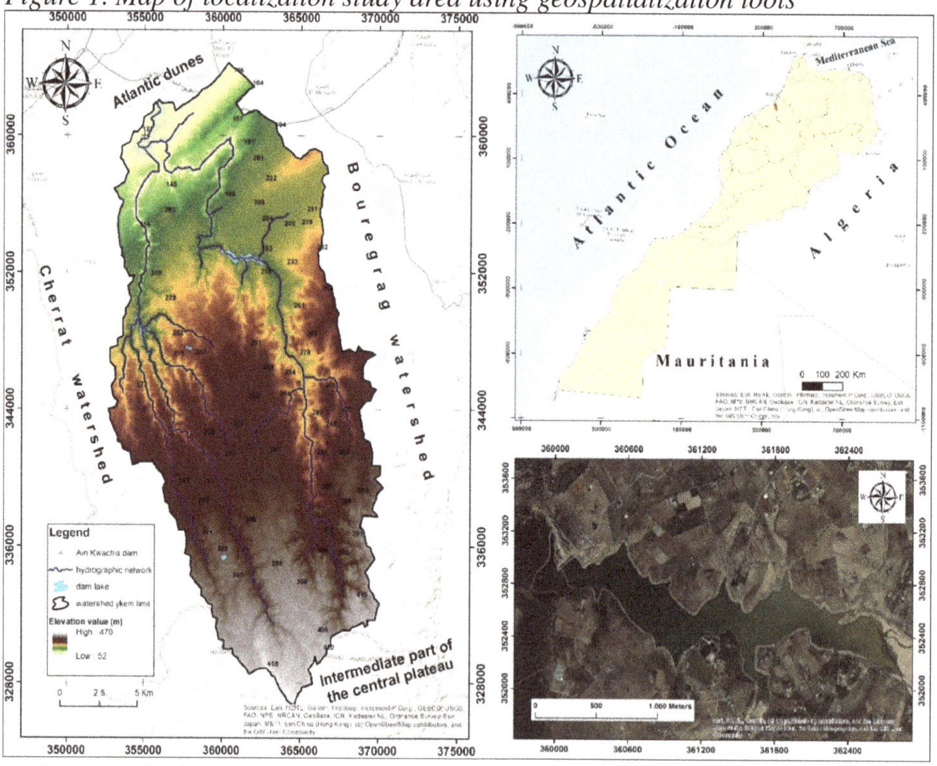

Source: *Realized using geospatialization tools*

The geographical aspect of the study area, in terms of altitude, has an influence on the climatic situation, as the topography remains a natural obstacle impacting on the reception of precipitation throughout the year (Mahmoud et al. 2020), and in terms of plateau, the area where the Ain Kwachia dam is located, does not present great altitudes, which makes the situation favorable, in addition, the nature of the climate on the central plateau is generally arid, which has an impact on rainfall throughout the year, even during the winter season. This observation calls for climatic adaptation, especially in terms of water resources during rainy days, in terms of catchment of water flowing via the hydrographic network in the Ykem watershed to the north-west of the central plateau.

2-2 Topography

The topographical nature of the Ykem watershed is characterized by heights that exceed 200 m over the entire surface of the basin, with the exception of the part overlooking the Atlantic Ocean, with a minimum height of 52 m located to the north, as well as the maximum height to the south towards the intermediate part of the central plateau, in addition to which the Ain Kwachia dam as a hydraulic structure is located on the khellata River at a height of over 200 m (Figure 1).

3-2 Climate

The type of climate in the lower plateau of the central plateau (Jaffal et al. 2022), where the Ykem watershed is located, is characterized by the dominance of a climate that tends towards aridity, with high rainfall in winter and a minimum in summer. The interannual average rainfall at the Cheikh reguig station on the river khellata in the Ykem basin is 400 mm, with a daily value not exceeding 200 mm as the maximum value recorded in 2014 Figure 2.

Figure 2. Daily Precipitation Corrected 2013/01/01 to 2022/12/31

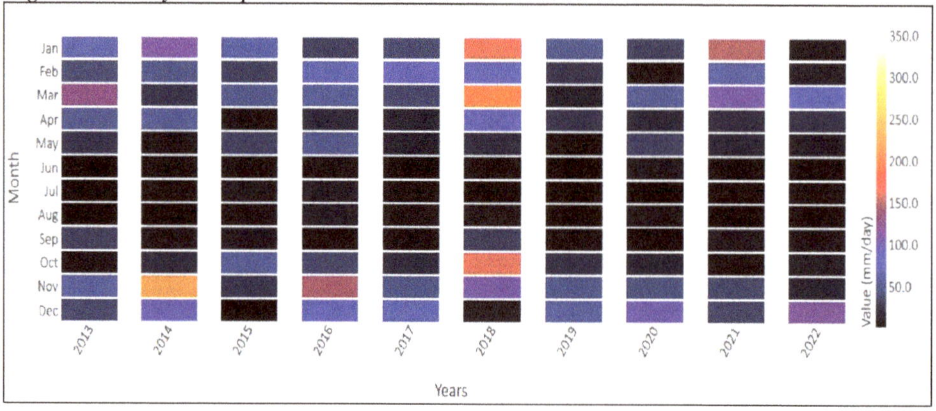

Source: *Data accessed from NASA POWER*

Temperatures in general average 18 °C throughout the year, rising to no more than 20 °C in summer, but in terms of maximum daily values, they exceed 24 °C, especially in July and August in summer, as shown in Figure 3 below:

Figure 3. Heatmap of Daily Temperature between 2013/01/01 to 2022/12/31

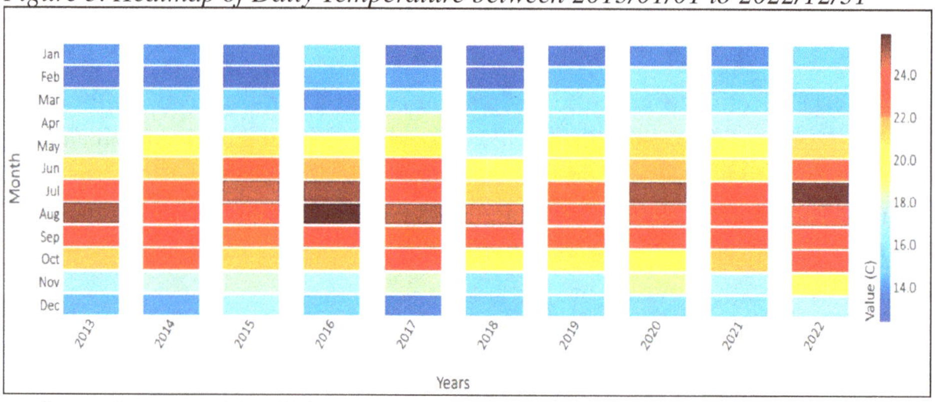

Source: *Data accessed from NASA POWER*

3. METHODS AND MATERIALS

In this part of the present chapter, the methodology adopted aims to monitor spatio-temporal variations in surface water resources, more specifically at the Ain Kwachia dam, over the period 2013 to 2023, using geospatialization techniques, namely remote sensing RS and the GIS geographic information system, with satellite images as the source of freely available data, for different periods, especially as climatic data requires purchase from the National Meteorological Directorate (DNM), which presents a constraint to research in terms of funding, as does satellite imagery in the official capacity of the Royal Center for Remote Sensing Space (CRTS), which concerns confidentiality for high-resolution images, justifying the choice of data falling under open access.

3-1 Methodology Framework

The methodology adopted to carry out this research is based on the use of geospatialization techniques, in terms of remote sensing (J. Li et al. 2023), (Zhao et al. 2023), (Martins et al. 2019) through the acquisition of satellite images, to quantify the spatio-temporal evolution of the Ain Kwachia dam impoundment, in addition to carrying out satellite image analyses via a GIS geographic information system, with the aim of obtaining processing results during the period between 2013 and 2023. As well as using Google Earth Pro satellite images (Hritz 2013), (Harrington et Cross 2015) for high resolution of the dam area, and calibrating them via GIS, with a view to monitoring the spatio-temporal evolution of the water surface at this hydraulic structure in terms of dam impoundment. The diagram below shows the methodological process involved in this research work (Figure 4).

Figure 4. Diagram of the working methodology

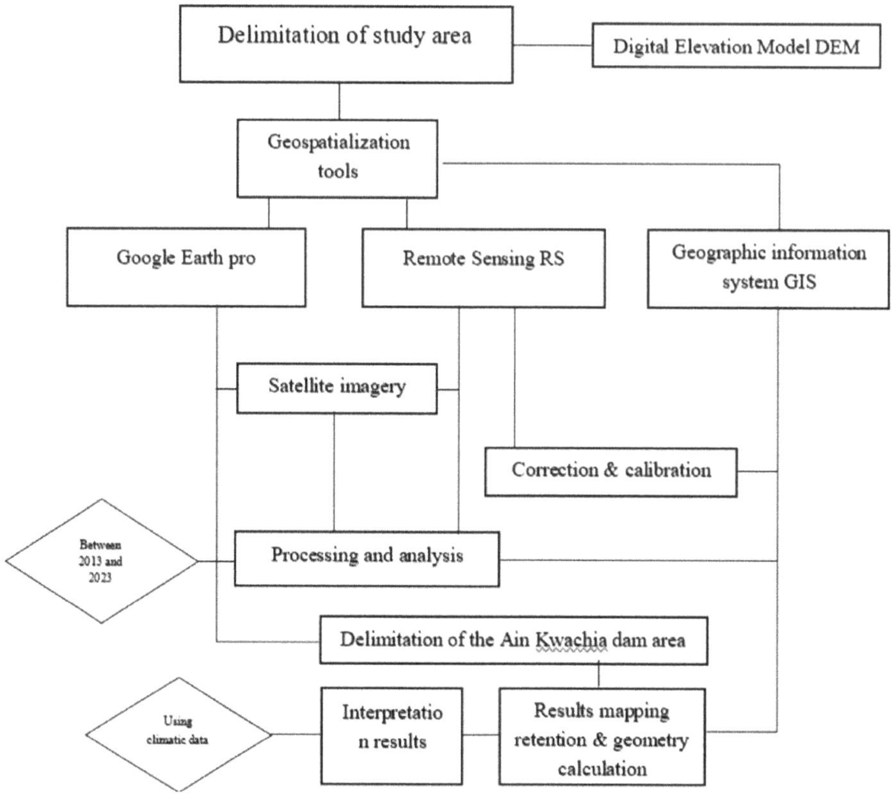

3-2 Satellite images

The use of geospatialization techniques to obtain satellite images (Z. Li et al. 2023) in terms of remote sensing, allows us to obtain images from the website of in terms of free access (https://www.usgs.gov/) with a resolution of 30 m, with a vast surface area of the Ykem watershed compared to that of the Ain Kwachia dam, for a more precise analysis of the dam's impoundment throughout the period between 2013 and 2023, which obliges us to also base ourselves on images extracted from Google Earth Pro, and this, following the availability of images with the exception of the satellite image for the year 2015 extracted from the Landsat satellite. As shown in Figure 5.

Figure 5. Satellite images used between 2013 and 2023

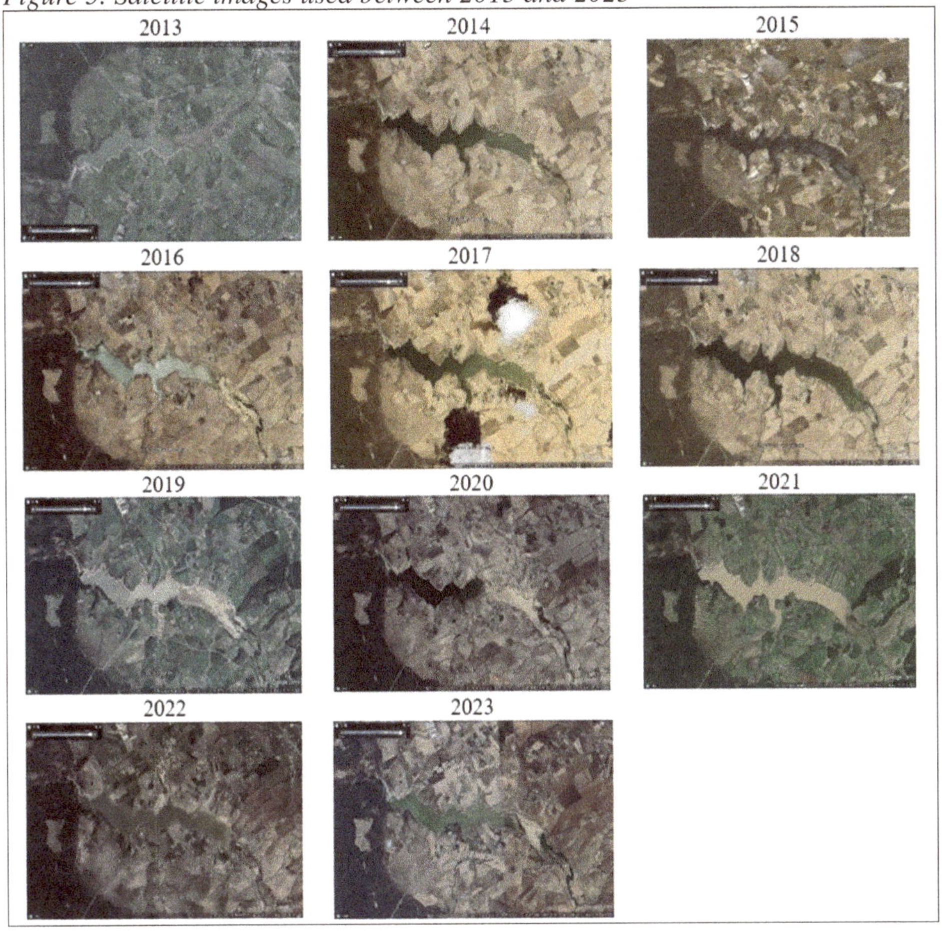

Source: *Google Earth Pro open access*

4. RESULTS

After carrying out the necessary analyses via the GIS geographic information system, based on satellite images, we obtained the results of the spatio-temporal evolution of the Ain Kwachia dam reservoir on an annual basis, during the period between 2013 and 2023, mentioned in Table 1, with minimum percentages of the dam reservoir ranging from 32,89% in 2020, also 34,11% recorded in 2013 and 2019, as well as a maximum percentage of 77,94% recorded in 2018. However, the rest of the years recorded a filling rate between 43,83% and 68,65%, as shown in Table 1.

Water storage data for the Ain Kwachia dam are presented in terms of reserves in million cubic meters (Mm^3), given that the maximum filling capacity of this hydraulic structure is 11 Mm^3, according to data from the Bouregreg and Chaouia hydraulic basin agency ABHBC.

Table 1. Ain Kwachia dam storage evolution in ha between 2013 and 2023

Source	Years	Reserve in Mm^3	Percentage filled (%)	Area (Ha)	Evolution in %
Google Earth	2013	4,28	38,88	61,4	34,11
Google Earth	2014	4,40	40,00	107	59,44
LANDSAT 8	2015	7,27	66,09	123,57	68,65
Google Earth	2016	5,20	47,23	84,3	46,83
Google Earth	2017	4,87	44,25	103	57,22
Google Earth	2018	8,25	75,00	140,3	77,94
Google Earth	2019	4,28	38,88	61,4	34,11
Google Earth	2020	2,69	24,47	59,2	32,89
Google Earth	2021	7,22	65,63	117,4	65,22
Google Earth	2022	7,21	65,52	112,8	62,67
Google Earth	2023	4,82	43,80	80,4	44,67

Source: Results based on GIS and ABHBC data

The results are mapped on the surface area of the Ain Kwachia dam reservoir, after processing satellite images from Google Earth Pro, via a GIS geographic information system, during the period between 20213 and 2023, using Landsat 8 satellite imagery, for reasons of image clarity, as shown in Table 1 on the one hand, and also represented in Figure 6 on the other, in order to visualize the evolution of the water surface at the dam during the period set for this research in both time and space.

Figure 6. Evolution mapping of retention of Ain kwachia dam between 2013 and 2023

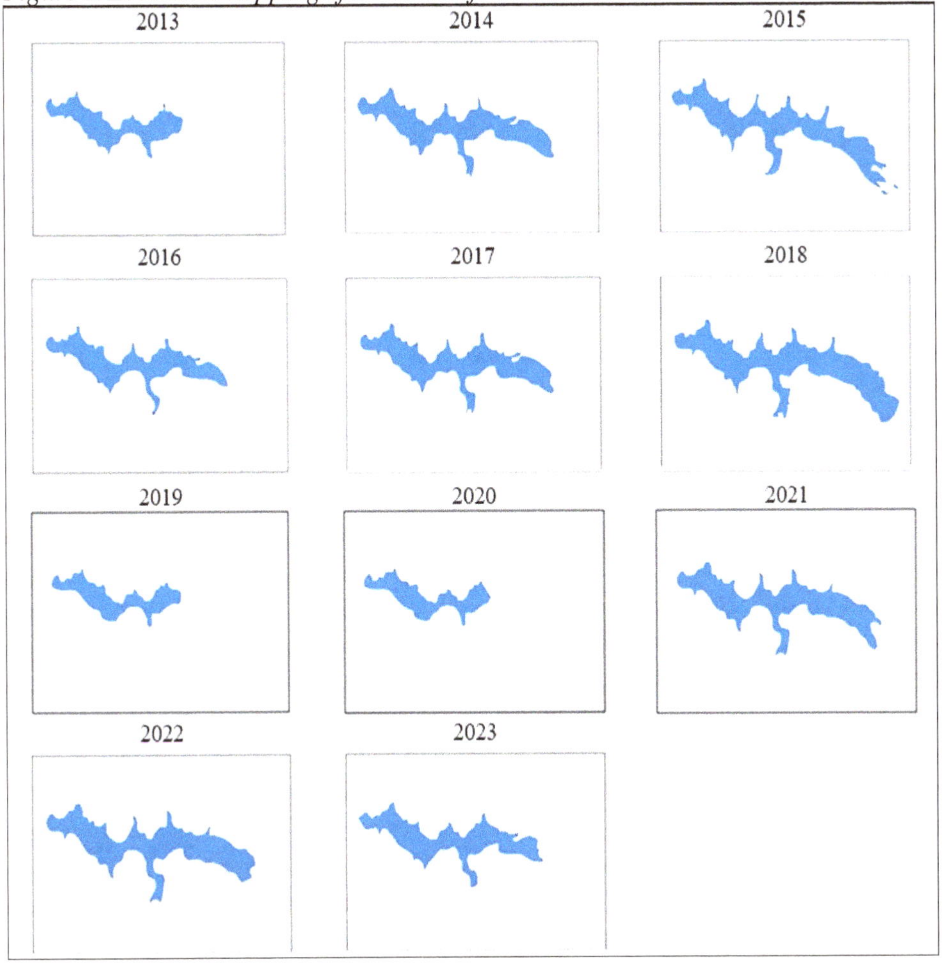

Source: *Analyses with GIS*

In terms of the Ain Kwachia dam reservoir's evolution, there was a rise from 61,4 ha to 123,57 ha in 2013 and 2015, a decrease to 84,3 ha in 2016, an increase to 140,3 ha in 2018 and a decrease to 59,2 ha in 2020. There was also notable variability in the dam reservoir's surface area between 2021 and 2023, measuring 117,4 ha and 80,4 ha, respectively. This is shown in Figure 7.

Figure 7. Diagram evolution the area of surface water of Ain kwachia dam between 2010 and 2023

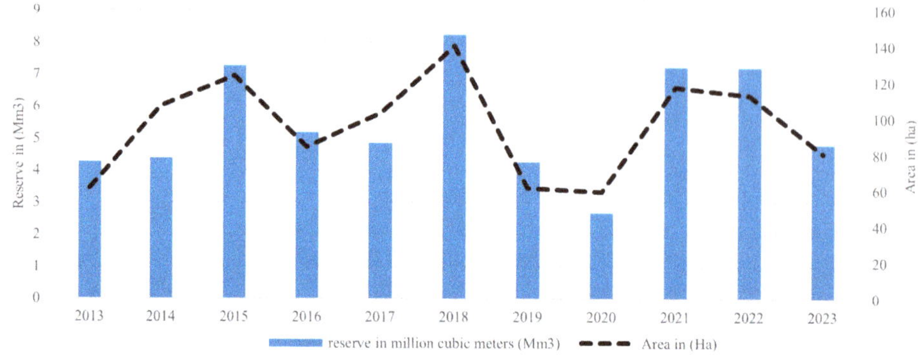

Source: *Data of ABHBC and analyses with GIS*

5. DISCUSSION

According to the methodology used for this research Figure 4, the Ain Kwachia dam reservoir's evolution of spatiotemporal variations has an impact on the dam's water storage. This is because, based on the surface area of the dam reservoir's values between 2013 and 2023, there is reservoir instability due to climatic variability with regard to interannual rainfall, with a minimum value of 190 mm. On the one hand, with a maximum value of 785 mm, and on the other, with average annual temperatures varying between 17,5 °C and 20 °C, this has repercussions on the water surface in the dam reservoir, through evaporation (El Bilali et al. 2022), (Tola et al. 2024).

This explains the effects of climate change on surface water resources and, of course, groundwater resources fed from the surface (El Khalki et al. 2021), (Hssaisoune et al. 2020). In particular, the last few years, specifically the period between 2020 and 2023, when rainfall is expected to decrease throughout the same period, the increase in temperature due to climate change, especially in the arid Central Plateau area, also causes an extreme situation in terms of surface water evaporation after a warming climate, as is the case in the Ykem watershed and of course the Ain Kwachia dam area. Seasonal temperature variations, especially in summer from June to August, have a major influence on the acceleration of surface water evaporation from lakes and dams, as in the case of the Ain Kwachia dam, the subject of this chapter. In the absence of real data from government agencies, we relied on open-access data using machine learning, based on Google Earth Engine functionalities. See Figure 8 for evaporation in area study for the period between 2013 and 2023.

Figure 8. the monthly evaporation values in the area study for the period between 2013 and 2023

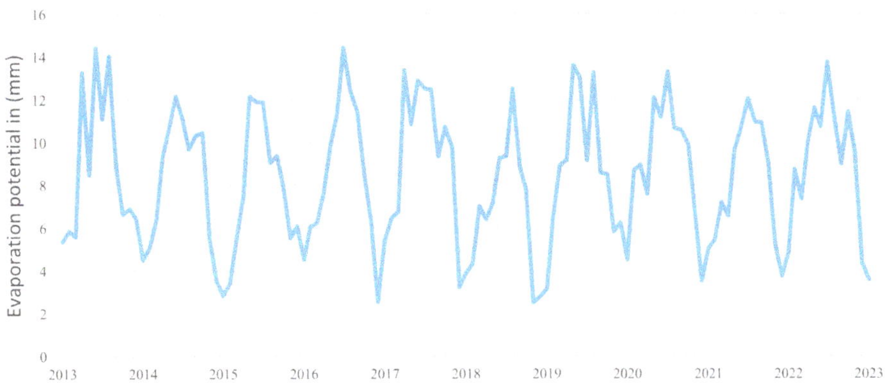

Source: *data using from Google Earth Engine*

Which contributes to the loss of surface water, and of course a reduction in surface water resources, as can be seen from the irregularity of precipitation, as well as temperatures in terms of wet and dry years, according to the calculation of the deviation from the mean of climatic values based on the following equation:

For deviation from mean precipitation (1

D_m=((P_a-P_m)/P_m)×100

For deviation from mean temperature (2

D_m=((T_a-T_m)/T_m)×100

D_m: Deviation from average
P_a: Annual precipitation
P_m: Average precipitation between 2013 and 2023
T_a: Average annual temperature
T_m: Average temperature between 2013 and 2023

We can create the following graph, which depicts wet and dry years in accordance with the graph in Figure 9, by computing the deviation from the mean of the climate elements, temperature and precipitation, respectively, based on the climate data series for the period between 2013 and 2023.

Figure 9. Deviation from average Precipitation and Temperature between 2013 and 2023

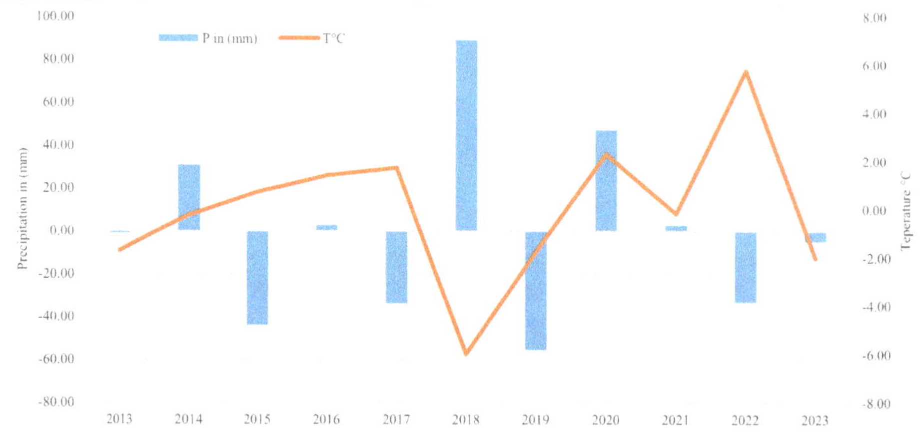

Source: *Use of precipitation and temperature data from Cheikh reguig station*

The increase in the water surface that the dam's reservoir occupies is primarily correlated with rainfall, that is, rainfall in wet years increases the dam's reservoir and, naturally, the water surface. This is the case for 2018, when the Ain Kwachia dam reservoir reached 140,3 ha due to a total annual rainfall of 785 mm; on the other hand, drought has an impact on the dam reservoir, which measured 34,11 ha in 2019 (Figure 10).

Figure 10. Annual of precipitation and temperature between 2013 and 2023

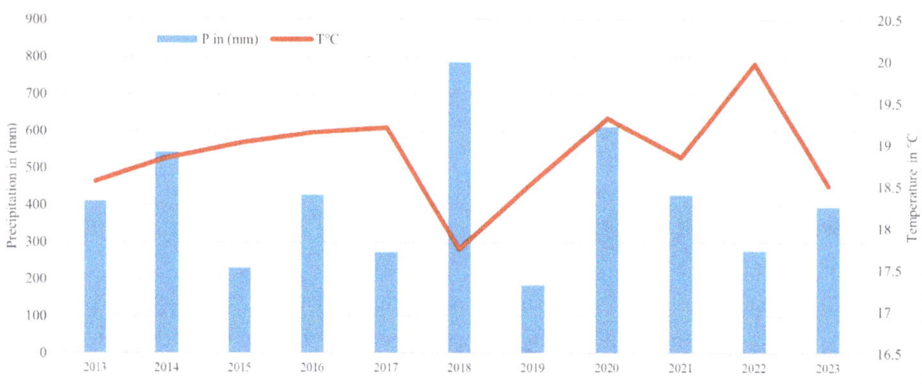

Source: *Annual of precipitation and temperature data from Cheikh reguig station*

With inter-annual variability during the monitoring period of water resources in the Ain Kwachia dam reservoir, surface water resources are linked to climatic conditions, particularly in terms of rainfall during the rainy period of the year, between October and April. Meanwhile, the hot period, between May and September, is characterized by rising temperatures and low rainfall. This specific climatic situation essentially contributes to the dam's reservoir's increase during rainy years and decrease during dry years. The quantification of the spatiotemporal evolution of the water in the Ain Kwachia dam reservoir indicates that hydraulic structures play a critical role in the storage and preservation of water resources, particularly during droughts. This helps to both mitigate (Hung, Van Trung, et Vo 2023) the aridity of the study area and ensure the sustainability of surface water for human use in the face of climate change (Figure 11).

Figure 11. Annual average, maximum and minimum precipitation and temperature values between 2013 and 2023

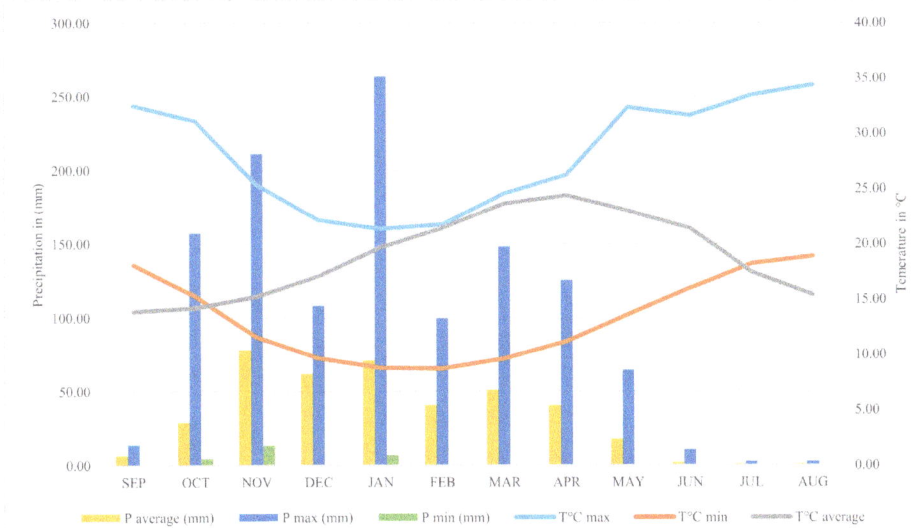

Source: *Average values of precipitation and temperature data from Cheikh reguig station*

It is important to emphasize water resource management strategies, at national and territorial level in Morocco, which are framed by legal texts aimed at water preservation, in addition to ensuring the proper management of surface and underground water resources, as is the case for law no. 36-15 relating to water (Ait Kadi et Ziyad 2018), and this, in a perspective of monitoring water sustainability, as well as adapting with the climatic context of Morocco, in of course all Moroccan zones characterized by the dominance of semi-arid and arid climatic stages. In this sense, the purpose of adopting a national water plan (NWP) (between 2020 and 2050 for 30years) is to manage water resources sustainably and equitably, taking into account climatic issues and challenges, such as managing periods of drought, coping with water scarcity, and meeting the needs of the Moroccan population in terms of daily uses and the requirements of economic activities, Also to be taken into consideration is the treatment of wastewater via wastewater treatment plants in several cities, for reuse of this water for watering green spaces, as well as the desalination of sea water to obtain non-conventional water, to combat water scarcity.

6. CONCLUSION AND RECOMMENDATION

In terms of concluding this research work, monitoring the evolution of surface water resources at the Ain Kwachia dam reservoir, in time and space, has highlighted the importance of hydraulic structures for storing surface water, during rainy periods, to serve human activities, especially agriculture in terms of farm irrigation in the Ykem watershed on the lower plateau of central Morocco, as well as the role of hydraulic engineering in coping with the impacts of climate change, particularly during periods of drought, especially between the years 2015 and 2017, in addition to those between 2019 and 2023, Both periods see an increase in temperature, which favors evaporation of surface water, as is the case for the Ain Kwachia dam reservoir.

In addition, the hydrological situation of the study area is particular, with the dominance of arid climate, the strategy of preserving surface water resources is necessary, because the Ykem watershed, where the Ain Kwachia dam is located, does not contain groundwater to have water supply, which makes it necessary to take this adaptation measure in terms of the dam, to ensure the sustainability of water resources in this important geographical location on the north-western part of the central plateau overlooking the coastal dunes of the Atlantic Ocean. Integrated water resource management, in terms of the construction of hydraulic structures, is one of the practical solutions for preserving water, especially in vulnerable areas such as the Ykem watershed, which is characterized by climatic aridity. In this sense, our recommendation, according to this research article, is to have other adaptation measures with the particular climate of the study area, through hydraulic developments with a view to water sustainability, of course, the results of this chapter are intended to help decision-makers and water resource managers take the necessary measures to cope with water scarcity, especially in similar locations in Morocco, via mitigation and adaptation strategies in view of the changes affecting certain areas of the country, as in the case of the Ykem watershed, where the Ain Kwachia dam is the subject of this research, especially the phenomenon of climate change, which is having repercussions on natural resources, namely surface water and, of course, groundwater, fed mainly from the surface by water penetration and snowmelt.

For this reason, hydraulic engineering is one of the practical solutions for preserving surface water resources, and we recommend building dams to capture the water that flows downstream, especially the main rivers, in addition, to avoid any kind of evacuation of water from rainy periods to the seas, to ensure water sustainability, in a perspective that aims to serve human activities in terms of irrigating farmland, supplying towns with drinking water and meeting all kinds of water demand.

ACKNOWLEDGMENT

We would like to thank the services of the Bouregreg and Chaouia water basin agency for providing us with all the data, as well as the open-access sites that enabled us to carry out this research work.

REFERENCES

Ait El Mekki, O., & Laftouhi, N. E. (2016). Combination of a geographical information system and remote sensing data to map groundwater recharge potential in arid to semi-arid areas: The Haouz Plain, Morocco. *Earth Science Informatics*, 9(4), 465–479. DOI: 10.1007/s12145-016-0268-0

Ait Kadi, M., & Ziyad, A. (201)8. Integrated Water Resources Management in Morocco. Springer. DOI: 10.1007/978-981-10-7913-9_6

Boutallaka, M., El Mazi, M., El Yazidi, M., El Hairchi, K., & Benabdelhadi, M. (2024). *Evolution of Bioclimatic Stages under climate change in the Ouergha Catchment (Northern Morocco)*. DOI: 10.1051/bioconf/202411503002

Chen, S., Liu, J., Ma, S., Fan, Y., Jia, J., Chen, J., & Chen, F. (2024). Holocene dust storm variations across northern monsoonal Asia and arid central Asia: Contrasting impacts of climate change. *Global and Planetary Change*, 240, 104524. DOI: 10.1016/j.gloplacha.2024.104524

El Aoula, R., Mhammdi, N., Dezileau, L., Mahe, G., & Kolker, A. S. (2021). Fluvial sediment transport degradation after dam construction in North Africa. *Journal of African Earth Sciences*, 182, 104255. DOI: 10.1016/j.jafrearsci.2021.104255

El Bilali, A., Taghi, Y., Briouel, O., Taleb, A., & Brouziyne, Y. (2022). A framework based on high-resolution imagery datasets and MCS for forecasting evaporation loss from small reservoirs in groundwater-based agriculture. *Agricultural Water Management*, 262, 107434. Advance online publication. DOI: 10.1016/j.agwat.2021.107434

El Khalki, E. M., Tramblay, Y., Hanich, L., Marchane, A., Boudhar, A., & Hakkani, B. (2021). Climate change impacts on surface water resources in the Oued El Abid basin, Morocco. *Hydrological Sciences Journal*, 66(15), 2132–2145. DOI: 10.1080/02626667.2021.1982137

Elfoul, & Ghachi. (2018). Impact Des Actions Anthropiques Sur Le Bilan Hydrologique Dans Le Bassin Versant De La Tassaout (Amont Du Barrage Moulay Youssef) : Modele Hydrologique Orchy Ii (1986-2010). *Journal of Water and Environmental Sciences*, 2(1), 297–304.

Garreaud, R. D., Vuille, M., Compagnucci, R., & Marengo, J. (2009). Present-day South American climate. *Palaeogeography, Palaeoclimatology, Palaeoecology*, 281(3), 180–195. DOI: 10.1016/j.palaeo.2007.10.032

Gómez-Fernández, D., López, R. S., Zabaleta-Santisteban, J. A., Medina-Medina, A. J., Goñas, M., Silva-López, J. O., Oliva-Cruz, M., & Rojas-Briceño, N. B. (2024). Landsat images and GIS techniques as key tools for historical analysis of landscape change and fragmentation. *Ecological Informatics*, 82, 102738. Advance online publication. DOI: 10.1016/j.ecoinf.2024.102738

Gramz, M., Batchi, M., El Bouziani, J., Azagane, M. H., El-Boukhari, A., Nafii, A., & Elharrari, Y. (2024). *Impact of climate change on surface water resources in arid zones - the case of the Ykem watershed in the central plateau of Morocco*. DOI: 10.1051/e3sconf/202448904008

Harrington, M., & Cross, M. (2015). GPS, GIS, and Google Earth. In Harrington, M., & Cross, M. (Eds.), *Google Earth Forensics* (pp. 35–47). Syngress. DOI: 10.1016/B978-0-12-800216-2.00003-9

Hekmatnia, M., Isanezhad, A., Ardakani, A. F., Ghojghar, M. A., & Ghaleno, N. D. (2023). An attempt to develop a policy framework for the global sustainability of freshwater resources in the virtual water trade. *Sustainable Production and Consumption*, 39, 311–325. DOI: 10.1016/j.spc.2023.05.022

Hritz, C. (2013). A malarial-ridden swamp: Using Google Earth Pro and Corona to access the southern Balikh valley, Syria. *Journal of Archaeological Science*, 40(4), 1975–1987. DOI: 10.1016/j.jas.2012.11.017

Hssaisoune, M., Bouchaou, L., Sifeddine, A., Bouimetarhan, I., & Chehbouni, A. (2020). Moroccan Groundwater Resources and Evolution with Global Climate Changes. *Geosciences*, 10(2), 81. DOI: 10.3390/geosciences10020081

Hung, P., Van Trung, L., & Vo, P. L. (2023). Decision Support Tool for Integrated Water Resources Management Based on GIS, Remote Sensing and SWAT Model: A Case Study in the Upper Part of Dong Nai River Basin, Vietnam. *Environmental Science and Engineering*, 1, 361–388. DOI: 10.1007/978-3-031-17808-5_23

Jaffal, M., Charbaoui, A., Kchikach, A., El Ghorfi, M., Khaldoun, A., Amine El Mahdi, S., Bodinier, J.-L., Yazami, O. K., Jourani, E.-S., & Manar, A. (2022). Gravity study of the Western Bahira Basin and the Gantour Phosphatic Plateau, central Morocco: Interpretation and hydrogeological implications. *Journal of African Earth Sciences*, 193, 104581. DOI: 10.1016/j.jafrearsci.2022.104581

Jiang, D., Jones, I., Liu, X., Simis, S. G. H., Cretaux, J.-F., Albergel, C., Tyler, A., & Spyrakos, E. (2024). Impacts of droughts and human activities on water quantity and quality: Remote sensing observations of Lake Qadisiyah, Iraq. *International Journal of Applied Earth Observation and Geoinformation*, 132, 104021. DOI: 10.1016/j.jag.2024.104021

Kana, J. D., Ahmad Diab, D., Hervé, D. G., N'kaya, G. D. M., Meli'i, J. L., Browdon, S. N., & Lebogo, S. P. K. N. (2022). A combined GIS-based remote sensing and wireline log data analysis for water resource management in the economic capital district of Cameroon. *Sustainable Water Resources Management*, 8(5), 129. Advance online publication. DOI: 10.1007/s40899-022-00714-0

Kumar, M., & Jha, R. (2022). Evaluating Water Resources Availability in Headwater Sub-catchments of Subarnarekha River Basin in a changing Environment using Remote Sensing and GIS Software. *Indian Journal of Environmental Protection*, 42(9), 1034–1041.

Li, J., Li, L., Song, Y., Chen, J., Wang, Z., Bao, Y., Zhang, W., & Meng, L. (2023). A robust large-scale surface water mapping framework with high spatiotemporal resolution based on the fusion of multi-source remote sensing data. *International Journal of Applied Earth Observation and Geoinformation*, 118, 103288. DOI: 10.1016/j.jag.2023.103288

Li, Z., Leong, W. J., Durand, M., Howat, I., Wadkowski, K., Yadav, B., & Moortgat, J. (2023). Super-resolution deep neural networks for water classification from free multispectral satellite imagery. *Journal of Hydrology (Amsterdam)*, 626, 130248. DOI: 10.1016/j.jhydrol.2023.130248

Mabrouki, Benchrifa, Ennouhi, Azoulay, Bencheikh, Rachiq, El-Moustaqim, Al-Jadabi, Azrour, Fatima-zahra, & El Hajjaji. (2023). Geographic Information System for the Study of Water Resources in Chaâba El Hamra, Mohammedia (Morocco). *Lecture Notes in Networks and Systems* 635 LNNS:469-74. .DOI: 10.1007/978-3-031-26254-8_67

Mahmoud, M. T., Mohammed, S. A., Hamouda, M. A., & Mohamed, M. M. (2020). Impact of Topography and Rainfall Intensity on the Accuracy of IMERG Precipitation Estimates in an Arid Region. *Remote Sensing (Basel)*, 13(1), 13. DOI: 10.3390/rs13010013

Martins, V. S., Kaleita, A., Barbosa, C. C. F., Fassoni-Andrade, A. C., Felipe de Lucia, L., & Novo, E. M. L. M. (2019). Remote sensing of large reservoir in the drought years: Implications on surface water change and turbidity variability of Sobradinho reservoir (Northeast Brazil). *Remote Sensing Applications: Society and Environment*, 13, 275–288. DOI: 10.1016/j.rsase.2018.11.006

Ongoma, V., Driouech, F., Brouziyne, Y., Chfadi, T., Epule Epule, T., Tanarhte, M., & Chehbouni, A. (2024). Morocco's climate change impacts, adaptation and mitigation—A stocktake. *Regional Environmental Change*, 24(1), 14. Advance online publication. DOI: 10.1007/s10113-023-02176-2

Ouharba, E. L. H., Mabrouki, J., & Triqui, Z. E. L. A. (2024). Assessment and Future Climate Dynamics in the Bouregreg Basin, Morocco – Impacts and Adaptation Alternatives. *Ecological Engineering and Environmental Technology*, 25(3), 51–63. DOI: 10.12912/27197050/177823

Rafik, A., Bahir, M., Beljadid, A., Chehbouni, A., Dhiba, D., & Ouhamdouch, S. (2022). The combination of the quality index, isotopic, and GIS techniques to assess water resources in a semi-arid context (Essaouira watershed in Morocco). *Groundwater for Sustainable Development*, 17, 100768. DOI: 10.1016/j.gsd.2022.100768

Sherif, M., Liaqat, M. U., Baig, F., & Al-Rashed, M. (2023). Water resources availability, sustainability and challenges in the GCC countries: An overview. *Heliyon*, 9(10), e20543. DOI: 10.1016/j.heliyon.2023.e20543 PMID: 37817990

Teixeira, A. M. D. A., Batista, L. V., Silva, R. M. D., Freitas, L. M. T., & Santos, C. A. G. (2024). Dynamic monitoring of surface area and water volume of reservoirs using satellite imagery, computer vision and deep learning. *Remote Sensing Applications: Society and Environment*, 35, 101205. Advance online publication. DOI: 10.1016/j.rsase.2024.101205

Thi, N. Q., Govind, A., Le, M.-H., Linh, N. T., Tran, T. M. A., Hai, N. K., & Ha, T. V. (2023). Spatiotemporal characterization of droughts and vegetation response in Northwest Africa from 1981 to 2020. *The Egyptian Journal of Remote Sensing and Space Sciences*, 26(3), 393–401. DOI: 10.1016/j.ejrs.2023.05.006

Tola, T. L., Zhang, K., Chukalla, A. D., Liu, L., Ezaz, G. T., Elsadek, E. A., & Abdallah, M. (2024). Spatiotemporal lakes surface area changes over 35 years and potential causes in the Central Rift Valley, Ethiopia. *Journal of Hydrology. Regional Studies*, 54, 101863. Advance online publication. DOI: 10.1016/j.ejrh.2024.101863

Villani, L., Castelli, G., Yimer, E. A., Chawanda, C. J., Nkwasa, A., Van Schaeybroeck, B., Penna, D., van Griensven, A., & Bresci, E. (2024). Impacts of climate change and vegetation response on future aridity in a Mediterranean catchment. *Agricultural Water Management*, 299, 108878. DOI: 10.1016/j.agwat.2024.108878

Wang, W., Shi, K., Wang, X., Zhang, Y., Qin, B., Zhang, Y., & Woolway, R. I. (2024). The impact of extreme heat on lake warming in China. *Nature Communications*, 15(1), 70. Advance online publication. DOI: 10.1038/s41467-023-44404-7 PMID: 38167519

Wu, F., Yang, X., Cui, Z., Ren, L., Jiang, S., Liu, Y., & Yuan, S. (2024). The impact of human activities on blue-green water resources and quantification of water resource scarcity in the Yangtze River Basin. *The Science of the Total Environment*, 909, 168550. DOI: 10.1016/j.scitotenv.2023.168550 PMID: 37979857

Zhao, B., Wu, J., Han, X., Tian, F., Liu, M., Chen, M., & Lin, J. (2023). An improved surface water extraction method by integrating multi-type priori information from remote sensing. *International Journal of Applied Earth Observation and Geoinformation*, 124, 103529. DOI: 10.1016/j.jag.2023.103529

Chapter 9
Assessing the Spatio-Temporal Evolution of Lake Aguelmam Azigza Using Remote Sensing and GIS: Middle Atlas – Morocco

Mohamed El Bouazzati
https://orcid.org/0009-0001-6971-2458
Faculty of Humanities and Social Sciences, Ibn Tofail University, Kenitra, Morocco

Mouhcine Batchi
https://orcid.org/0000-0002-0915-6592
Faculty of Humanities and Social Sciences, Ibn Tofail University, Kenitra, Morocco

Faycal Fatah
https://orcid.org/0000-0002-7014-4312
Faculty of Humanities and Social Sciences, Ibn Tofail University, Kenitra, Morocco

Amina Moumane
https://orcid.org/0009-0001-9188-6964
Faculty of Humanities and Social Sciences, Ibn Tofail University, Kenitra, Morocco

Ibtissam Motib

DOI: 10.4018/979-8-3693-9651-3.ch009

Faculty of Humanities and Social Sciences, Ibn Tofail University, Kenitra, Morocco

ABSTRACT

This study investigates the spatio-temporal evolution of Lake Aguelmam Azigza from 1977 to 2024 using remote sensing and GIS techniques. Utilizing Landsat satellite imagery and the modified normalized difference water index (MNDWI), the study reveals a significant fluctuation in the lake's surface area over the analyzed period, with a concerning overall decline of 66.32%. The lake's surface area decreased from 49.11 hectares in 1977 to 16.54 hectares in 2024. This decline is likely attributed to climate change-induced factors such as drought and variations in precipitation, highlighting the vulnerability of Moroccan water resources. The research emphasizes the need for proactive measures, including resource monitoring and drought mitigation strategies, to ensure the sustainability of water resources in the region. Despite limitations related to data availability, the study provides valuable insights into the dynamics of Lake Aguelmam Azigza and underscores the urgency of addressing the challenges posed by climate change on water resources in Morocco and similar regions.

1. INTRODUCTION

The Mediterranean region, particularly the Moroccan Middle Atlas, is recognized as being highly susceptible to the impacts of climate change. This vulnerability is well-documented in the Fifth Assessment Report of the Intergovernmental Panel on Climate Change (IPCC, AR5, 2013), which emphasizes the profound implications for water resources and ecosystems. Projections for the period 2070-2099 indicate a drastic reduction in annual rainfall, potentially decreasing by as much as 30%. This decline is expected to lead to a corresponding reduction in water resources, estimated to be between 30% and 50% (Milano, 2012). The Moroccan Middle Atlas, situated at the intersection of arid and temperate climatic zones, exemplifies the complexity of climatic interactions in the Mediterranean. This region is influenced by a convergence of mid-latitude and subtropical processes, further shaped by the Atlantic Ocean, the Mediterranean Sea, and the Sahara Desert. Such diverse influences contribute to substantial spatial and temporal variability in precipitation patterns, a phenomenon extensively discussed in the works of Giorgi and Lionello (2008) and Raible et al. (2010).

The focus of this study is Lake Aguelmam Azigza, located within a tectono-karstic depression in the Middle Atlas. The lake has been intermittently studied since the 1940s, with research consistently underscoring its relatively pristine environment and the minimal human impact within its catchment area. The climate of this region is characterized by Mediterranean sub-humid conditions, marked by wet winters and dry summers. Precipitation variability in the Middle Atlas is a key factor driving changes in the surface area of lakes, including Aguelmam Azigza, as evidenced by historical data and more recent monitoring efforts.

In this context, our research aims to monitor the spatio-temporal evolution of lakes in the Moroccan Middle Atlas, with a particular focus on Lake Aguelmam Azigza, over the period from 1977 to 2024. We employ advanced remote sensing techniques, including the analysis of satellite images and Digital Terrain Models (DTMs), as well as Geographic Information System (GIS) tools, to gain a deeper understanding of the variability in the lake's surface area. This approach allows for a comprehensive assessment of how climatic factors influence the lake's dynamics over time.

2. STUDY AREA

Lake Aguelmam Azigza, known as “the green lake” in Berber, is a natural body of water of tectono-karstic origin, situated approximately 30 km east of the town of Khénifra. It is part of the upper basin of the Oum Er Rbiaa River, nestled in the heart of the Middle Atlas mountain range at an elevation of 1,546 meters (Adlal, 2019). The lake is geographically positioned at 32°58’N latitude and 5°26’W longitude.

The surrounding landscape is characterized by limestone hills with steep slopes, predominantly covered with cedar and holm oak forests. The lake's hydrology is primarily sustained by precipitation, including both rainfall and snowmelt, as well as surface runoff and underground water sources. The waters of Lake Aguelmam Azigza are noted for their clarity and softness, with a slightly basic pH and a high carbonate content, indicative of the surrounding geological formations (Khénifra National Park Development and Management Plan, 2018). See Figure 1 for a detailed visual representation of the lake and its surrounding environment.

Figure 1. Geographical location map of the Aguelmam Azigza lake

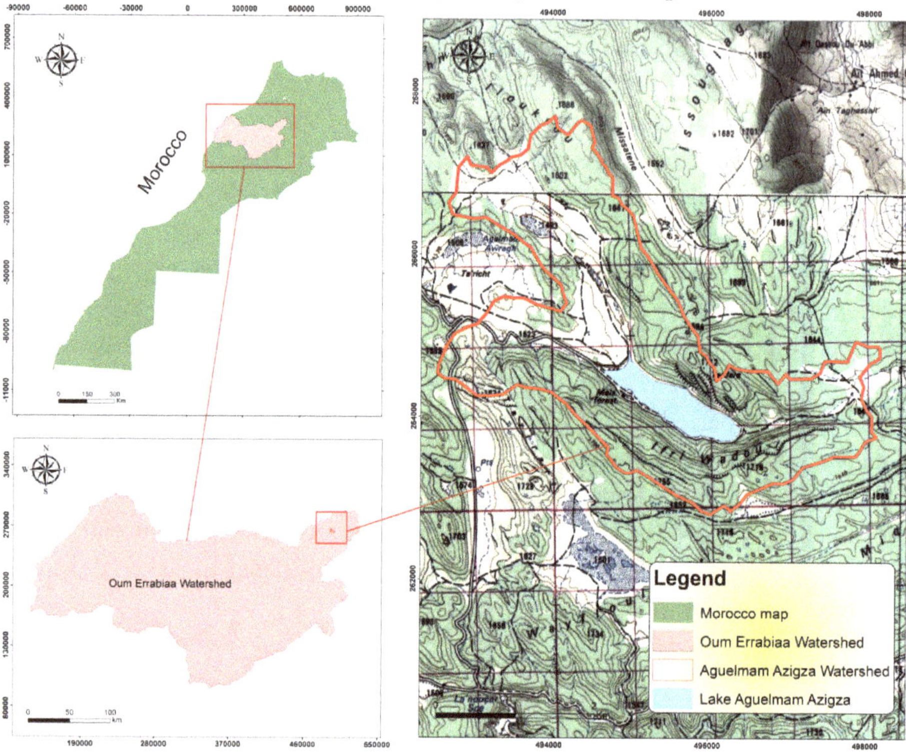

2.1 Geology and Geomorphology

According to Martin (1981), Lake Aguelmam Azigza is situated within the structural domain of the Middle Atlas Causse, a region characterized by classic high limestone plateaux landscapes. These plateaux are primarily composed of neritic carbonates from the Lower and Middle Lias, which rest on underlying Triassic argillites (Adllal, 2019). The lake itself is a collapse doline, formed within the carbonate rock (limestone and dolomite) of the Lias. It is surrounded by steep, fissured walls, which feature large detachment cracks and sections of the wall that have collapsed (El Khalki, 2001).

2.2 Hydrology and Hydrogeology

Lake Aguelmam Azigza is intermittently fed by waters from the surrounding polje, despite the absence of visible outlets. The lack of an outlet, coupled with the lake's relative persistence, indicates a significant connection with the water table, which forms at the interface between the Lias and Trias formations (El Khalki et al., 2001). The surface hydrographic network of the region is both discontinuous and irregular, with valleys generally remaining dry. Large-scale basins in the area are either permanently or temporarily filled with water, forming lakes (Adallal, 2019).

2.3 Climate Data

The Middle Atlas region experiences a cold, humid climate of the Mediterranean mountain type, influenced by its altitude, geographical location, and exposure to marine weather patterns (Martin, 1981). Clouds originating from the west contribute to heavy precipitation, including rain and snow, in the Middle Atlas.

Table 1. Location of climatic stations (ABHS, 2005)

Name	N° IRE	X	Y	Z
Tarhat	4875	476400	267500	1036
Ait Aissa	6232	530740	285300	-

Two climatic stations in the region provide precipitation data (see Table 1). The average rainfall, based on available data (figure 3), indicates the following patterns:

- The rainy season extends from November to April.
- Maximum rainfall occurs in December at the Tarhat station (78.6 mm) and in March at the Ait Aissa station (71.4 mm).
- Minimum rainfall is observed in July, with 8.9 mm at the Tarhat station and 15.4 mm at the Ait Aissa station.
- The average rainfall values are 49 mm and 40.7 mm at the Tarhat and Ait Aissa stations, respectively.

Figure 2. Monthly Rainfall Data for Climate Stations (ABHS,2021)

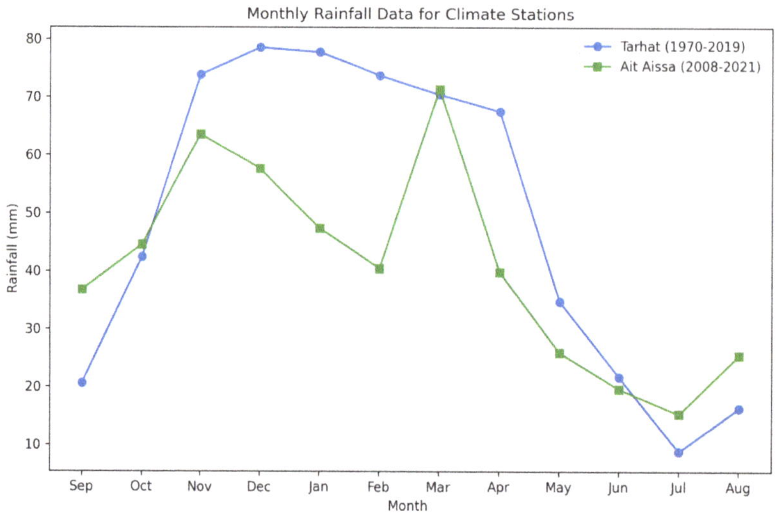

It is crucial to consider that the last decade has been marked by prolonged drought cycles, which can alter averages depending on the reference periods selected.

- Temperature:

  The temperature data indicate that July is the hottest month, with a maximum temperature of 43.65°C, while January is the coldest, with a minimum temperature of -5°C.

- Evaporation:

Evaporation data from 1986 to 2019 reveal significant fluctuations, with notable peaks in 1998, 1999, 2000, and 2001, where values often exceeded 3,000 mm. In contrast, the years 1996, 2010, 2011, and 2018 show lower evaporation rates, often below 2,500 mm. Overall, there is an upward trend in evaporation over the years, with considerable annual variations likely influenced by specific climatic factors.

3. MATERIALS AND METHOD

3.1 Satellite Data

This study utilizes satellite images from various Landsat missions to analyze the spatio-temporal evolution of Lake Aguelmam Azigza. The selected images span from 1977 to 2024, covering nearly five decades of data, and provide insights into changes in the lake's surface area over time. This extensive temporal coverage is crucial for understanding long-term trends and the impact of both natural and anthropogenic factors on this water resource. This table 2 provides a concise overview of the Landsat missions used in the study, including details about the sensors, acquisition dates, and the specific spectral bands utilized for analyzing the lake's surface changes.

Table 2. **Summary of Landsat Satellite Images and Sensor Specifications Used for Analyzing Lake Aguelmam Azigza**

Satellite	Sensor(s)	Acquisition Dates	Ground Resolution	Spectral Bands	Details
Landsat 2	MSS	May 1977	80 m	- Band 4: Visible Green (0.5 to 0.6 µm) - Band 5: Visible Red (0.6 to 0.7 µm) - Band 6: Near-Infrared (0.7 to 0.8 µm) - Band 7: Near-Infrared (0.8 to 1.1 µm)	Line-scanning device with six scan lines per spectral band.

continued on following page

Table 2. Continued

Satellite	Sensor(s)	Acquisition Dates	Ground Resolution	Spectral Bands	Details
Landsat 5	MSS, TM	April 1986, May 1996, May 2007	30 m (reflective), 120 m (thermal)	- Band 1: Visible Blue (0.45 - 0.52 μm) - Band 2: Visible Green (0.52 - 0.60 μm) - Band 3: Visible Red (0.63 - 0.69 μm) - Band 4: Near-Infrared (0.76 - 0.90 μm) - Band 5: Near-Infrared (1.55 - 1.75 μm) - Band 6: Thermal (10.40 - 12.50 μm) - Band 7: Mid-Infrared (2.08 - 2.35 μm)	MSS data acquisitions ended in 1999. TM sensor includes mid-range infrared.
Landsat 8	OLI, TIRS	April 2015	30 m (most bands), 15 m (Pan)	- Band 1: Coastal Aerosol (0.43 - 0.45 μm) - Band 2: Blue (0.450 - 0.51 μm) - Band 3: Green (0.53 - 0.59 μm) - Band 4: Red (0.64 - 0.67 μm) - Band 5: Near-Infrared (0.85 - 0.88 μm) - Band 6: SWIR 1 (1.57 - 1.65 μm) - Band 7: SWIR 2 (2.11 - 2.29 μm) - Band 8: Panchromatic (0.50 - 0.68 μm) - Band 9: Cirrus (1.36 - 1.38 μm)	Improved radiometric precision with a 12-bit dynamic range. 4096 grey levels vs. 256 in Landsat 1-7.
Landsat 9	OLI, TIRS	April 2024	30 m (most bands), 15 m (Pan)	Same as Landsat 8	Continued from Landsat 8 with similar capabilities.

3.2 Satellite Image Classifications

To detect spatio-temporal changes in Lake Aguelmam Azigza between 1977 and 2024, we extracted the lake's water surface from Landsat satellite images taken in 1977, 1986, 1996, 2006, 2015 and 2024. For each image, we calculated the Normalized Difference Water Index (NDWI) and the Modified Normalized Difference Water Index (MNDWI).

In simple terms the NDWI and MNDWI is a tool used in remote sensing to identify and highlight water bodies like lakes and rivers in satellite images.

3.2.1 NDWI

The NDWI (Formula 1) was employed to monitor changes in the water surface of Lake Aguelmam Azigza. Originally proposed by Gao, this index is designed to detect variations in vegetation liquid water content and is also used for managing water bodies and assessing soil moisture. NDWI values close to +1 indicate higher moisture levels, while values approaching zero or negative values signify drier conditions or the presence of terrestrial vegetation and soil (Karmaoui et al., 2023).

$$\boldsymbol{NDWI = \frac{Green\ band - NIR\ band}{Green\ band + NIR\ band}} \quad (1)$$

The NDWI for each satellite is calculated from the table 3:

Table 3. The bands used in the calculation of the NDWI index for each satellite

Landsat 02 MSS	Landsat 05 TM	Landsat 08 OLI	Landsat 09
$MNDWI = \frac{B4 - B6}{B4 + B6}$	$MNDWI = \frac{B2 - B4}{B2 + B4}$	$MNDWI = \frac{B3 - B5}{B3 + B5}$	$MNDWI = \frac{B3 - B5}{B3 + B5}$

3.2.2 MNDWI

Xu (2006) introduced a modified version of the NDWI by replacing near-infrared (NIR) reflectance with mid-infrared (MIR) reflectance, resulting in the (MNDWI) (Formula 2). Xu's research demonstrated that MNDWI offers improved accuracy in delineating wetlands compared to the original NDWI. This modification enhances the ability to distinguish water bodies by effectively eliminating non-water pixels, such as those representing vegetation, soil, and urban areas

$$\mathbf{MNDWI = \frac{Green\ band - MIR\ band}{Green\ band + MIR\ band}} \quad (2)$$

The MNDWI for each satellite is calculated from the table 4:

Table 4. The bands used in the calculation of the MNDWI index for each satellite

Landsat 02 MSS	Landsat 05 TM	Landsat 08 OLI	Landsat 09
$MNDWI = \frac{B4 - B7}{B4 + B7}$	$MNDWI = \frac{B2 - B5}{B2 + B5}$	$MNDWI = \frac{B3 - B6}{B3 + B6}$	$MNDWI = \frac{B3 - B6}{B3 + B6}$

The classification scheme for the MNDWI is as follows: values ranging from 0 to +1 are classified as water bodies, indicating the presence of water, while values ranging from -1 to 0 are classified as other types of land cover, including vegetation, soil, and urban areas. This classification effectively distinguishes between water bodies and non-water areas based on their MNDWI index values. (Computing MNDWI using Sentinel on GEE, 2024)

3.3 Classification Accuracy

Accuracy assessment is essential for effective decision-making and ecosystem management, ensuring that satellite and classified images accurately reflect actual land cover. This process involves validating how well pixel classifications match the ground truth data. In this study, a Google Earth Pro image from April 2024—corresponding to the period of the satellite images used for NDWI and MNDWI calculations—served as reference data for comparison. We created 3,000 Accuracy Assessment Points, comparing the classes assigned by the NDWI and MNDWI maps with the actual classes observed in Google Earth Pro. The results of this comparison are presented in a confusion matrix (table5).

Table 5. confusion matrix (Ragan,2018)

	Predicted Water	Predicted Non-Water
Actual Water	a (true positive)	b (false negative)
Actual Non-Water	c (false positive)	d (true negative)

In a simple word:

- a: Number of pixels correctly classified as 'water'.
- b: Number of pixels incorrectly classified as 'non-water' when they are actually water.
- c: Number of pixels incorrectly classified as 'water' when they are not water.
- d: Number of pixels correctly classified as 'non-water'.

Then we calculate overall accuracy, which is the proportion of correctly classified pixels out of the total pixels, using the formula (3):

$$\textbf{Overall Accuracy} = \frac{\mathbf{a + d}}{\mathbf{a + b + c + d}} \tag{3}$$

Next, we determine class-specific accuracy for:

• Water

$$\textbf{Water Accuracy} = \frac{\mathbf{a}}{\mathbf{a + c}} \tag{4}$$

• Non-Water:

$$\textbf{No} - \textbf{Water Accuracy} = \frac{\mathbf{d}}{\mathbf{b + d}} \tag{5}$$

Additionally, we calculate the Kappa coefficient, a discrete multivariate technique used in accuracy assessments (Karmaoui et al., 2023), to measure the agreement between NDWI and MNDWI classifications and Google Earth imagery, accounting for random chance.

3.4 Calculation of Lake Water Surfaces

After computing the NDWI and MNDWI indices, the resulting raster files were imported into ArcGIS 10.4. We began by converting these raster datasets into vector polygons through the raster-to-vector conversion process. This transformation allowed us to delineate the lake's water surface more accurately. Next, we performed a detailed analysis to calculate the water surface area of Lake Aguelmam Azigza based on the polygons created from the indices. This involved extracting the area measurements of the polygons representing water bodies to determine the extent of the lake's surface for the study period.

3.5 Mapping With ArcGIS 10.4

Following the water surface calculations, we proceeded with the map layout process in ArcGIS 10.4. This step included designing the final map presentation, which involved several key components:

- **Geographical Coordinates**: Adding coordinate grids and labels to provide geographic reference points.
- **Legends**: Creating a legend to explain the symbols and colors used on the map for different land cover types.
- **Directional Indicators**: Including a north arrow to orient the map for better spatial understanding.

- **Scale Bar**: Incorporating a scale bar to indicate distances and ensure accurate interpretation of map features.
- **Additional Cartographic Elements**: Adding other essential elements such as title, source information, and any necessary annotations to enhance the map's clarity and usefulness.

Each of these elements was carefully integrated to ensure the final maps were both accurate and visually informative, supporting the overall goals of the study.

4. RESULTS

4.1 Water Surfaces of Lake Aguelmam Azigza

Figure 3. Variation of Lake Aguelmam Azigza Surface Area Over Time using NDWI

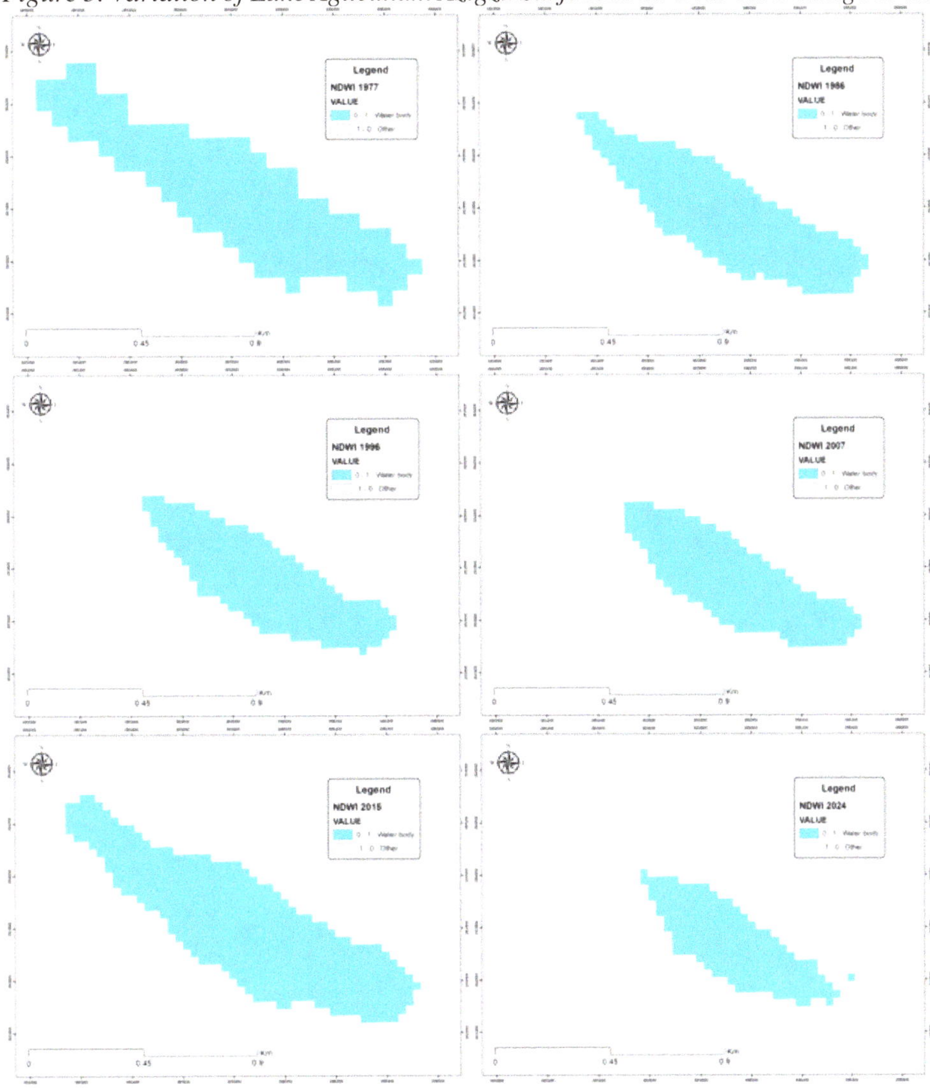

Figure 4. Variation of Lake Aguelmam Azigza Surface Area Over Time using MNDWI

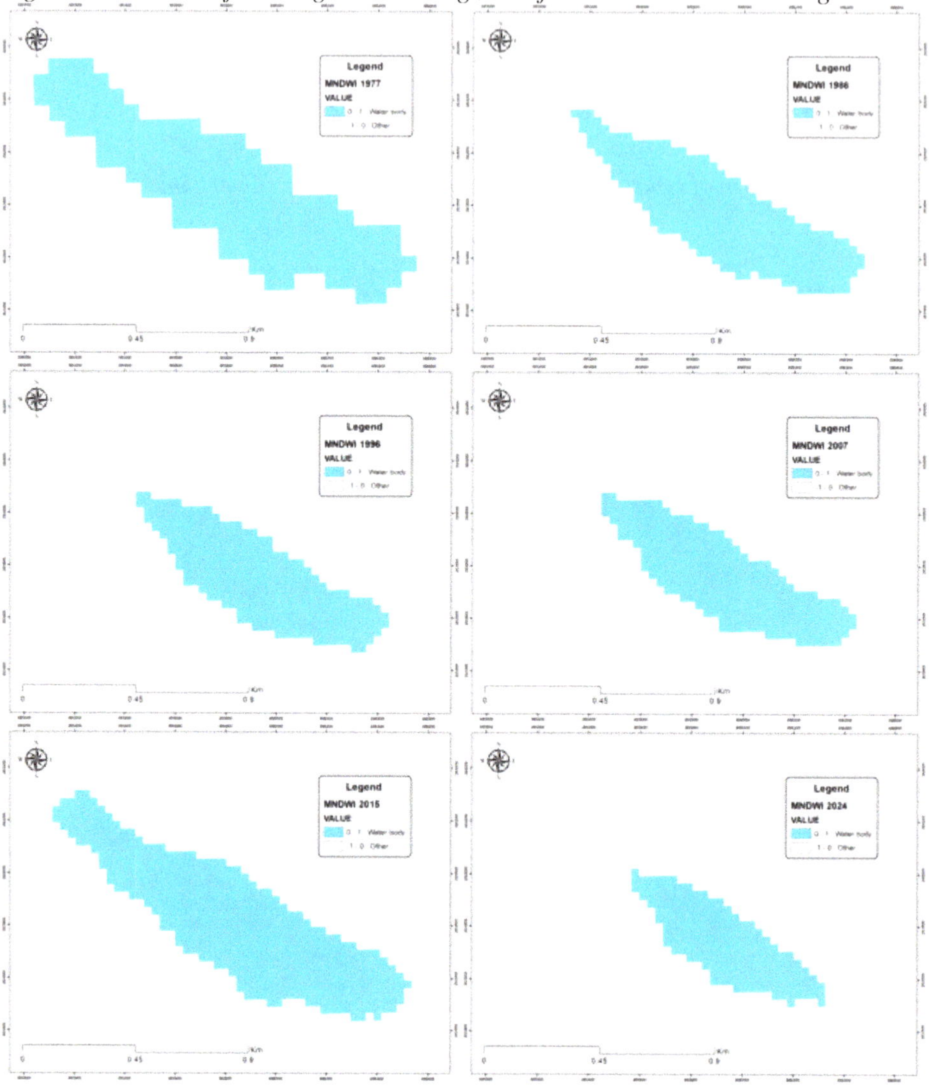

Figure 5. Variation of Lake Aguelmam Azigza Surface Area Over Time

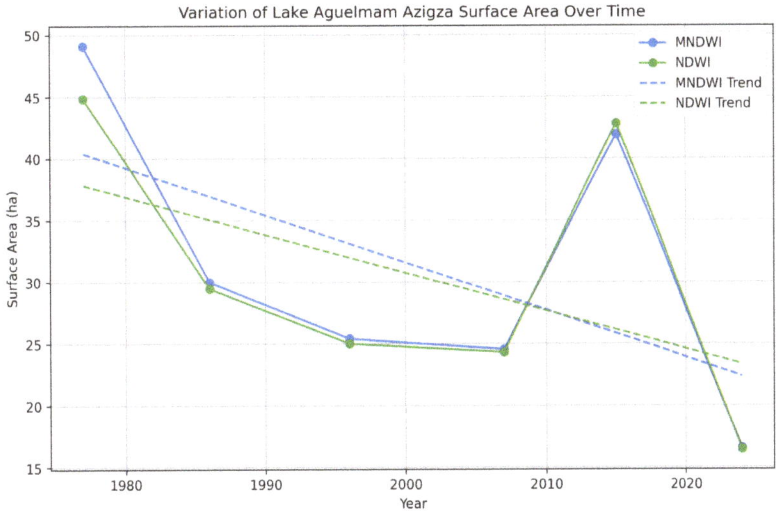

Figures 4 and 5 present surface area data for Lake Aguelmam Azigza across different periods from 1977 to 2024, measured using two indices: The MNDWI and the NDWI. These indices are commonly used in remote sensing to identify water bodies and monitor their surface changes over time. The data provided are in hectares (ha).

Both MNDWI and NDWI indicate a fluctuating trend in the surface area of Lake Aguelmam Azigza over the observed periods.

There is an overall decline in the lake's surface area from 1977 to 2024, although the surface area fluctuates significantly during these years.

- Initial Period (1977-1986):

In 1977, the lake's surface area was relatively large, with MNDWI showing 49.1 ha and NDWI showing 44.86 ha.

By 1986, there was a significant reduction in the surface area, with MNDWI decreasing to 29.97 ha and NDWI to 29.44 ha. This represents a decrease of approximately 39% and 34% respectively.

- Mid-Period (1986-2007):

The downward trend continues through 1996, with the surface area shrinking to 25.4 ha (MNDWI) and 25.0 ha (NDWI).

By 2007, the lake's surface area further decreases slightly to 24.53 ha (MNDWI) and 24.29 ha (NDWI), indicating a period of relative stability but at a much-reduced size compared to 1977.

- Recent Period (2007-2024):

Between 2007 and 2015, there is a noticeable recovery in the lake's surface area. By 2015, the surface area increases to 41.94 ha (MNDWI) and 42.84 ha (NDWI). This recovery might be attributed to a period of increased precipitation or other climatic factors favoring water retention.

However, the surface area dramatically decreases again by 2024, with MNDWI showing 16.54 ha and NDWI showing 16.44 ha. This marks a sharp decline, with the lake's surface area falling by more than 60% compared to 2015.

The MNDWI and NDWI values are closely aligned throughout the periods, which confirms the reliability of these indices in measuring water bodies.

The slight variations between the indices across all periods are negligible, indicating consistent results in the surface area estimation.

The data suggest that Lake Aguelmam Azigza has experienced significant fluctuations in its surface area over the past few decades, which could be influenced by various factors, including climate variability, seasonal precipitation patterns, and possibly human activities. The dramatic reduction in surface area in the most recent period (2024) could indicate severe drought conditions or increased evaporation rates, further exacerbating the vulnerability of the lake under changing climatic conditions.

This analysis underscores the importance of continuous monitoring of water bodies in the Middle Atlas region. The fluctuating trends in the lake's surface area are critical indicators of environmental change, providing insights into broader hydrological dynamics and the impacts of climate change in this sensitive region

The water index (NDWI) presented in table 6 was calculated in order to determine the water surface of Lake Aguelmam Azigza between the years 1977 and 2024.

4.2 Validation of NDWI and MNDWI Results Using Google Earth Imagery

• Confusion Matrix for NDWI Validation

To validate the NDWI analysis results, we constructed a confusion matrix using the data from 3,000 assessment points. The matrix compares the NDWI classifications with the Google Earth image from April 2024, as follows:

Presence of Water in NDWI and Google Earth Image: 157

No Water in NDWI but Water in Google Earth Image: 21

No Water in NDWI and No Water in Google Earth Image: 2821

Presence of Water in NDWI but Absence of Water in Google Earth Image: 1

The confusion matrix provides a clear view of the NDWI's classification accuracy (table 6).

Table 6. confusion matrix of NDWI

	Presence of water (Google Earth)	Absence of water (Google Earth)	Total
Presence of water (NDWI)	157	1	158
Absence of water (NDWI)	21	2821	2842
Total	178	2822	3000

The validation of NDWI results using Google Earth imagery from April 2024 demonstrates a high level of accuracy. The overall accuracy of the classification was 99.27%, indicating that the vast majority of the pixels were correctly classified. Specifically, the precision for identifying the presence of water was 88.3%, while the precision for identifying the absence of water was exceptionally high at 99.96%. The Kappa coefficient, calculated to be 0.935, further confirms the strong agreement between the NDWI results and the reference data, showing that the match extends beyond what would be expected by chance. These results (represented in table 7) underscore the reliability of the NDWI in accurately mapping water and non-water areas .

Table 7. Accuracy Statistics for NDWI Validation

Accuracy Statistics	Value
Overall Accuracy	99.27%
Precision (Presence of Water)	88.3%

continued on following page

Table 7. Continued

Accuracy Statistics	Value
Precision (Absence of Water)	99.96%
Kappa Coefficient	0.935

- **Confusion Matrix for NDWI Validation**

The confusion matrix for validating the MNDWI against the Google Earth image from April 2024 is summarized as follows:

Presence of Water in MNDWI and Google Earth Image: 153

No Water in MNDWI but Water in Google Earth Image: 22

No Water in MNDWI and No Water in Google Earth Image: 2825

Presence of Water in MNDWI but Absence of Water in Google Earth Image: 0

The matrix (table 8) clearly illustrates the classification performance of MNDWI, showing a high rate of correct identification for both water and non-water areas:

Table 8. confusion matrix of MNDWI

	Presence of water (Google Earth)	Absence of water (Google Earth)	Total
Presence of water (MNDWI)	153	0	153
Absence of water (MNDWI)	22	2825	2847
Total	175	2825	3000

- **Accuracy Assessment Results for MNDWI**

The validation of the MNDWI against the Google Earth image from April 2024 showcases a high degree of accuracy. The overall accuracy of the classification is 99.27%, indicating that the vast majority of the pixels were accurately classified. The precision for identifying the presence of water is 87.4%, while the precision for identifying the absence of water is perfect at 100%. The Kappa coefficient, calculated at 0.936, further emphasizes the excellent agreement between the MNDWI results and the reference data, demonstrating that the classification accuracy significantly surpasses what would be expected by chance. These findings confirm the reliability of the MNDWI in distinguishing between water and non-water areas, with no false positives detected for water areas (see table 9).

Table 9. Accuracy Statistics for MNDWI Validation

Accuracy Statistics (MNDWI)	Value
Overall Accuracy	99.27%
Precision (Presence of Water)	87.4%
Precision (Absence of Water)	100%
Kappa Coefficient	0.936

5. DISCUSSION

The MNDWI and the NDWI both demonstrated strong performance in detecting water bodies, with their values showing close agreement. However, the NDWI exhibited slightly better accuracy in detecting the presence of water, while the MNDWI proved to be perfectly accurate in identifying the absence of water. When considering the Kappa coefficient, the MNDWI showed a marginally superior performance, indicating a very high level of agreement with the reference imagery from Google Earth. This suggests that while both indices are highly accurate, the MNDWI offers a slight advantage in overall agreement (Kappa) and in identifying areas devoid of water, whereas the NDWI is marginally better at detecting water presence. These findings align with previous research that has underscored the effectiveness of MNDWI. For instance, Singh et al. (2015) found that MNDWI is particularly adept at delineating water features that are mixed with vegetation, outperforming NDWI in such scenarios. Moreover, MNDWI has been successfully employed in other studies, such as in the mapping of recent area changes in Lake Chad, where it was utilized to estimate the lake's area fluctuations from 2003 to 2016 (Buma et al., 2018). The data presented reflect the varying conditions of Lake Aguelmam Azigza over a period of forty-seven years, illustrating significant fluctuations in the lake's surface area. During dry periods, the surface area was reduced to as little as 16.54 hectares, while in wet seasons, it expanded to 49.11 hectares. Specifically:

- Between 1977 and 1986, the surface area of Lake Aguelmam Azigza decreased by 38.97%, corresponding to a reduction of 19.14 hectares.
- From 1986 to 1996, the lake lost an additional 4.55 hectares of its surface area.
- The decline continued between 1996 and 2007, albeit at a slower rate, with the lake shrinking by approximately 3.5% (0.89 hectares).
- Between 2007 and 2015, the lake's surface area saw a significant increase of over 41.51%, adding 70.97 hectares.

- However, from 2015 to 2024, the lake's surface area is projected to decrease by 60.56%, equating to a loss of 25.4 hectares.
- Overall, from 1977 to 2024, Lake Aguelmam Azigza has lost 66.32% of its surface area.

The reduction in the area of Aguelmam Azigza is a multifaceted phenomenon driven by # climatic, geological, and anthropogenic factors. The lake, located in a semi-arid region highly susceptible to climate change, undergoes significant fluctuations in its water levels, chemistry, and biological conditions in response to regional hydrological balances (Adallal et al., 2023). This sensitivity is further exacerbated by the increasing frequency and intensity of droughts in Morocco, a trend that has been observed since the 1970s and is closely linked to the North Atlantic Oscillation (Sebbar et al., 2019). These droughts, coupled with declining annual rainfall, fewer wet days, and an increase in dry years since the 1980s (El Orfi, 2023), have had a severe impact on the nation's water resources, leading to diminished surface water and reduced groundwater recharge (FAO, 2023). Additionally, the dissolution of carbonate geological formations contributes to changes in the lake's area (Boutsougame et al., 2022). This trend is clearly reflected in Lake Aguelmam Azigza, where a general decline in lake volume has been observed since 2012 (El Orfi, 2023), coinciding with significant annual rainfall fluctuations. For instance, there was a dramatic decrease in annual rainfall from 844 mm in 2009-10 to just 13 mm in 2019-20 (Boutsougame et al., 2022). Variations in lake water temperature, influenced by both climatic conditions and underground water inputs (Adallal et al., 2014; Gayral and Panouse, 1954), further contribute to the lake's vulnerability. Temperature measurements have shown a maximum of 24.8°C in July and a minimum of 7.8°C in January (Adallal et al., 2014). The impact of these stressors is not confined to Aguelmam Azigza; similar patterns of drought-induced water level reductions, sometimes leading to complete desiccation for several years, have been observed in other Moroccan lakes like Dayet Aoua, underscoring the widespread vulnerability of these ecosystems (Talbi, 2024) Fluctuations in the lake's shoreline are primarily driven by precipitation and evapotranspiration (Boutsougame et al., 2022).

In Aguelmam Sidi Ali, multiple regression analysis ($r = 0.87$) has linked annual water level variations to climatic conditions and the initial lake level (Sayad et al., 2011). Periods of drought over the past three decades have lowered the water levels in these lakes by nearly 7 meters, although abundant rainfall during 2008-09 and 2009-10 led to a sharp rise in water levels (Sayad et al., 2011). Drought sequences identified using the standardized precipitation index reveal that several regions of Morocco have experienced a series of dry spells since the early 1980s (Driouech, 2010). Projections indicate a continued decline in average rainfall (16% to 23% by 2050 and 21% to 45% by 2100) and a rise in average temperatures (2.5°C to 2.8°C by 2050 and 3.4°C to 6.1°C by 2100), further intensifying water stress in the region (El Orfi, 2023; UNESCO, 2020). Field investigations and satellite data suggest that intensified agricultural activities also threaten the sustainability and water potential of the Middle Atlas hydrosystems (El Orfi, 2023). The shrinking of Aguelmam Azigza is therefore a stark indicator of the escalating pressures on water resources in Morocco, underscoring the urgent need for sustainable water management practices this study, however, is subject to several important limitations. Firstly, challenges were encountered in the collection of precipitation data; the available data are incomplete and may be inaccurate, particularly for historical periods, which could affect the accuracy of the analyses.

6. CONCLUSION

Climate change has significant consequences for ecosystems and societies globally, and Morocco, located in the Mediterranean zone, is highly vulnerable. The Moroccan Middle Atlas, known as the country's water tower due to its major rivers and tectono-karstic and volcanic lakes, is experiencing severe impacts from climate change, including a significant drop in lake levels due to recurring droughts.

This study analyzed the spatio-temporal evolution of Lake Aguelmam Azigza's surface area from 1977 to 2024 using remote sensing and GIS techniques. By applying the MNDWI to Landsat 02 MSS, Landsat 05 TM, Landsat 08 OLI, and Landsat 09 satellite images, we observed a decline in the lake's water surface over time, particularly in recent years. This decline is attributed to climate change.

The findings underscore the urgent need for proactive measures to mitigate climate change effects and ensure the sustainability of vital water resources. Recommended strategies include:

- Enhanced Monitoring: Continuously track water levels and quality using remote sensing technologies.
- Sustainable Water Management: Implement practices that promote efficient and responsible water use.
- Community Involvement**: Launch awareness campaigns to educate the public on the importance of water conservation.
- Solid Policy Frameworks: Develop and enforce robust policies to support sustainable water resource management.

To protect Lake Aguelmam Azigza and other critical water resources, collective action and a commitment to responsible management are crucial. Encouraging scientific research can be achieved through:

- Facilitating Data Access: Provide researchers with easy access to climate and hydrological data to enhance understanding and predictions of droughts.
- Increasing Research Funding: Allocate more resources to studies on water management and climate change adaptation.
- Encouraging International Collaboration: Foster partnerships with international institutions to share knowledge and technologies.

Effective water resource management in Morocco also requires leveraging GIS and remote sensing technologies. These tools enable:

- Mapping and Monitoring Water Bodies: Identify areas vulnerable to drought.
- Optimizing Irrigation Practices: Improve water efficiency and reduce wastage.
- Forecasting Drought Periods: Integrate climate and hydrological data into GIS models to anticipate and mitigate drought impacts.
- Analyzing Land-Use Change: Understand how land-use changes affect water resources and develop conservation strategies.

By adopting these technologies and strategies, Morocco can make informed decisions for sustainable water management and enhance resilience against drought.

REFERENCES

Abhatoo: "QUEL AVENIR DE L'EAU AU MAROC?": RAPPORT DE SYNTHESE DES TRAVAUX DE LA JOURNEE SCIENTIFIQUE DU 17 MARS 2022. (n.d.). Retrieved August 18, 2024, from http://www.abhatoo.net.ma/maalama-textuelle/developpement-economique-et-social/developpement-economique/agriculture/ressources-en-eau-et-lutte-contre-la-secheresse/quel-avenir-de-l-eau-au-maroc-rapport-de-synthese-des-travaux-de-la-journee-scientifique-du-17-mars-2022

Adallal, R. (2019). *Variabilité environnementale des lacs du moyen atlas marocain : Fonctionnement hydrogéochimique, hydrologique et réponse au changement climatique.* 212.

Adallal, R., Hanane, I., Benkaddour, A., Vallet-coulomb, C., Rhoujjati, A., Sonzogni, C., & Vidal, L. (2022). Hydrogeochemical Processes of the Azigza Lake System (Middle Atlas, Morocco) Inferred from Monthly Monitoring. *Aquatic Geochemistry*, 29(1), 25–47. Advance online publication. DOI: 10.1007/s10498-022-09409-6

Boutsougame, A., El Youssfi, L., Aassine, H., Khuili, S., Cherkaoui, S. I., Ouazzani, H., & Alaoui, M. (2022). Impact of geology and climate change on wetlands: Case of Lake Aguelmam Azegza (Middle Atlas, Morocco). *E3S Web of Conferences, 337*, 01007. DOI: 10.1051/e3sconf/202233701007

Buma, W., Lee, S.-I., & Seo, J. Y. (2018). Recent Surface Water Extent of Lake Chad from Multispectral Sensors and GRACE. *Sensors (Basel)*, 18(7), 2082. DOI: 10.3390/s18072082 PMID: 29958481

Chardon, M. (1982). Martin (J.).—Le Moyen Atlas Central. Etude géomorphologique. *Revue de Geographie Alpine*, 70(4), 444–444.

Computing MNDWI using Sentinel on GEE | Life in GIS. (n.d.). Retrieved August 31, 2024, from https://www.lifeingis.com/computation-of-mndwi-in-google-earth-engine-using-sentinel-2/

Daki, Y., Zahour, G., Lachgar, R., & El Hadi, H. (2016). Caractérisation De La Sécheresse Climatique Du Bassin Versant D'oum Er Rbia (Maroc) Par Le Biais De L'indice De Précipitation Standardisé (SPI). *European Scientific Journal*, 12(14), 198. DOI: 10.19044/esj.2016.v12n14p198

Damnati, B. (2000). The climate of North Africa during the last interglacial-glacial cycle: A synthesis. *Journal of Archaeological Science*, 27(12), 1085–1089.

El- Bouhali, A., Amyay, M., & El Ouazanı Ech- Chahdi, K. (2024). Changes in water surface area of the Middle Atlas-Morocco lakes: A response to climate and human effects. *International Journal of Engineering and Geosciences*, 9(2), 221–232. DOI: 10.26833/ijeg.1391957

El Khalki, Y., & Akdim, B. (2001). Les dolines d'effondrement et les dolines-lacs des Causses du SW du Moyen Atlas (Maroc). *Karstologia : revue de karstologie et de spéléologie physique, 38*(1), 19-24. DOI: 10.3406/karst.2001.2481

El Khalki, Y., & Hafid, A. (2002). Turbidité, indicateur du fonctionnement perturbé du géosystème karstique de l'Atlas de Beni Mellal (Moyen Atlas méridional, Maroc). *Karstologia*, 40(1), 39–44. DOI: 10.3406/karst.2002.2506

El Orfi, T. (2023). *Quantification, modélisation et usages des ressources en eau dans le Haut Bassin de l'Oum Er Rbia (Amont du barrage Ahmed El Hansali, Maroc)* [These de doctorat, Université de Lorraine]. https://theses.fr/2023LORR0265

Etebaai, I. (2009). *L'environnement actuel et le fonctionnement hydroclimatique de quelques systèmes lacustres dans le Moyen Atlas marocain : Cas des lacs Ifrah, Iffer et Afourgagh*. 344.

Gasse, F. (1991). Hydrological changes in the African tropics since the Last Glacial Maximum. *Quaternary Science Reviews*, 10(4), 379–386.

Giorgi, F., & Lionello, P. (2008). Climate change projections for the Mediterranean region. *Global and Planetary Change*, 63(2-3), 90–104. DOI: 10.1016/j.gloplacha.2007.09.005

IPCC, AR5. (2013). Climate Change 2013: The Physical Science Basis. Contribution of Working Group I to the Fifth Assessment Report of the Intergovernmental Panel on Climate Change.

Jellali, M. (1997). *Développement des ressources en eau au Maroc*. Www.Water .Gov.Ma/Semide/Fr/Themes/Ppne.Htm. http://www.albacharia.ma/xmlui/handle/123456789/31203

Karmaoui, A., Moumane, A., El Jaafari, S., Menouni, A., Al Karkouri, J., Yacoubi, M., & Hajji, L. (2023). Thirty Years of Change in the Land Use and Land Cover of the Ziz Oases (Pre-Sahara of Morocco) Combining Remote Sensing, GIS, and Field Observations. *Land (Basel)*, 12(12), 12. Advance online publication. DOI: 10.3390/land12122127

La sécheresse au Maghreb: Diagnostic, impacts et perspectives pour le renforcement de la résilience du secteur agricole. (2023). FAO. DOI: 10.4060/cc7126fr

Le Réseau ADUM. (n.d.). Retrieved August 18, 2024, from https://adum.parisnanterre.fr/as/ed/cv.pl?mat=104963&site=adumfr

Magny, M. (2013). Holocene climate changes in the central Mediterranean as recorded by lake-level fluctuations at Lake Accesa (Tuscany, Italy). *Quaternary Science Reviews*, 66, 66–79.

Milano, M. (2012). *The Mediterranean Region under Climate Change*. IRD Éditions.

Orfi, T. E. (n.d.). *Quantification, modélisation et usages des ressources en eau dans le Haut Bassin de l'Oum Er Rbia (Amont du barrage Ahmed El Hansali, Maroc).*

Raible, C. C. (2010). Mediterranean climate variability over the last centuries: A review. *Climate of the Past*, 6(1), 99–114.

Sayad, A., Chakiri, S., Martin, C., Bejjaji, Z., & Echarfaoui, H. (2011). Effet des conditions climatiques sur le niveau du lac Sidi Ali (Moyen Atlas, Maroc). *Physio-Géo*, 5(Volume 5), 251–268. DOI: 10.4000/physio-geo.2145

Sebbar, A., Ilmen, R., Mohammed, H., Hassan, F., & Badri, W. (2019, December 12). *Caractérisation de la sécheresse climatique au Maroc.*

Singh, K. V., Setia, R., Sahoo, S., Prasad, A., & Pateriya, B. (2015). Evaluation of NDWI and MNDWI for assessment of waterlogging by integrating digital elevation model and groundwater level. *Geocarto International*, 30(6), 650–661. DOI: 10.1080/10106049.2014.965757

Talbi, M. (n.d.). *Impact Des Stress Hydriques Sur Les Potentiels De Dayet Aoua (Moyen Atlas - Maroc)*. Retrieved August 18, 2024, from https://www.academia.edu/20886804/IMPACT_DES_STRESS_HYDRIQUES_SUR_LES_POTENTIELS_DE_DAYET_AOUA_MOYEN_ATLAS_MAROC_

Tramblay, Y. (2012). Impact of climate change on extreme rainfall in Morocco: New uncertainties and new vulnerabilities. *Journal of Hydrology (Amsterdam)*, 479, 1–19.

UN World Water Development Report 2020 – Water and Climate Change | UNESCO. (n.d.). Retrieved August 18, 2024, from https://www.unesco.org/en/wwap/wwdr/2020

Chapter 10

PAP/CAR-Based Water Erosion Analysis in the Oued Lakhdar Watershed:

A GIS and Remote Sensing Approach

Rachid Laaouidi
Faculté des Sciences Humaines et Sociales, Université Ibn Tofail, Morocco

Siham Roudani
https://orcid.org/0009-0001-9798-6368
Faculté des Sciences Humaines et Sociales, Université Ibn Tofail, Morocco

Mouhcine Batchi
https://orcid.org/0000-0002-0915-6592
Faculté des Sciences Humaines et Sociales, Université Ibn Tofail, Morocco

Zakariae Amhani
https://orcid.org/0000-0002-9917-6449
Faculté des Sciences Humaines et Sociales, Université Ibn Tofail, Morocco

ABSTRACT

This work is the result of a qualitative assessment of water erosion in the Oued Lakhdar watershed, located within the larger Oum Erbia watershed. The PAP/CAR model and geomatic tools were used for this evaluation. The Oum Erbia watershed, spanning 1664 km^2, experiences severe erosion due to factors such as heavy and aggressive rainfall, steep slopes, sparse vegetation, and the predominance of soft rocks. Human activities, including deforestation, vegetation degradation, and in-

DOI: 10.4018/979-8-3693-9651-3.ch010

appropriate agricultural practices, have further exacerbated the erosion problem in the Oued Lakhdar watershed. The PAP/CAR method allowed for a qualitative assessment of erosion throughout the entire watershed and the identification of priority erosion zones. Data analysis using the PAP/CAR method revealed five main erosion states: extreme erosion (20.57%), strong erosion (18.56%), moderate erosion (23.02%), notable erosion (27.02%), and weak erosion (10.84%). In total, 62.15% of the watershed experiences moderate to extreme erosion.

1. INTRODUCTION

In the context of ongoing climate change and land use land cover changes, fertile soils are vital for sustaining human populations (Raj et al., 2019). However, these soils are increasingly threatened by water erosion, a significant global challenge (Osman, 2014). Currently, many soils are in a state of deterioration due to accelerated erosion processes. The Mediterranean region, known for its climatic extremes, faces particularly severe erosion issues (Ferreira et al., 2022). Deforestation for agricultural purposes represents a significant environmental issue, both locally and globally (Thornes & Wainwright, 2004).

In Mediterranean mountainous areas, the expansion of agricultural land through the removal of natural vegetation has amplified erosion vulnerability (López-Bermúdez et al., 1998). These regions face heightened erosion risks not only from the climate but also from increased human intervention and land conversion. Ineffective farming methods, especially in arid, high-altitude regions with steep slopes and shallow soils, coupled with occasional intense rainfall, have significantly accelerated erosion.

In Morocco, particularly in semi-arid regions, soil and environmental degradation is predominantly driven by water erosion (Bou-imajjane et al., 2020; Mazigh et al., 2022; Simonneaux et al., 2015), impacting approximately 40% of the country's land to varying extents (Ayt Ougougdal et al., 2020). The average annual soil erosion rate in Morocco ranges between 23 and 55 t ha−1 yr−1, with extreme values between 115 and 524 t ha−1 yr−1 (Acharki et al., 2022). These variations are influenced by topographic, meteorological, geological, and anthropogenic factors.

The Oued Lakhdar watershed, which is the focus of this study, is part of the larger Oum Rbia basin, located entirely within the Central High Atlas region. This area exhibits physical, climatic, and socio-economic conditions that are conducive to soil degradation, highlighting the need to assess erosion to mitigate its negative effects.

Today, advances in geomatics, such as remote sensing and geographic information systems (GIS), have become essential tools for processing, mapping, managing, and manipulating geographic data. These technologies enable the updating and overlaying of maps to produce an accurate representation of the erosive state of a region (Arabameri et al., 2018; Pijl et al., 2020; Jain et al., 2001; Wei et al., 2022). Applying these technologies to the study of the Oued Lakhdar watershed allows for the identification of the most erosion-prone areas, analysis of surface phenomena, and informed decision-making for managing soil degradation.

This study focuses on the Oued Lakhdar watershed, aiming to assess erosion trends, states, and causal risk factors using the widely used PAP/RAC method (Edd-efli et al., 2023; Elbadaoui et al., 2023; Ouallali et al., 2016). The goal is to evaluate erosion risk by analyzing the distribution, interactions, and influence of factors such as soil types, slope length, rainfall erosivity, slope exposure, soil moisture index, and land surface temperature.

2. STUDY AREA: OUED LAKHDAR WATERSHED

The Oued Lakhdar watershed is a significant component of the hydraulic system of the Oum Erbia River, one of the primary rivers in Morocco. As one of the 10 sub-basins that make up the Haouz-Mejjate basin, it spans an area of 3120.9 km^2. Located in the High Atlas region, specifically in the central limestone High Atlas, it is upstream of the Hassan Premier dam. The watershed is bordered by:

- North, Northeast, and East: The Oued El Abid watershed and its tributary Oued Ahançal.
- South: The Oued Tassaouet watershed.
- Southwest and West: The Oued "Assif n'Ghazzaf" watershed, a tributary of Oued Lakhdar downstream of the Hassan 1st dam.

Geographically, the watershed is situated south of Azilal, east of Demnate, and west of Zawyat Ahançal. The center of Ait Mhamed is located in its northern part. The terrain includes high mountains (over 2000 meters) upstream, mid-altitude mountains (1500 to 2000 meters) in the central part, and low mountains (900 to 1500 meters) near the Hassan 1st dam downstream (Layati & El Ghachi, 2024; Qalmoun et al., 2022).

The climate of the Oued Lakhdar basin varies from semi-arid to arid, characterized by hot, dry summers and cold, wet winters. The average annual precipitation is around 560 mm. Temperature variations are significant, with maximum temperatures averaging 28.22°C and minimum temperatures averaging 18.4°C. Rainfall is highly variable both temporally and spatially, influenced by topographic exposure (Layati & El Ghachi, 2023).

The geological composition of the basin is highly varied, predominantly consisting of Jurassic dolomitic limestone (Lias and Dogger periods), with occasional limestone or calcareous sandstone. These formations significantly impact the hydrology and soil characteristics of the region (Layati & El Ghachi, 2023).

Figure 1. Geographic location of the basin within the context of Morocco's major watersheds

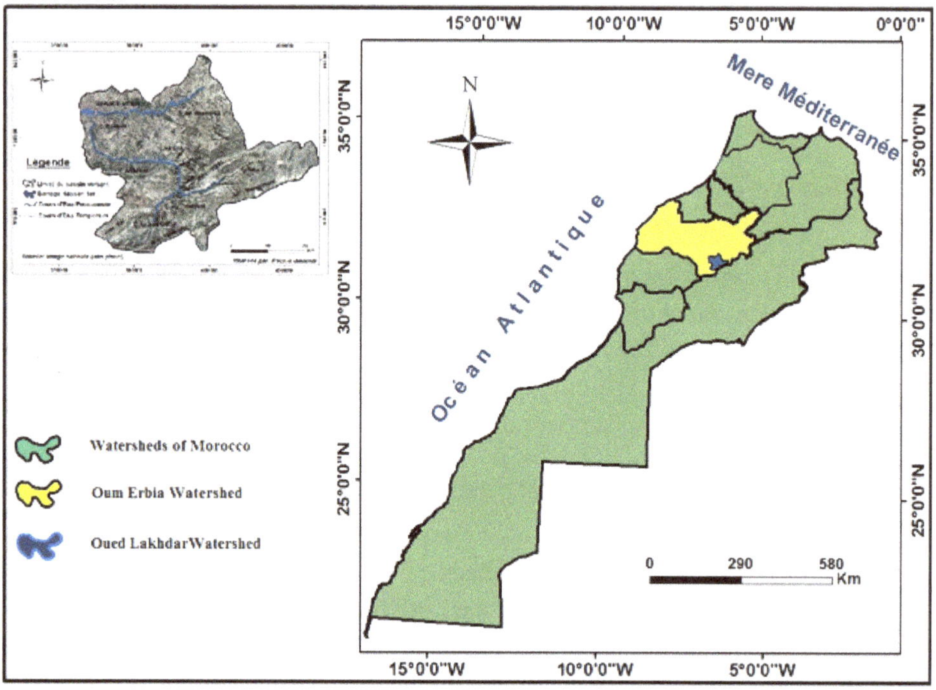

Figure 2. Geographic location of the Oued Lakhdar watershed within the context of the province

3. METHODOLOGY

The PAP/CAR method (Tahouri et al., 2022), which focuses on natural factors influencing hydric erosion (slope, vegetation cover, and lithology) (Driss et al., 2023). This method is determined through three phases:

3.1 Predictive Approach

The predictive approach involves the identification, evaluation, and integration of key factors such as slope, lithology, soil types, land use, and vegetation cover to form initial hypotheses about erosion risks (Tahouri et al. 2022). This method produces maps of homogeneous erosive units, which serve as the foundation for understanding potential erosion trends. The process follows seven distinct steps:

- S 1 and 2: Creation of slope and lithology maps.
- 3: Generation of the erodibility map by overlaying slope and lithology data.

- 4 and 5: Development of land use and vegetation cover maps.
- 6: Creation of the soil protection map by combining land use and vegetation cover data.
- 7: Production of the erosive states map by integrating the erodibility and soil protection maps.

3.2 Descriptive Approach

focuses on identifying, describing, and qualitatively assessing ongoing erosion processes at the study site, including varying degrees of erosion exposure and evolving trends. This method relies on direct field observation and verification, using the predictive erosive states map as a reference framework (Ouallali et al. 2016). It aims to uncover site-specific erosion constraints, providing ground-truth validation that helps confirm or adjust earlier assessments of both active and potential erosion.

3.3. Integration Approach

The integration approach yields the final map, which depicts the susceptibility of water erosion to degradation or improvement. This map results from merging qualitative data from the predictive and descriptive phases (Diani et al. 2023). The process involves creating an erodibility map by overlaying slope and lithofacies maps, followed by the soil protection map, which combines land use and vegetation cover density. The final erosive states map emerges from integrating the erodibility and soil protection maps.

4. RESULTS

4.1. Development of the Erodibility Map

4.1.1 Slope Map

The slope is considered one of the primary factors in the erosion process (Fu et al. 2011). Its influence on erosion is more significant than on runoff, especially when the soil is not fully covered. For soils with the same crops, runoff increases with the slope. Steep slopes with rapid flow are generally the cause of excessive erosion, the importance of which depends on the geology, soil, and protection of the vegetation cover.

The slope map was generated from the Digital Elevation Model (DEM). It was created in ArcGIS using the Spatial Analyst Tools extension: Surface/Slope (Fig.4).

Table 1. Slope Classes According to PAP/CAR

PAP/CAR Code	Slope Class
1	None to low (0-3%)
2	• Moderate (3-12 **Predictive Approach** %)
3	Steep (12-20%)
4	Very steep (20-35%)
5	Extreme (>35%)

Based on the analysis of the slope map database, we created the following table, which represents the areas corresponding to each slope class, as well as the cumulative areas of the different slope classes (Table1).

Table 2. Areas, Frequencies, and Cumulative Frequencies Based on Slope Classes

Classes	Code	Area (Km^2)	Cumulative Frequency	% of Areas	% Cumulative
< 3	1	376.71	376.71	22	22
3-12	2	570.68	947.39	34	56
12-20	3	251.66	1199.05	15	71
20-35	4	363.8	1562.85	22	93
> 35	5	119.44	1682.29	7	100

The results show percentages of 15%, 22%, and 7% for classes 3, 4, and 5, respectively. Low and moderate slopes represent 22% for class 1 and 34% for class 2, indicating that 56% of the study area falls within slope classes 3, 4, and 5 (12-35%) and 44% within slope classes 1 and 2 (0-12%).

Figure 3. Frequency Histogram of Slope Areas (Km²) in the Oued Lakhdar Watershed

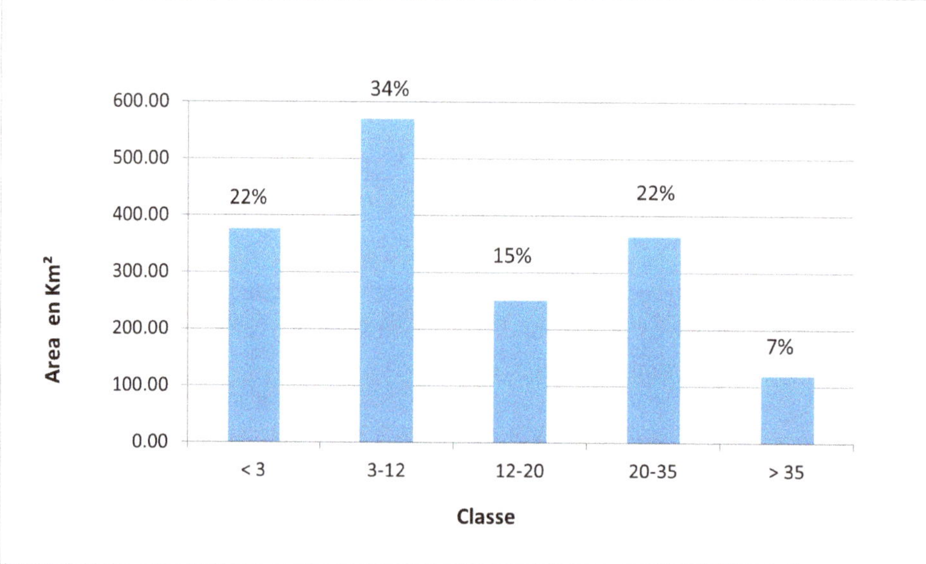

More than 56% of the study area has steep slopes. Therefore, our basin demonstrates a very rugged terrain, which will intensify runoff activity. Erosion will significantly outweigh accumulation, given that slopes of less than 3% occupy only 22% of the watershed, while 78% of the basin area has slopes greater than 22%.

The slope class map provides information on the variation in slope steepness, which plays a crucial role as it can influence the degree of erosion. (On gentle slopes, soils are less vulnerable to erosion.) This map shows that the steepest slopes, ranging from very steep to steep, are concentrated in the southwestern, intermediate, and southeastern parts of the basin. These areas are characterized by rugged terrain, especially in the mountainous massifs oriented southeast and southwest. Moderate slopes occupy the northeastern and southwestern parts of the basin, formed by glacis and slopes with varying morphology. The gentle to very gentle slopes correspond to the downstream part of the basin.

Figure 4. Slope map according to PAP/CAR classes

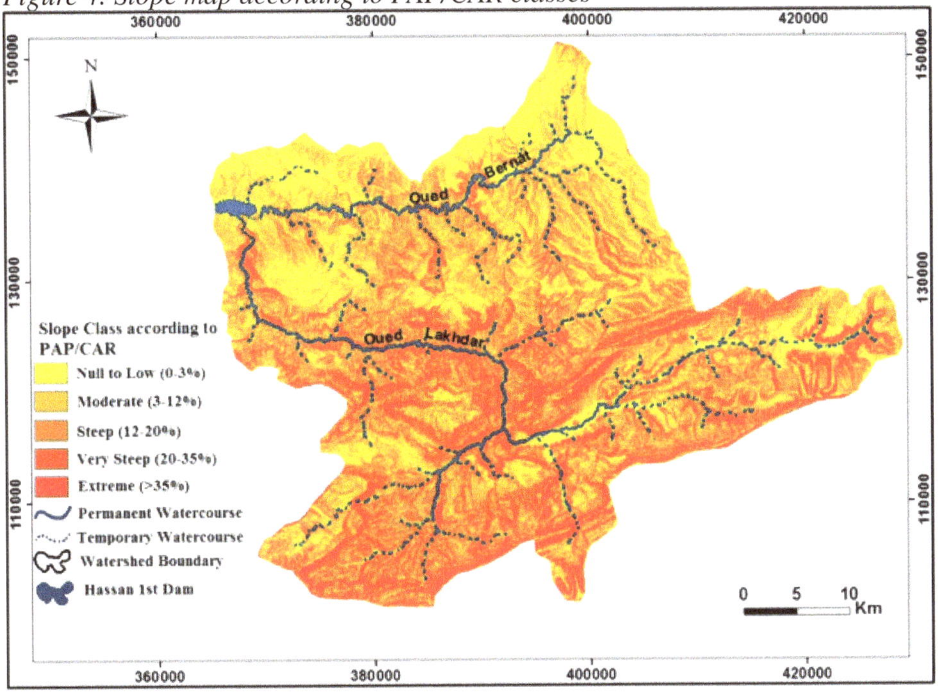

4.1.2 Lithofacies Map

The lithofacies (lithology) n map (Fig. 5) was created based on the geological map of the Oued Lakhdar watershed. This map is prepared from geological, lithological, and pedological data, identifying the different types of rocks or surface sediments/ soils, classified according to their relative degree of cohesion and mechanical resistance to erosion (Tab.3).

Based on the substrate's vulnerability to erodibility, this map shows the distribution of terrains according to their degree of friability. The various outcrops have been divided into 4 classes, with each degree of friability assigned a code: 1 for hard rocks, and progressively 2, 3, and 4 for increasingly less hard rocks according to the table (Tab.4).

Table 3. Lithofacies classes according to PAP/CAR

Lithofacies Class	Material Type
1 (a)	Unaltered compact rocks, strongly cemented conglomerates, crusts, outcrops of ferruginous sandstone (massive limestones, highly rocky soils, igneous or eruptive rocks, locally crusted soils).
2 (b)	Moderately weathered fractured rocks or very compact and cohesive soils.
3 (c)	Weakly or moderately compacted sedimentary rocks or soils (slate, schist, marl, etc.).
4 (d)	Weakly resistant rocks and/or deeply weathered rocks or soils (marl, gypsum, clayey slate, etc.).
5 (e)	Loose, non-cohesive sediment or soil, and detrital material.

The processing of the database allowed the creation of Table 3, which lists the percentages of areas and cumulative areas of the different lithofacies classes.

Table 4. Lithofacies classes based on areas, percentages, and cumulative percentages

Lithofacies Class	Degree of Soil Cohesion	Area (Km^2)	Cumulative Frequency (Km^2)	% of Areas	% Cumulative
1	Unaltered compact rocks	158.14	158.14	9.40	9.40
2	Moderately weathered rocks	130.82	288.96	7.78	17.18
3	Weakly compacted rocks	668.81	957.77	39.75	56.93
4	Weakly resistant or loose non-cohesive rocks	724.56	1682.3	43.07	100.00

The histogram of area frequencies (Fig. 6) shows that class (4) of weakly resistant or loose non-cohesive rocks and class (2) of moderately weathered rocks represent 39.75% and 43.07% of the Oued Lakhdar watershed, respectively. Class 1 does not exceed 10%. The class representing the smallest area is that of moderately weathered fractured rocks or very compact and cohesive soils (class 2). All these data indicate that our basin has a lithology that will not withstand natural, and especially anthropogenic, stresses for long.

Figure 5. Lithofacies map

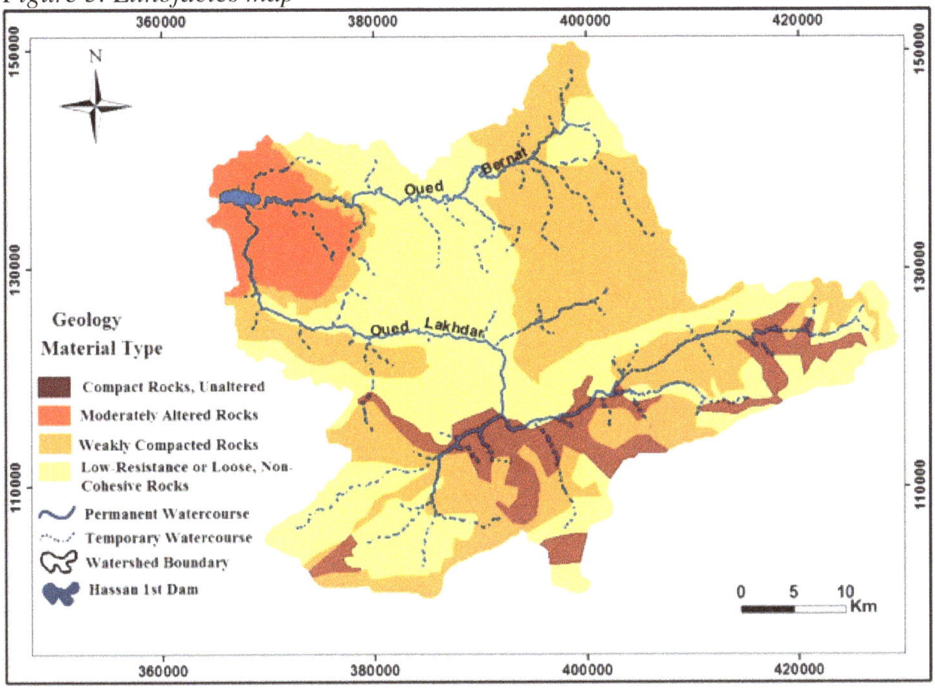

The lithology map reveals that the majority of the basin is composed of weakly or moderately compacted rocks and poorly resistant or unconsolidated, non-cohesive rocks, representing more than 82%. These are located throughout the northern and southern parts of the watershed where the altitude is high and the slope is steep. They also occupy the downstream part of the watershed. The strong slope inclination in the upstream part is likely the origin of these poorly developed soils or weakly to moderately compacted or unconsolidated, non-cohesive rocks. On steep slopes, alteration products are constantly removed by erosion, leaving only the C horizon of the soil, consisting of fractured rock, coarse rock fragments, and few fine particles trapped in pockets protected from gravity and flow or fixed by the root system of plants. In contrast, unaltered, compact rocks are located in the southern part, with a percentage of 9.40%. Moderately altered, fractured rocks or very compact and cohesive soils (class 2) occupy only 7.78%, mainly in the northern part.

Figure 6. Histogram of surface frequencies according to lithological classes

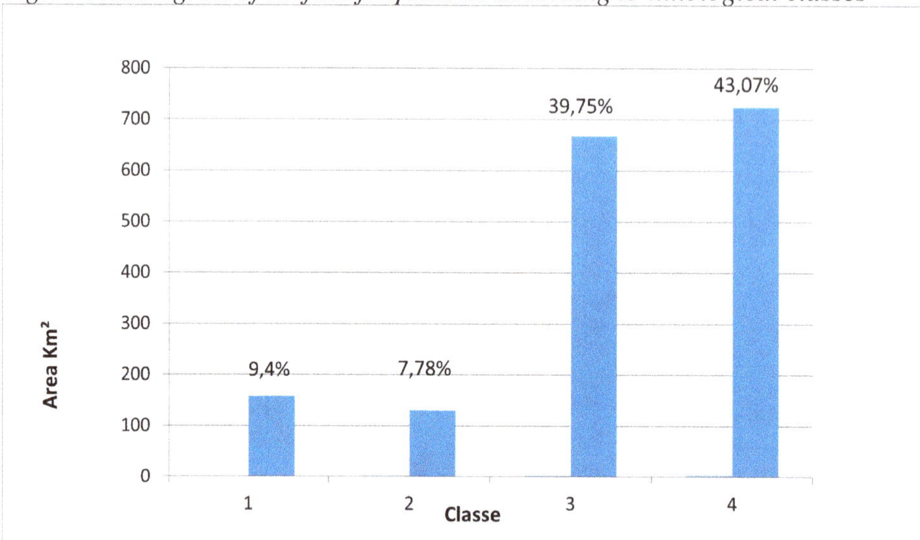

4.1.3 Erodibility Map

The erodibility map (Fig. 7) is derived from the combination of the slope map and lithofacies map (Okou et al. 2016). The resulting polygons from this overlay are categorized based on a matrix presented in Table 5, which ranks the terrain according to varying degrees of erodibility. This classification system provides a framework for assessing the susceptibility of different areas to erosion.

Table 5. Erodibility degrees according to PAP/CAR

Slope Classes	Lithofacies Classes
	1(a)
1	1(EN)
2	1(EN)
3	2(EB)
4	3(EM)
5	4(EA)

Table 6. Erodibility Classes

Class	Erodibility Level
1	Low (EN)
2	Moderate (EB)
3	Medium (EM)
4	High (EA)
5	Extreme (EX)

The result of processing the erodibility database (Tab. 7) provided several insights, particularly regarding the areas relative to each class.

Table 7. Erodibility Data Table

Erodibility Class	Area (Km^2)	% of Areas	Cumulative % of Areas
1 (EN)	449.64	27.36	27.36
2 (EB)	271.29	16.51	43.87
3 (EM)	384.03	23.37	67.24
4 (EA)	266.25	16.20	83.44
5 (EX)	272.00	16.55	100

The histogram of area frequencies according to erodibility classes shows that the most representative areas belong to the low and medium erodibility classes 1 (EN) and 2 (EB), with areas of 449.64 km^2 and 384.03 km^2, respectively. For the extreme erodibility class 5 (EX), it does not exceed 272 km^2. The moderate (2 (EB)) and high (4 (EA)) erodibility classes are less representative, with areas of 271.29 km^2 and 266.25 km^2, respectively. This will have serious long-term repercussions if the quality of protection is low.

Figure 7. Histogram of surface frequencies according to erodibility classes

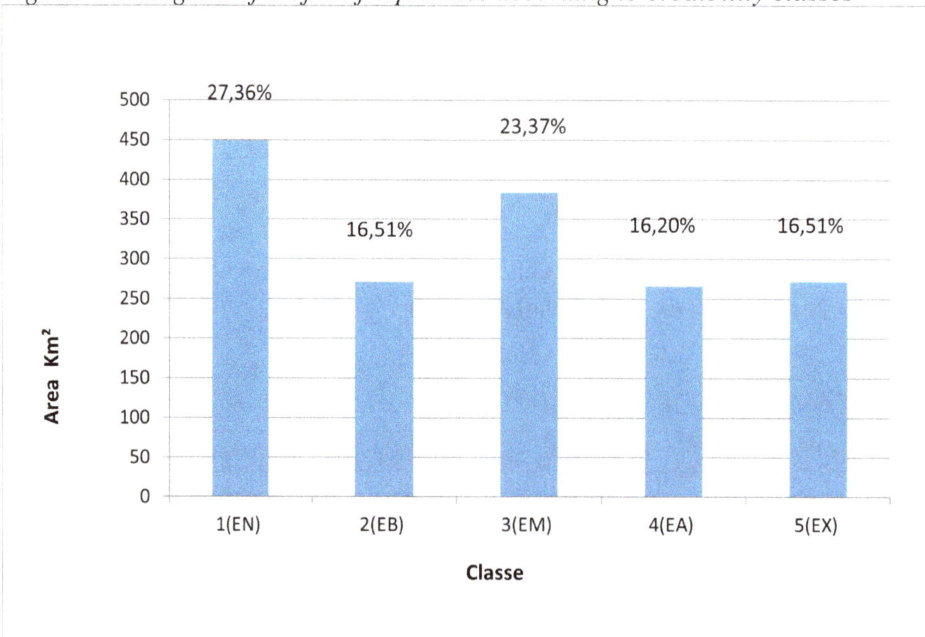

Figure 8. Soil Erodibility Map

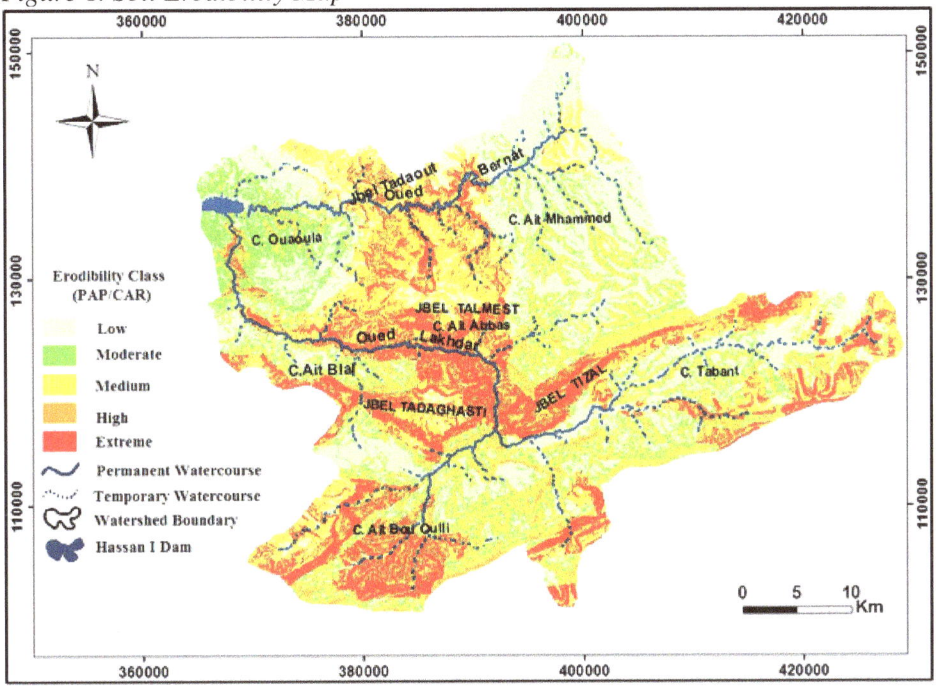

4.2. Development of the Soil Protection Map

The soil protection map is created by overlaying the land use map and the vegetation cover density map (Driss et al. 2023). The first map divides the area into polygons based on vegetation type (e.g., sparse or dense shrublands, forests, reforestation) or land use (e.g., orchards, dry farming, intensive vegetable cultivation). The second map considers the vegetation cover density, aiming to refine the land use map. The matrix used for this overlay is specific to the Mediterranean bioclimatic conditions and was developed through years of correlation studies and workshops.

4.2.1 Land Use Map

The land use map (Fig.9) was created based on a satellite image downloaded using an application called SASPLANET, with a spatial resolution of 1 meter. After converting the coordinate system from WG84 to Lambert, polygons were digitized based on vegetation type through visual interpretation and field visits. The resulting map identifies different types of vegetation cover.

Table 8. Land Use Classes

Class	Land Use Type
1	Bare or grassy land
2	Orchards
3	Sparse shrublands
4	Moderately dense forest
5	Dense forest

The database processing allowed for the creation of the following table, which lists the percentages of areas and cumulative areas of different land use classes.

Table 9. Area Classes and Percentages of Land Use Types

Class	Land Use Type	Area (km^2)	% Area
1	Bare or grassy land	857.53	51.06
2	Orchards	41.65	2.48
3	Sparse shrublands	513.22	30.56
4	Moderately dense forest	182.38	10.86
5	Dense forest	84.68	5.04
Total		**1679.47**	**100**

The data table provides insights into area percentages (Tab.9). The most common class is bare or grassy land, covering 857.53 km^2, accounting for 51.06% of the total area. Sparse shrublands are moderately represented, covering 513.22 km^2 or 30.56% of the total area. Moderately dense forest covers 182.38 km^2, representing only 10.86% of the total area. Dense forests and orchards account for less than 5.04%, with areas of 84.68 km^2 and 41.65 km^2, respectively.

Figure 9. Histogram of Land Use Classes by Area (km²)

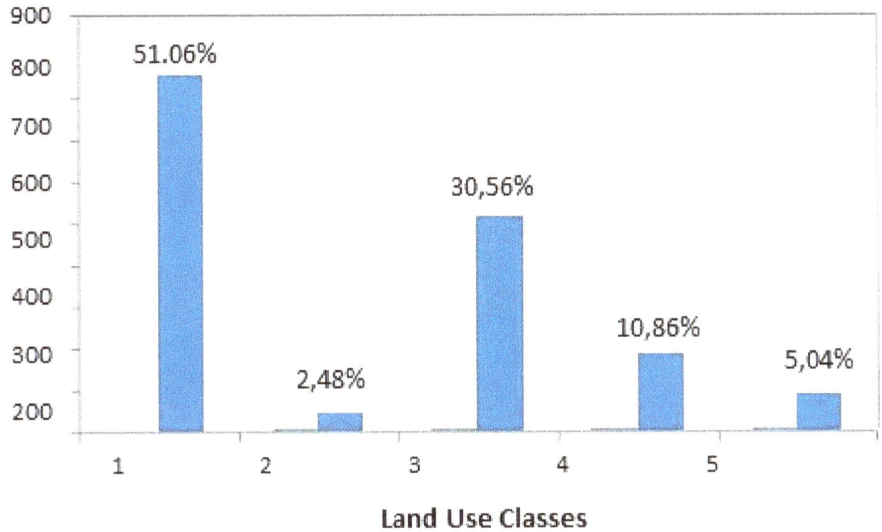

The distribution of different land use types (Fig.10) shows that the upstream part includes almost all land use types, with moderately dense forests and sparse shrublands being the most represented. This can be explained by the amount of water this area receives. There is also a presence of bare or grassy land and sparsely represented orchards, indicating deforestation in favor of cultivation and grazing.

Figure 10. Land Use Map

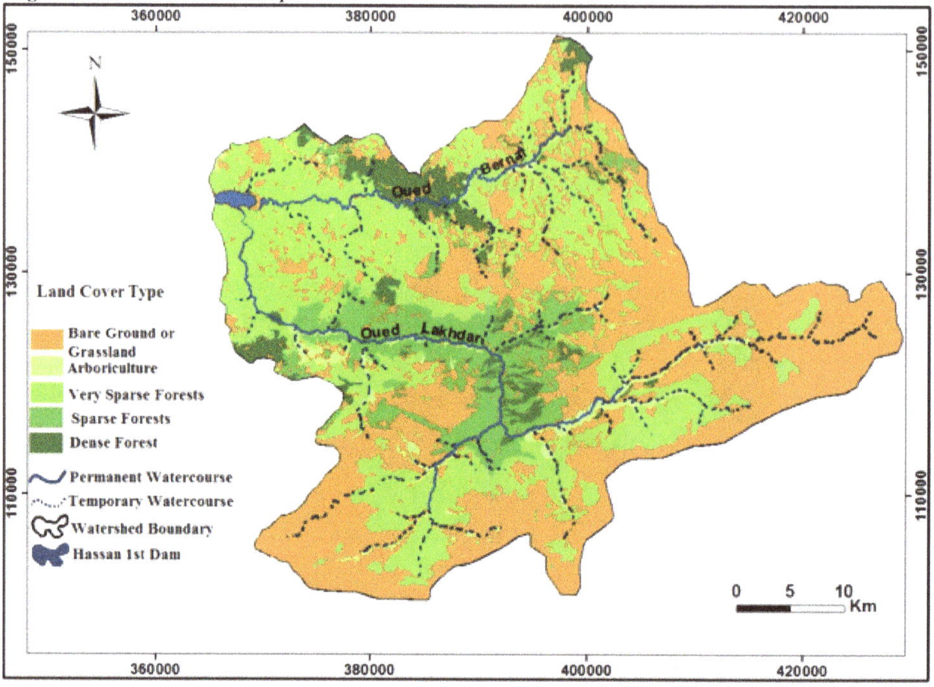

In the midsection, moderately dense forests, sparse shrublands, and orchards are present. Dense forests are less represented due to deforestation, fires, and grazing. The downstream part is characterized by three types of cover: mainly sparse shrublands, followed by orchards within the sparse shrublands (e.g., olive trees, other fruit trees, and vineyards), and bare or grassy land.

From this distribution, it can be concluded that the midsection and downstream areas are better protected by the marked presence of natural forests (dense to moderately dense) compared to the upstream part. Further verification will be done once the overlay with the vegetation cover density map is complete.

4.2.2 Vegetation Cover Density Map

The vegetation cover density map (Fig.12) was created using a satellite navigation tool that allows for the in-situ digitization of various themes. The result is a mapping of the Oued Lakhdar watershed into polygons based on cover density. This map serves two purposes: to refine the land use map and to consider a key factor in soil resistance, as denser vegetation provides stronger soil protection, reducing the impact of rainfall. Table shows the results of the database processing related to vegetation cover density.

Table 10. Area Classes and Percentages of Vegetation Cover Density

Class	Vegetation Cover Density	Area (km²)	% Area	% Cumulative Area
1	Less than 25%	380.45	22.60	22.60
2	25% – 50%	808.46	48.03	70.63
3	50% – 75%	431.74	25.65	96.28
4	Greater than 75%	62.55	3.72	100.00
Total		**38719.73**	**100.00**	**100.00**

Figure 11. Histogram of Vegetation Cover Classes by Area

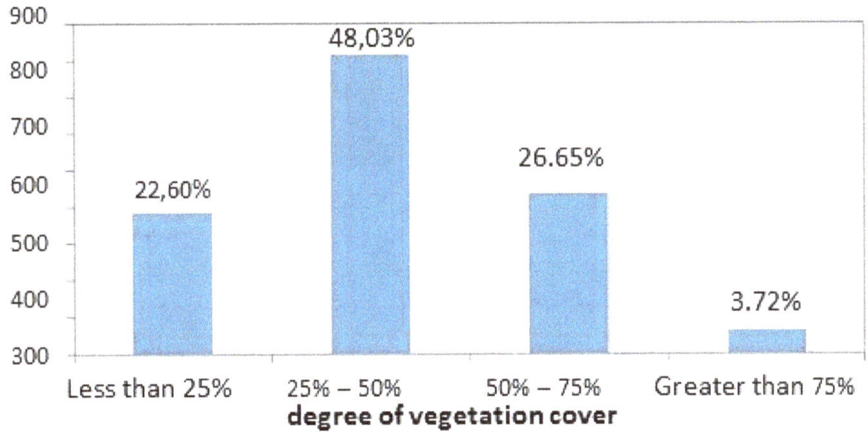

Figure 12. Vegetation Cover Density Map

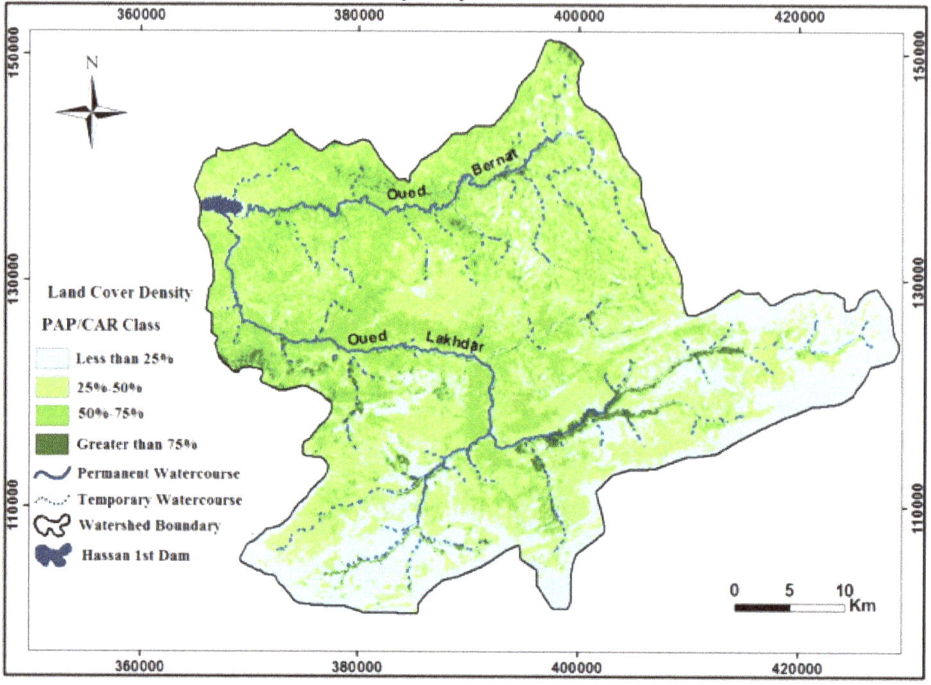

The most frequent class is Class 2 (25% - 50%) with 48.03% of the total area. Classes 1 and 3 do not exceed 25.65%, with areas of 380.45 km^2 and 431.74 km^2, respectively. The least represented class is Class 4 (greater than 75%) with a percentage of 3.72%. The cumulative frequency curve confirms that more than 70% of the area has low vegetation cover density, making the soil highly vulnerable. Another constraint is that the remaining dense cover is severely degraded, particularly due to agricultural expansion and grazing.

The distribution of these classes within the Oued Lakhdar watershed (Fig.10) shows that low densities are present in almost the entire southern area of the basin, while the upstream part shows a small area of high-density cover. According to the land use map, this area shows moderate density, mainly due to deforestation, fires, and overgrazing. In contrast, medium and dense densities, with some very dense vegetation areas, are present throughout the northeast and middle sections of the basin.

The downstream area shows the presence of dense and medium classes (Classes 2 and 3), with some sparse vegetation areas (Class 1). Based on this information, considering the vegetation cover density factor, it can be concluded that the Oued Lakhdar watershed has low to moderate protection, making it highly vulnerable to erosion.

4.2.3 Soil Protection Map

The soil protection degree map, as shown in Figure 14, The purpose of this map is to prioritize the Oued Lakhdar watershed into polygons based on protection degree. This step is essential for identifying areas with low or high protection and, consequently, "black zones" that require urgent and rapid intervention through reforestation to stabilize the soil.

Table 11. Land Use vs. Vegetation Cover Density Matrix

Land Use Type	Vegetation Cover Density
1	5 (Very Low)
2	5 (Very Low)
3	3 (Medium)
4	4 (Low)
5	5 (Very Low)

Table 12. Soil Protection Degree Legend

Protection Class	Soil Protection Degree
1	Very High (MA)
2	High (A)
3	Medium (M)
4	Low (B)
5	Very Low (MB)

The database processing allowed the derivation of Table 12, from which the graph (Fig.13.) was created. It shows that the Oued Lakhdar watershed is poorly protected. The most frequent class is Class 5, with an area of 744.16 km^2, representing 44.16% of the total basin area. For other classes, the area values decrease from the lowest to the highest protection degree, indicating that in the Oued Lakhdar watershed, the larger the area, the weaker the protection.

Figure 13. Histogram of the frequencies of the protection degree classes as a function of surface areas

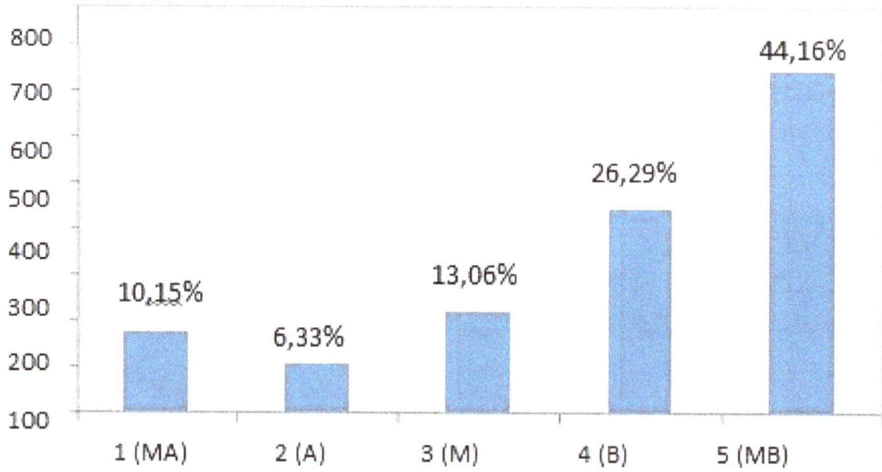

Table 13. Data Table on Protection Degree

Class	Protection Degree	Area (km²)	% Area	% Cumulative Area
1 (MA)	Very High (MA)	171	10.15	10.15
2 (A)	High (A)	257	15.31	25.46
3 (M)	Medium (M)	507	30.19	55.65
4 (B)	Low (B)	500	29.77	85.42
5 (MB)	Very Low (MB)	744.16	44.16	100.00

Figure 14. Soil Protection Degree Map

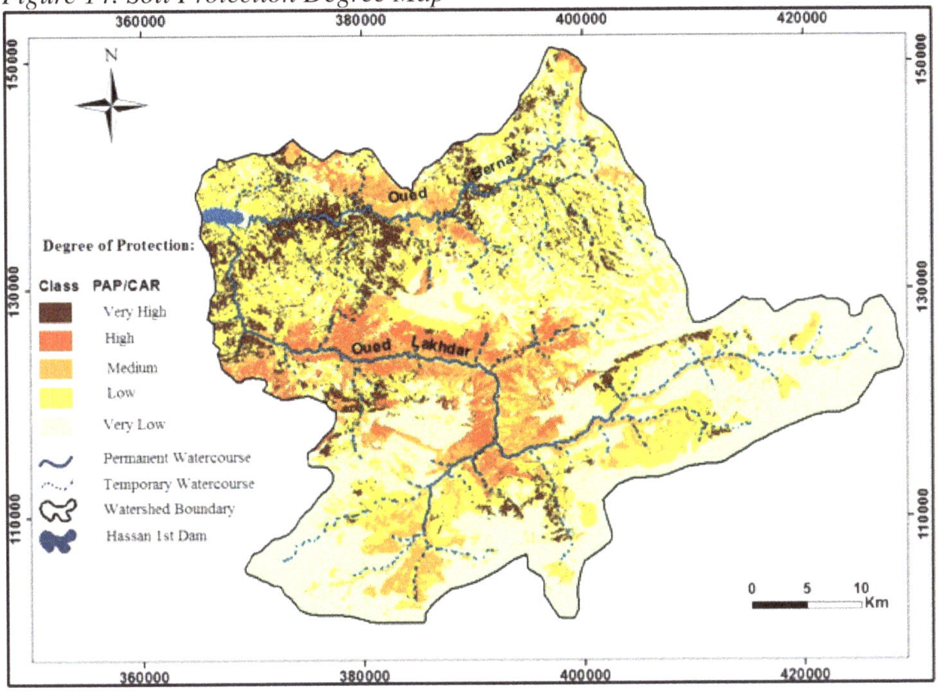

The distribution of protection degrees (Fig. 14) shows that the south and northeast parts of the basin have high soil protection degrees. The central section shows low protection degrees, making it the most vulnerable to erosion. Consequently, urgent reforestation and soil stabilization measures are required in the central section to prevent further soil erosion. The upstream section shows a medium degree of protection, due to mixed land uses and varied vegetation densities.

4.3 Development of the Erosion Status:

Map of the erosion status map is created by overlaying the erodibility map and the soil protection map (Okou et al. 2016). The resulting polygons are then classified according to the matrix (Table 14).

Table 14. Soil Protection Level and Erodibility Level Soil Erosion Status Matrix

Degré de protection	Degré d'érodibilité				
des sols	1(EN)	2(EB)	3(EM)	4(EA)	5(EX)
1(MA)	1	1	1	2	2
2(A)	1	1	2	3	4
3(M)	1	2	3	4	4
4(B)	2	3	3	5	5
5(MB)	2	3	4	5	5

Table 15. Codification of Erosive States

Class	Erosion status levels
1	Very low erosion
2	Low erosion
3	Notable erosion
4	High erosion
5	Very high erosion

The database processing led to the creation of Table 16, which presents the classes according to their areas, percentage of total area, and cumulative areas. Based on this table, a frequency histogram was generated (Fig 15). This histogram shows that the most representative classes are those of low erosion with 444.21 km^2, notable erosion with 378.39 km^2, very high erosion with 338.63 km^2, and high erosion with 305.28 km^2, representing 27%, 23%, 20.59%, and 18.56% respectively. Regarding the last three erosion classes (from notable to very high), they represent a total area of 1022.3 km^2, which is more than 62% of the total basin area. The very low erosion class does not exceed 10.85% with an area of 178.43 km^2. This indicates that the Oued Lakhdar watershed has a large area that is highly vulnerable to erosion. This will have a significant impact on soil productivity and, consequently, on the quality of life of the region's population.

*Table 16. Erosion Degree Data Table" **or** Erosion Degree*

Class	Erosion Degree	areas in km Km^2	% the area	% Cumulative percentage of the area
1	Very low erosion	178,43	10,85	10,85
2	Low erosion	444,21	27,00	37,85
3	Notable erosion	378,39	23,00	60,85

continued on following page

Table 16. Continued

Class	Erosion Degree	areas in km Km²	% the area	% Cumulative percentage of the area
4	High erosion	305,28	18,56	79,41
5	Very high erosion	338,63	20,59	100,00
	Total	1644,94	100,00	

Figure 15. Histogram of erosion degree classes as a function of areas

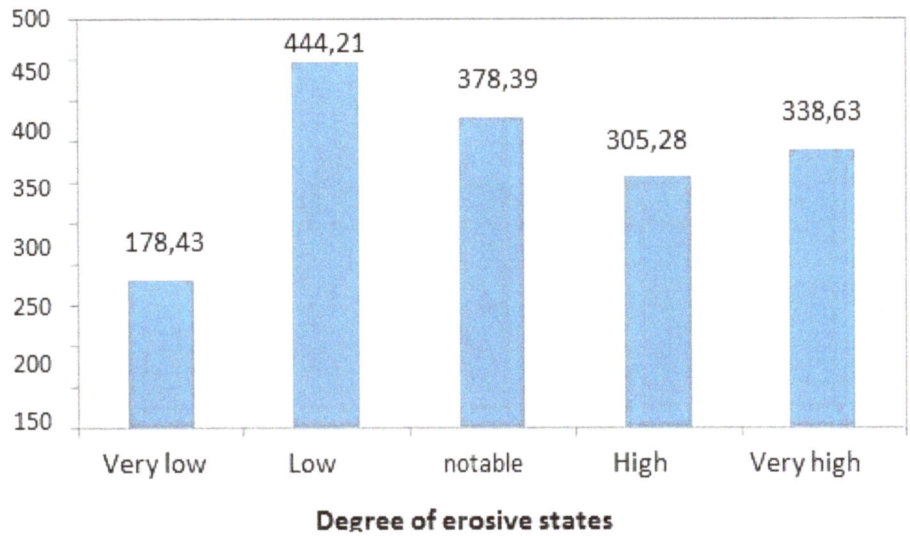

The distribution of these areas on the erosion status map (Fig. 16) shows that the most eroded zones are located in the upstream part of the basin, particularly in the southern area, specifically at Jbel Rat, Jbel Aguensou-n-Issis, and Jbel Tifdaniwine, situated in the commune of Ait Bou Oulli, as well as at Jbel Waougoulzat and Jbel Tizal within the commune of Tabant. This is explained by the absence of dense vegetation cover in the more eroded areas with steep slopes. Elsewhere, very high erosion follows the pattern drawn by the hydrographic network with differences, particularly in the upstream regions, for example, at Oued Bernat and Oued Lakhdar.

It is also noted that the areas that escape very high erosion are affected by high erosion, located mainly in the intermediate part, especially at Jbel Tadaout and in some areas at Jbel Tadaghasti and the commune of Ait Bou Oulil. In the commune of Ait Mhammed, erosion is currently notable. However, considering the human factor, it is also observed that where populations live coincides with areas of high erosion, with small localities developing, taking into account the soil component. Regarding low and very low erosion, they occur in reduced areas, particularly in the downstream parts, especially more frequently in the commune of Ouaoula. This is partly explained by the presence of very high erosion protection, due to the presence of vegetation cover (density between 50% and 75%) in the less eroded areas, and partly by the very low erodibility, explained by the presence of moderately altered fractured rocks or very compacted or cohesive soils with a low to moderate slope (0-3%) or moderate (3-12%)

Figure 16. Erosion status map

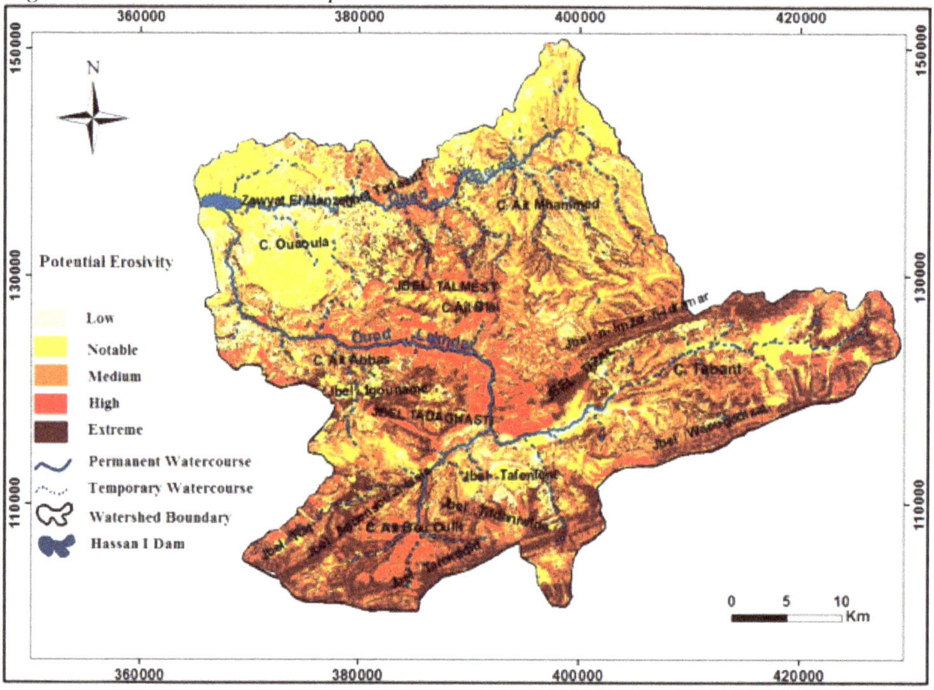

4.4. Descriptive Approach

This section involved a field trip aimed at observing and studying the various forms and processes of water erosion in the Oued Lakhdar watershed. The objective was also to verify the mapped data derived from satellite images and maps during the predictive phase.

The different forms of water erosion that could significantly affect our basin include:

- **Sheet Erosion**

This is the initial stage of soil degradation due to erosion. It results in soil degradation across the entire surface, manifesting as a diffuse form of erosion (Dlamini et al. 2011). In our study area, diffuse runoff processes are almost ubiquitous. Runoff develops on strictly clayey or marl formations. Diffuse runoff merely strips away the soil or superficial formations. The signs of erosion include changes in soil color, thinning of the superficial layer, and the downward migration of fine particles from the top to the bottom of the slope (Fig.17). These particles accumulate at the bottom of the slope, creating a distinguishable tint on aerial photographs.

Figure 17. Example of sheet erosion

- **Rill, Gully, and Ravine Erosion**

Linear forms of erosion are the most commonly encountered in the Oued Lakhdar watershed. Figure 18, as an example, shows that rill erosion has become more pronounced and evolved into gully erosion at Jbel Tizal. Ait Tabant, upstream of Jbel Tizal, gully erosion is becoming increasingly significant. These forms develop when rainfall intensity exceeds the soil surface's infiltration capacity. Initially, puddles form, which then connect via small streams. When these streams reach a certain velocity, they gain the energy to carve incisions, concentrating runoff and creating erosion limited by flow lines. This energy is no longer dispersed across the entire soil surface but is concentrated along steeper slope lines. Linear erosion indicates that runoff has organized, increased in speed, and acquired energy capable of cutting into the soil and carrying increasingly larger particles—not just clays and silts like in sheet erosion, but also gravels, stones, and blocks when organized into ravines.

Figure 18. Rills and ravines at Tabant-Jbel Tizal

- **Generalized Ravine Erosion (Badlands)**

The term "badlands" refers to a ruiniform landscape of marl or clay terrain, heavily eroded by runoff on steep slopes. These lands are generally unsuitable for agriculture (Fig.19). This is a highly advanced stage of the erosion process, where basins converge, creating highly fragmented zones with severely degraded soils. In the study area, badlands constitute a significant portion, approximately 6.35%. Topographically, they correspond to areas with very steep slopes, exceeding 25%.

Figure 19. Evolution of badlands on steep slopes (douar.abachko-commune ait bouli)

- **Mass Movements**

Mass movements refer to the downslope movement of soil, rock, or debris due to gravity (Martinsen 1994), often triggered by water infiltration. Solifluction occurs when water saturates layers between sandstone and marl, reducing cohesion and causing material to slide. These movements range from small surface slides to large rotational slides, where thousands of cubic meters of material shift. Superficial slides involve slow movement of the topsoil, while rotational slides are more dramatic, structured by faults and triggered by water, slope, tectonics, and human activities. Mudflows, a dense mixture of water and surface material, occur during prolonged rainfall or human exploitation of steep slopes.

4.5 Erosion Forms Map

This phase is based on the cartographic delineation of erosion forms characterizing the Oued Lakhdar basin according to PAP/CAR guidelines. The erosion forms map was created using satellite imagery and field observations.

Table 17 and Figure 20 summarize the spatial analysis of the erosion forms map database, with the first column showing the descriptive abbreviations of the forms used for PAP/CAR guidelines. Weak erosion, limited vegetation cover, and sheet or diffuse erosion account for 1634.06 km^2 (75.44% of the total basin area), while deep ravines with bank collapse occupy the second place with 250.8 km^2. Badlands and moderately deep ravines do not exceed 226.34 km^2, and the rest are highly localized forms, not exceeding 54.72 km^2.

Table 17. Relative Areas of Erosion Forms

Erosion Forms	Code	Area in km^2	% of Total Area
Weak erosion, limited vegetation cover, sheet or diffuse erosion	L-2-3	1123.07	75.44
Rills and superficial gullies	C2	44.22	2.05
Moderately deep ravines	C3	88.43	4.10
Deep ravines with bank collapse	C4	250.8	11.58
Badlands, generalized ravine erosion	CX	137.32	6.35
Periodically flooded and/or alluvial zones	W1	0.22	0.24
Localized solifluxion/gravity slides	M1	0.02	0.23

Figure 20. Frequency Histogram Based on Relative Areas

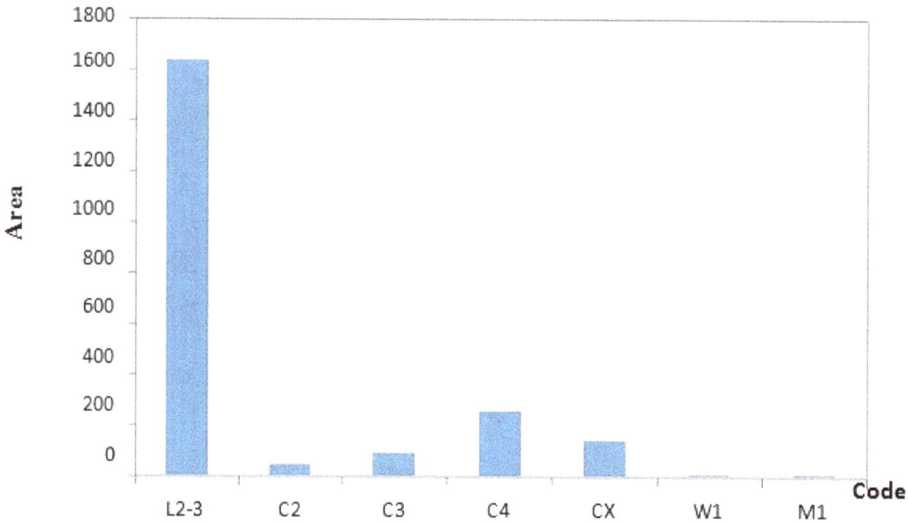

The distribution of these forms in the Oued Lakhdar watershed (Fig 21) shows that the most represented form in the basin is weak erosion, limited vegetation cover, and sheet or diffuse erosion. This can be attributed to the steep slopes in this area, promoting sheet or diffuse erosion, and to moderate protection provided by vegetation cover, leading to weak erosion in some areas of the basin. In the southern upstream part of the basin, at Jbel Waougoulzat and Tarekddid, the most represented forms are badlands, rills, and ravine erosion, explained by the terrain's vulnerability due to factors such as aggressive rainfall, snow, and human practices exacerbating these conditions.

Figure 21. Map of Actual Erosion in the Oued Lakhdar Watershed

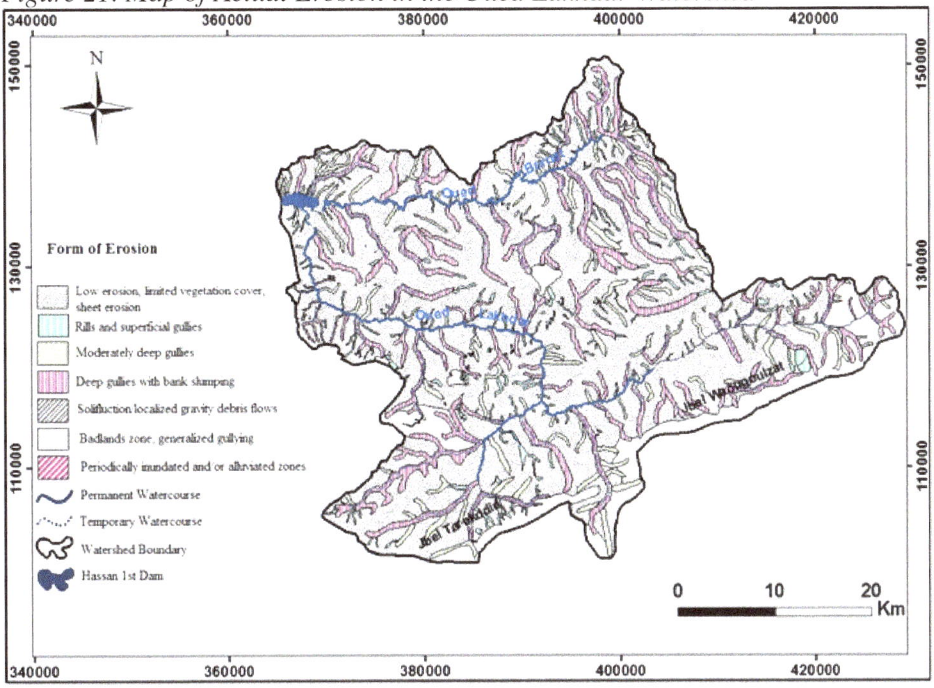

4.6 Integration Approach

This phase is the outcome of the two preceding steps. It involves complementing the results obtained during the predictive phase with descriptive data on erosion processes mapped during the second phase. Combining these two results provides a highly accurate cartographic product that reflects the actual state of soil degradation and the future evolution of erosion.

It results in the final cartographic product, identifying and assessing both potential erosion (erosive status) and current erosion in its various forms, intensities, and evolutionary trends.

4.6.1. Creation of the Consolidated PAP/CAR Erosion Map

This consolidated map was created by combining the potential erosion map and the erosion forms map (Fig 22) (Labbaci et al. 2020). The spatial analysis of the combined map reveals the following:

The basin area of Oued Lakhdar covers an area of 232.02 km^2. It is characterized by low potential erosion with a proportion of 27.85% of the total area and a low real erosion proportion of 75.44% (Table 18). The low potential erosion is primarily due to terrain, vegetation cover, and soil conditions, which help mitigate erosion processes. Meanwhile, the relatively high percentage of low actual erosion indicates that the basin is subject to moderate erosion processes, likely due to intense agricultural activities and natural vegetation cover.

The combination of moderate to high potential and real erosion is found in the upper part of the basin, which includes more severe erosion forms such as badlands, gullies, and bank scouring, accounting for 28.95% of the basin area.

Table 18. Summary of Combined Potential and Actual Erosion in the Oued Lakhdar Basin

Combined Erosion	Code	Area in km²	% of Total Area
Low Potential Erosion/Low Actual Erosion	L-1-2	127.85	27.85
Low Potential Erosion/Moderate Actual Erosion	L-1-3	34.56	12.16
Moderate Potential Erosion/Moderate Actual Erosion	M-1-2	68.09	28.95
Moderate Potential Erosion/High Actual Erosion	M-2-3	54.72	16.20

4.6.2. Final Summary Map of Erosion Intensity

The final summary map was created based on the above results, highlighting the distribution of erosion intensities across the Oued Lakhdar basin. The spatial analysis shows that weak erosion covers 39.68% of the basin, moderate erosion covers 40.87%, and high erosion covers 19.45% of the basin area (Table.19)

Table 19. Summary of Erosion Intensity in the Oued Lakhdar Basin

Erosion Intensity	Code	Area in km²	% of Total Area
Weak Erosion	L-1	128.85	39.68
Moderate Erosion	M-1	68.09	40.87
High Erosion	H-1	54.72	19.45

Figure 22. Final Summary Map of Erosion Intensity in the Oued Lakhdar Watershed

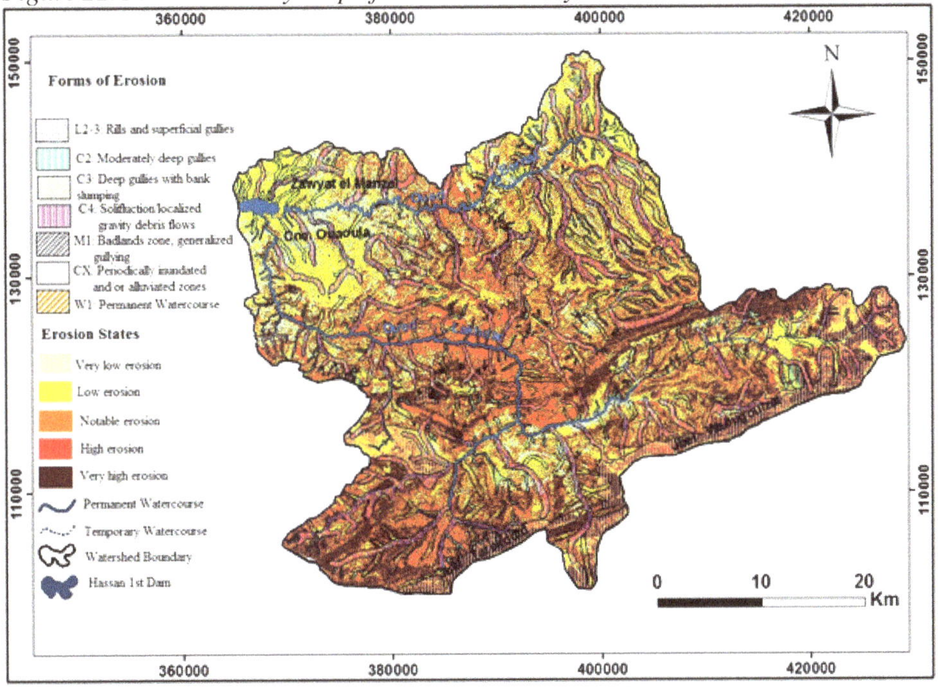

5. CONCLUSION

This study highlights the importance of integrating both predictive and descriptive approaches in understanding and mapping erosion processes in the Oued Lakhdar basin. The predictive phase provided a valuable initial assessment of erosion risks, while the descriptive phase offered ground-truthing and further refinement of the mapped data.

The integration of these two approaches resulted in a highly accurate and comprehensive erosion map that reflects both the current state of soil degradation and future trends in the basin. The final product serves as a critical tool for land management and conservation efforts, helping to identify areas most at risk and prioritize interventions.

The findings also emphasize the need for continued monitoring and management of erosion processes in the Oued Lakhdar basin, particularly in areas with high erosion intensity. Effective erosion control measures, such as reforestation, sustainable agricultural practices, and infrastructure improvements, are crucial to mitigating soil degradation and ensuring the long-term sustainability of the basin's natural resources.

Addressing the challenges in this project requires collaboration between civil society, the regional directorate of water and forests, and farmers to mitigate damage in the sub-basin. Key objectives include combating soil erosion, restoring degraded forests, improving living conditions, conserving natural resources, and boosting land productivity. Initiatives already underway focus on reforestation, forest restoration, fire prevention, dry stone terraces, gabion dams, and water, biomass, and soil conservation, with farmer participation playing a crucial role. Additionally, local development actions such as fruit tree cultivation, water harvesting, infrastructure improvement, and economic diversification (beekeeping, dairy cooperatives, and crafts) aim to reduce reliance on natural resources, thus helping curb erosion and promote sustainable land use.

REFERENCES

Acharki, S., El Qorchi, F., Arjdal, Y., Amharref, M., Bernoussi, A. S., & Ben Aissa, H. (2022). Soil erosion assessment in Northwestern Morocco. *Remote Sensing Applications: Society and Environment*, 25, 100663. DOI: 10.1016/j.rsase.2021.100663

Arabameri, A., Pradhan, B., Pourghasemi, H. R., Rezaei, K., & Kerle, N. (2018). Spatial Modelling of Gully Erosion Using GIS and R Programing : A Comparison among Three Data Mining Algorithms. *Applied Sciences (Basel, Switzerland)*, 8(8), 1369. DOI: 10.3390/app8081369

Ayt Ougougdal, H., Khebiza, M. Y., Messouli, M., Bounoua, L., & Karmaoui, A. (2020). Delineation of vulnerable areas to water erosion in a mountain region using SDR-InVEST model : A case study of the Ourika watershed, Morocco. *Scientific African*, 10, e00646. DOI: 10.1016/j.sciaf.2020.e00646

Bou-imajjane, L., Belfoul, M. A., Elkadiri, R., & Stokes, M. (2020). Soil erosion assessment in a semi-arid environment : A case study from the Argana Corridor, Morocco. *Environmental Earth Sciences*, 79(18), 409. DOI: 10.1007/s12665-020-09127-8

Diani, K., Ettazarini, S., & Hahou, Y. (2023). Identification of soil erosion sites in semiarid zones: Using GIS, remote sensing, and PAP/RAC model. In *Handbook of Hydroinformatics* (pp. 169–183). Elsevier. DOI: 10.1016/B978-0-12-821961-4.00003-8

Dlamini, P., Orchard, C., Jewitt, G., Lorentz, S., Titshall, L., & Chaplot, V. (2011). Controlling factors of sheet erosion under degraded grasslands in the sloping lands of KwaZulu-Natal, South Africa. *Agricultural Water Management*, 98(11), 1711–1718. DOI: 10.1016/j.agwat.2010.07.016

Driss, S., Brahim, B., Afaf, A., Habiba, A., Ferdaouss, L., & Abdel-Ilah, M. (2023). Mapping of Zones Vulnerable to Water Erosion Using PAP / CAR and GIS Method Upper the Aït Moulay Hmad Dam, Northeast of the Central Massif, Morocco. *Iraqi Geological Journal*, 56(1B), 181–197. DOI: 10.46717/igj.56.1B.14ms-2023-2-22

Eddefli, F., Tayebi, M., Hajaj, S., Khaddari, A., Ouakil, A., & El Harti, A. (2023). Hydric Erosion Mapping Enhancement in Korifla Sub-Watershed (Central Morocco). *Journal of Landscape Ecology*, 16(3), 54–75. DOI: 10.2478/jlecol-2023-0017

Elbadaoui, K., Mansour, S., Ikirri, M., Abdelrahman, K., Abu-Alam, T., & Abioui, M. (2023). Integrating Erosion Potential Model (EPM) and PAP/RAC Guidelines for Water Erosion Mapping and Detection of Vulnerable Areas in the Toudgha River Watershed of the Central High Atlas, Morocco. *Land (Basel)*, 12(4), 837. DOI: 10.3390/land12040837

Ferreira, C. S. S., Seifollahi-Aghmiuni, S., Destouni, G., Ghajarnia, N., & Kalantari, Z. (2022). Soil degradation in the European Mediterranean region : Processes, status and consequences. *The Science of the Total Environment*, 805, 150106. DOI: 10.1016/j.scitotenv.2021.150106 PMID: 34537691

Fu, S., Liu, B., Liu, H., & Xu, L. (2011). The effect of slope on interrill erosion at short slopes. *Catena*, 84(1-2), 29–34. DOI: 10.1016/j.catena.2010.08.013

Jain, S. K., Kumar, S., & Varghese, J. (2001). Estimation of Soil Erosion for a Himalayan Watershed Using GIS Technique. *Water Resources Management*, 15(1), 41–54. DOI: 10.1023/A:1012246029263

Labbaci, A., Marghadi, S., Laaribya, S., & Moukrim, S. (2020). Integrating Sentinel-2 Data and PAPCAR Model to Map Water Erosion: Case of Beni Boufrah Watershed. *RJESTE*, 3(1). Advance online publication. DOI: 10.4314/rjeste.v3i1.4S

Layati, E., & El Ghachi, M. (2023). Quantitative analysis of the morphometric parameters of the Oued Lakhdar watershed (Morocco) : Statistical and GIS approaches. In *1st International Conference On Water And Earth Sciences (ICWES'22)* (p. 47-58). Sciendo. https://doi.org/DOI: 10.2478/9788367405270-007

Layati, E., & El Ghachi, M. (2024). Contribution of Geomatics in the Quantification of Water Erosion : Application in the Oued Lakhdar (Morocco). *Photonirvachak (Dehra Dun)*, 52(6), 1315–1327. DOI: 10.1007/s12524-024-01880-8

López-Bermúdez, F., Romero-Díaz, A., Martínez-Fernandez, J., & Martínez-Fernandez, J. (1998). Vegetation and soil erosion under a semi-arid Mediterranean climate : A case study from Murcia (Spain). *Geomorphology*, 24(1), 51–58. DOI: 10.1016/S0169-555X(97)00100-1

Martinsen, O. (1994). Mass movements. In Maltman, A. (Ed.), *The Geological Deformation of Sediments* (pp. 127–165). Springer Netherlands. DOI: 10.1007/978-94-011-0731-0_5

Mazigh, N., Taleb, A., Bilali, A. E., & Ballah, A. (2022). The Effect of Erosion Control Practices on the Vulnerability of Soil Degradation in Oued EL Malleh Catchment using the USLE Model Integrated into GIS, Morocco. *Trends in Sciences*, 19(2), 2059. DOI: 10.48048/tis.2022.2059

Okou, F. A. Y., Tente, B., Bachmann, Y., & Sinsin, B. (2016). Regional erosion risk mapping for decision support: A case study from West Africa. *Land Use Policy*, 56, 27–37. DOI: 10.1016/j.landusepol.2016.04.036

Osman, K. T. (2014). Soil Erosion by Water. In K. T. Osman, *Soil Degradation, Conservation and Remediation* (p. 69-101). Springer Netherlands. DOI: 10.1007/978-94-007-7590-9_3

Ouallali, A., Moukhchane, M., Aassoumi, H., Berrad, F., & Dakir, I. (2016). The Mapping of the Soils' Degradation State by Adaptation the PAP/RAC Guidelines in the Watershed of Wadi Arbaa Ayacha, Western Rif, Morocco. *Journal of Geoscience and Environment Protection*, 04(07), 77–88. DOI: 10.4236/gep.2016.47009

Pijl, A., Reuter, L. E. H., Quarella, E., Vogel, T. A., & Tarolli, P. (2020). GIS-based soil erosion modelling under various steep-slope vineyard practices. *Catena*, 193, 104604. DOI: 10.1016/j.catena.2020.104604

Qalmoun, A., Bouzrarf, K., & Belqat, B. (2022). Assessment of the ecological status of the Oum Er-rabie River basin (Central Morocco) through physicochemical, bacteriological parameters and biotic indices. *Biologia*, 77(9), 2533–2547. DOI: 10.1007/s11756-022-01128-1

Raj, A., Jhariya, M. K., Yadav, D. K., Banerjee, A., & Meena, R. S. (2019). Soil for Sustainable Environment and Ecosystems Management. In Jhariya, M. K., Banerjee, A., Meena, R. S., & Yadav, D. K. (Eds.), *Sustainable Agriculture, Forest and Environmental Management* (pp. 189–221). Springer Singapore. DOI: 10.1007/978-981-13-6830-1_6

Simonneaux, V., Cheggour, A., Deschamps, C., Mouillot, F., Cerdan, O., & Le Bissonnais, Y. (2015). Land use and climate change effects on soil erosion in a semi-arid mountainous watershed (High Atlas, Morocco). *Journal of Arid Environments*, 122, 64–75. DOI: 10.1016/j.jaridenv.2015.06.002

Tahouri, J., Sadiki, A., Karrat, L., Johnson, V. C., Chan, N. W., Fei, Z., & Kung, H. T. (2022). Using a modified PAP/RAC model and GIS-for mapping water erosion and causal risk factors : Case study of the Asfalou watershed, Morocco. *International Soil and Water Conservation Research*, 10(2), 254–272. DOI: 10.1016/j.iswcr.2021.07.003

Thornes, J. B., & Wainwright, J. (2004). *Environmental Issues in the Mediterranean*. Routledge. DOI: 10.4324/9780203495490

Wei, Y., Liu, Z., Zhang, Y., Cui, T., Guo, Z., Cai, C., & Li, Z. (2022). Analysis of gully erosion susceptibility and spatial modelling using a GIS-based approach. *Geoderma*, 420, 115869. DOI: 10.1016/j.geoderma.2022.115869

Compilation of References

Boutsougame, A., El Youssfi, L., Aassine, H., Khuili, S., Cherkaoui, S. I., Ouazzani, H., & Alaoui, M. (2022). Impact of geology and climate change on wetlands: Case of Lake Aguelmam Azegza (Middle Atlas, Morocco). *E3S Web of Conferences, 337*, 01007. DOI: 10.1051/e3sconf/202233701007

Aawar, T., & Khare, D. (2020). Assessment of climate change impacts on streamflow through hydrological model using SWAT model: A case study of Afghanistan. *Modeling Earth Systems and Environment*, 6(3), 1427–1437. DOI: 10.1007/s40808-020-00759-0

Abdelmajid, S., Mukhtar, A., Baig, M. B., & Reed, M. R. (2021). Climate change, agricultural policy and food security in Morocco. Emerging Challenges to Food Production and Security in Asia, Middle East, and Africa: Climate Risks and Resource Scarcity, 171–196. DOI: 10.1007/978-3-030-72987-5_7

Abhatoo: "QUEL AVENIR DE L'EAU AU MAROC?": RAPPORT DE SYNTHESE DES TRAVAUX DE LA JOURNEE SCIENTIFIQUE DU 17 MARS 2022. (n.d.). Retrieved August 18, 2024, from http://www.abhatoo.net.ma/maalama-textuelle/developpement-economique-et-social/developpement-economique/agriculture/ressources-en-eau-et-lutte-contre-la-secheresse/quel-avenir-de-l-eau-au-maroc-rapport-de-synthese-des-travaux-de-la-journee-scientifique-du-17-mars-2022

Acharki S, Arjdal Y, El Mansouri B (2024) Meteorological drought forecasting via soft computing models in Gharb perimeter (Northwest Morocco). E3S Web Conf 489:04015. DOI: 10.1051/e3sconf/202448904015

Acharki, S., El Qorchi, F., Arjdal, Y., Amharref, M., Bernoussi, A. S., & Ben Aissa, H. (2022). Soil erosion assessment in Northwestern Morocco. *Remote Sensing Applications: Society and Environment*, 25, 100663. DOI: 10.1016/j.rsase.2021.100663

Achite, M., Elshaboury, N., Jehanzaib, M., Vishwakarma, D., Pham, Q., Anh, D., Abdelkader, E., & Elbeltagi, A. (2023). Performance of Machine Learning Techniques for Meteorological Drought Forecasting in the Wadi Mina Basin, Algeria. *Water (Basel)*, 15(4), 765. https://doi.org/. DOI: 10.3390/w15040765

Adallal, R. (2019). *Variabilité environnementale des lacs du moyen atlas marocain : Fonctionnement hydrogéochimique, hydrologique et réponse au changement climatique*. 212.

Adallal, R., Hanane, I., Benkaddour, A., Vallet-coulomb, C., Rhoujjati, A., Sonzogni, C., & Vidal, L. (2022). Hydrogeochemical Processes of the Azigza Lake System (Middle Atlas, Morocco) Inferred from Monthly Monitoring. *Aquatic Geochemistry*, 29(1), 25–47. Advance online publication. DOI: 10.1007/s10498-022-09409-6

Adham, A., Sayl, K. N., Abed, R., Abdeladhim, M. A., Wesseling, J. G., Riksen, M., Fleskens, L., Karim, U., & Ritsema, C. J. (2018). A GIS-based approach for identifying potential sites for harvesting rainwater in the Western Desert of Iraq. *International Soil and Water Conservation Research*, 6(4), 297–304. DOI: 10.1016/j.iswcr.2018.07.003

Aguedai, H., Jelbi, M., Lahlou, F., & Abdelaziz, M. (2022). Hydrochemical and geophysical characterization of the Mnasra coastal aquifers (Rharb basin NW Morocco). *Arabian Journal of Geosciences*, 15(18), 1480. DOI: 10.1007/s12517-022-10738-7

Aguedai, H., Lahlou, F., & El Mansouri, B. (2020). Intrusion Zones Identification in the Mnasra Aquifer (Morocco) Using the Seawater Intrusion Models and the Geophysics Data. In Jarar Oulidi, H., Fadil, A., & Semane, N. E. (Eds.), *Geospatial Technology* (pp. 51–68). Springer International Publishing. DOI: 10.1007/978-3-030-24974-8_4

Ahmed, H. (2019). Applications of Remote Sensing in Oceanographic Research. International Journal of Oceanography & Aquaculture. DOI: 10.23880/ijoac-16000159

Ahmed, F., & Kilic, K. (2018). Fuzzy Analytic Hierarchy Process : A performance analysis of various algorithms CO. *Fuzzy Sets and Systems*, 1, 1–19. DOI: 10.1016/j.fss.2018.08.009

Ahmed, N., Wang, G., Booij, M. J., Xiangyang, S., Hussain, F., & Nabi, G. (2022). Separation of the impact of landuse/landcover change and climate change on runoff in the upstream area of the Yangtze River, China. *Water Resources Management*, 36(1), 1–21. DOI: 10.1007/s11269-021-03021-z

Ait El Mekki, O., & Laftouhi, N. E. (2016). Combination of a geographical information system and remote sensing data to map groundwater recharge potential in arid to semi-arid areas: The Haouz Plain, Morocco. *Earth Science Informatics*, 9(4), 465–479. DOI: 10.1007/s12145-016-0268-0

Ait Kadi, M., & Ziyad, A. (201)8. Integrated Water Resources Management in Morocco. Springer. DOI: 10.1007/978-981-10-7913-9_6

Aïtelghazi, A., Rhinane, H., Bensalmia, A., & Giuliani, G. (2019). Using the Landsat-7 data to study the correlation between the surface temperature and phytoplankton turbidity Case study: Al Massira Lake (Settat - Morocco). *Materials Today: Proceedings*, 13, 496–504. DOI: 10.1016/j.matpr.2019.04.006

Akpejiori, I. J. (2018). Assessing the impacts of land management practices on surface runoff and soil erosion in Ham-Chun Watershed, Northern Thailand (Master's thesis, University of Twente). 2018.

Alabjah, B., Amraoui, F., Chibout, M., & Slimani, M. (2018). Assessment of saltwater contamination extent in the coastal aquifers of Chaouia (Morocco) using the electric recognition. *Journal of Hydrology (Amsterdam)*, 566, 363–376. DOI: 10.1016/j.jhydrol.2018.09.003

Al-Adamat, R., AlAyyash, S., Al-Amoush, H., Al-Meshan, O., Rawajfih, Z., Shdeifat, A., Al-Harahsheh, A., & Al-Farajat, M. (2012). The combination of indigenous knowledge and geo-informatics for water harvesting siting in the Jordanian Badia. DOI: 10.4236/jgis.2012.44042

Al-Adamat, R. (2008). GIS as a decision support system for siting water harvesting ponds in the Basalt Aquifer/NE Jordan. *Journal of Environmental Assessment Policy and Management*, 10(02), 189–206. DOI: 10.1142/S1464333208003020

Aly, Z., Bonn, F. J., & Magagi, R. (2007). Analysis of the backscattering coefficient of salt-affected soils using modeling and RADARSAT-1 SAR data. *IEEE Transactions on Geoscience and Remote Sensing*, 45(2), 332–341. DOI: 10.1109/TGRS.2006.887163

Amiri, V., Nakhaei, M., Lak, R., & Kholghi, M. (2016). Geophysical, isotopic, and hydrogeochemical tools to identify potential impacts on coastal groundwater resources from Urmia hypersaline Lake, NW Iran. *Environmental Science and Pollution Research International*, 23(16), 16738–16760. DOI: 10.1007/s11356-016-6859-y PMID: 27184149

Arabameri, A., Pradhan, B., Pourghasemi, H. R., Rezaei, K., & Kerle, N. (2018). Spatial Modelling of Gully Erosion Using GIS and R Programing : A Comparison among Three Data Mining Algorithms. *Applied Sciences (Basel, Switzerland)*, 8(8), 1369. DOI: 10.3390/app8081369

Arnell, N. W. (1999). Climate change and global water resources. *Global Environmental Change*, 9, S31–S49. DOI: 10.1016/S0959-3780(99)00017-5

Arsanjani, J. J., Helbich, M., Kainz, W., & Boloorani, A. D. (2013). Integration of logistic regression, Markov chain and cellular automata models to simulate urban expansion. *International Journal of Applied Earth Observation and Geoinformation*, 21, 265–275. DOI: 10.1016/j.jag.2011.12.014

Arunrat, N., Wang, C., Pumijumnong, N., Sereenonchai, S., & Cai, W. (2017). 'Farmers' intention and decision to adapt to climate change: A case study in the Yom and Nan basins, Phichit province of Thailand'. *Journal of Cleaner Production*, 143, 672–685. DOI: 10.1016/j.jclepro.2016.12.058

Ashtekar, A. S., Mohammed-Aslam, M. A., & Moosvi, A. R. (2019). Utility of normalized difference water index and GIS for mapping surface water dynamics in sub-upper Krishna Basin. *Photonirvachak (Dehra Dun)*, 47(8), 1431–1442. DOI: 10.1007/s12524-019-01013-6

Ayalew, A. D., Wagner, P. D., Sahlu, D., & Fohrer, N. (2022). Land use change and climate dynamics in the Rift Valley Lake Basin, Ethiopia. *Environmental Monitoring and Assessment*, 194(11), 791. DOI: 10.1007/s10661-022-10393-1 PMID: 36107274

Ayalew, A. D., Wagner, P. D., Tigabu, T. B., Sahlu, D., & Fohrer, N. (2023). Hydrological responses to land use and land cover change and climate dynamics in the Rift Valley Lakes Basin, Ethiopia. *Journal of Water and Climate Change*, 14(8), 2788–2807. DOI: 10.2166/wcc.2023.138

Ayt Ougougdal, H., Khebiza, M. Y., Messouli, M., Bounoua, L., & Karmaoui, A. (2020). Delineation of vulnerable areas to water erosion in a mountain region using SDR-InVEST model : A case study of the Ourika watershed, Morocco. *Scientific African*, 10, e00646. DOI: 10.1016/j.sciaf.2020.e00646

Aziane, N., Larif, M., & Khaddari, A. (2020) State of nitric pollution of the Mnasra aquifer, coastal zone of the Gharb plain (Morocco). Moroccan Journal of Chemistry 8. https://doi.org/DOI: 10.48317/IMIST.PRSM/MORJCHEM-V8I4.17149

Baban, S. M. J., & Wan-Yusof, K. (2003). Modelling optimum sites for locating reservoirs in tropical environments. *Water Resources Management*, 17(1), 1–17. DOI: 10.1023/A:1023066705226

Bachta, M. S., & Ben Nasr, J. (2020). Water Policy in Tunisia. In Zekri, S. (Ed.), *Water Policies in MENA Countries* (pp. 161–184). Springer International Publishing. DOI: 10.1007/978-3-030-29274-4_8

Badon-Ghyben, W. (1889). Nota in verband met de voogenomen put boring nabij Amsterdam (Notes on the probable results of the proposed well drilling near Amsterdam). *Tijdschr. Kon. Inst. Ing.*, 8-22.

Bahija, Abdelmoumen, & Souad. (2023). Analysis of the Hydrological Functioning of the Oued Reghaya Watershed in the Context of Climate Change (Tensift, Morocco). African Journal of Advanced Pure and Applied Sciences, 160–65.

Bahir, M., Mennani, A., Jalal, M., & Youbi, N. (2000). Ressources hydriques du bassin synclinal d'Essaouira (Maroc). *Estudios Geológicos (Madrid)*, 56(3-4), 185–195. DOI: 10.3989/egeol.00563-4150

Bahir, M., Ouhamdouch, S., Ouazar, D., & El Moçayd, N. (2020). Climate change effect on groundwater characteristics within semi-arid zones from western Morocco. *Groundwater for Sustainable Development*, 11, 100380. DOI: 10.1016/j.gsd.2020.100380

Bajjali, W. (2023). Spatial Interpolation. In *ArcGIS Pro and ArcGIS Online* (pp. 223–242). Springer International Publishing. DOI: 10.1007/978-3-031-42227-0_11

Banasik, K., & Woodward, D. (2010). Empirical Determination of Runoff Curve Number for a Small Agricultural Watershed in Poland. 2nd Joint Federal Interagency Conference, Las Vegas, NV. DOI: 10.3390/w6051118

Banholzer, S., Kossin, J., & Donner, S. (2014). The Impact of Climate Change on Natural Disasters. In Singh, A., & Zommers, Z. (Eds.), *Reducing Disaster: Early Warning Systems For Climate Change*. Springer. DOI: 10.1007/978-94-017-8598-3_2

Batchi M, Karkouri JA, Fenjiro I, & Maaqili ME (2017) Étude comparative de deux modèles (DRASTIC et SI) pour l'évaluation de la sensibilité de la nappe phréatique de Mnasra (Maroc nord-occidental) à la pollution d'origine agricole. physio-geo 43–64. DOI: 10.4000/physio-geo.5213

Bayer, A. D., Fuchs, R., Mey, R., Krause, A., Verburg, P. H., Anthoni, P., & Arneth, A. (2021). Diverging land-use projections cause large variability in their impacts on ecosystems and related indicators for ecosystem services. *Earth System Dynamics: ESD*, 12(1), 327–351. DOI: 10.5194/esd-12-327-2021

Bear, J., Cheng, A. H. D., Sorek, S., Ouazar, D., & Herrera, I. (Eds.). (1999). *Seawater intrusion in coastal aquifers: concepts, methods and practices* (Vol. 14). Springer Science & Business Media. DOI: 10.1007/978-94-017-2969-7

Belihu, M., Tekleab, S., Abate, B., & Bewket, W. (2020). Hydrologic response to land use land cover change in the Upper Gidabo Watershed, Rift Valley Lakes Basin, Ethiopia. *HydroResearch*, 3, 85–94. DOI: 10.1016/j.hydres.2020.07.001

Ben Khélifa, W., & Mosbahi, M. (2022). Modeling of rainfall-runoff process using HEC-HMS model for an urban ungauged watershed in Tunisia. *Modeling Earth Systems and Environment*, 8(2), 1749–1758. DOI: 10.1007/s40808-021-01177-6

Benabderrazik, K., Kopainsky, B., Tazi, L., Joerin, J., & Six, J. (2021). Agricultural intensification can no longer ignore water conservation – A systemic modelling approach to the case of tomato producers in Morocco. *Agricultural Water Management*, 256, 107082. DOI: 10.1016/j.agwat.2021.107082

Benini, L., Antonellini, M., Laghi, M., & Mollema, P. N. (2016). Assessment of water resources availability and groundwater salinization in future climate and land use change scenarios: A case study from a coastal drainage basin in Italy. *Water Resources Management*, 30(2), 731–745. DOI: 10.1007/s11269-015-1187-4

Benseddik, B., El Mrabet, E., El Mansouri, B., Chao, J., & Kili, M. (2017). Delineation of artificial recharge zones in Mnasra Aquifer (NW, Morocco). *Modeling Earth Systems and Environment*, 3(1), 10. DOI: 10.1007/s40808-017-0267-2

Berhail, S., & Katipoğlu, O. M. (2023). Spatiotemporal detection of abrupt change in trends of rainfall and dry and wet periods at different time scales: The case of the Medjerda basin in northeast Algeria. *Acta Geophysica*, 71(5), 2497–2516. DOI: 10.1007/s11600-023-01097-2

Boudhar, A. (2009). *Télédétection Du Manteau Neigeux et Modélisation de La Contribution Des Eaux de Fonte Des Neiges Aux Débits Des Oueds Du Haut Atlas de Marrakech.* Université Cadi Ayyad, Faculté Des Sciences et Techniques. DOI: 10.1016/j.jag.2009.09.008

Bou-imajjane, L., Belfoul, M. A., Elkadiri, R., & Stokes, M. (2020). Soil erosion assessment in a semi-arid environment : A case study from the Argana Corridor, Morocco. *Environmental Earth Sciences*, 79(18), 409. DOI: 10.1007/s12665-020-09127-8

Bounif, M., Rahimi, A., & Bouassria, A. (2023) Characterization of the temporal variability of climate and climatic drought in the Doukkala irrigated perimeter (Morocco). https://doi.org/DOI: 10.5281/ZENODO.10184704

Boutallaka, M., El Mazi, M., El Yazidi, M., El Hairchi, K., & Benabdelhadi, M. (2024). *Evolution of Bioclimatic Stages under climate change in the Ouergha Catchment (Northern Morocco).* DOI: 10.1051/bioconf/202411503002

Bo, X. I. A. O., Qing-Hai, W. A. N. G., Jun, F. A. N., Feng-Peng, H. A. N., & Quan-Hou, D. A. I. (2011). Application of the SCS-CN Model to Runoff Estimation in a Small Watershed with High Spatial Heterogeneity. *Pedosphere*, 21(6), 738–749. DOI: 10.1016/S1002-0160(11)60177-X

Bradshaw, Sodhi, Peh, & Brook. (2007). Global evidence that deforestation amplifies flood risk and severity in the developing world. *Global Change Biology*, *13*, 2379-95.

Brooks, K. N., Ffolliott, P. F., Gregersen, H. M., & DeBano, L. F. (2003). *Hydrology and the Management of Watersheds*. Iowa State University Press. DOI: 10.5923/j.re.20140402.04

Buma, W., Lee, S.-I., & Seo, J. Y. (2018). Recent Surface Water Extent of Lake Chad from Multispectral Sensors and GRACE. *Sensors (Basel)*, 18(7), 2082. DOI: 10.3390/s18072082 PMID: 29958481

Burke, L., Kura, Y., Kassem, K., Revenga, C., Spalding, M., McAllister, D., & Caddy, J. (2001). *Coastal Ecosystems*. World Resources Institute.

Burningham, H., & French, J. (2017). Understanding coastal change using shoreline trend analysis supported by cluster-based segmentation. *Geomorphology*, 282, 131–149. DOI: 10.1016/j.geomorph.2016.12.029

Bzioui, M. (2004). Rapport National 2004 Sur Les Ressources En Eau Au Maroc. *UN Water-Africa*, 94. Advance online publication. DOI: 10.4314/aga.v19i3.1723

Caballero, I., Ruiz, J., & Navarro, G. (2019). Sentinel-2 satellites provide near-real time evaluation of catastrophic floods in the west mediterranean. *Water (Basel)*, 11(12), 2499. DOI: 10.3390/w11122499

Campo, A. G. D. (2012). GIS in Environmental Assessment: A Review of Current Issues and Future Needs of current issues and future needs. Journal of Environmental Assessment Policy and Management, 14(1), 1-23. .DOI: 10.1142/S146433321250007X

Cao, H., Tian, Y., Liu, Y., & Wang, R. (2024). Water body extraction from high spatial resolution remote sensing images based on enhanced u-net and multi-scale information fusion. *Scientific Reports*, 14(1), 16132. DOI: 10.1038/s41598-024-67113-7 PMID: 38997473

Cardoso De Matos, A. (2022). The Spread of Scientific Knowledge and Technology Transfer: André Coyne (1891–1960) and the Construction of Dams in 20th Century Portugal. In Borgato, M. T., & Phili, C. (Eds.), *In Foreign Lands: The Migration of Scientists for Political or Economic Reasons* (pp. 203–219). Springer International Publishing. DOI: 10.1007/978-3-030-80249-3_9

Chachadi A.G., & Lobo-Ferreira J.P. (2005). Assessing aquifer vulnerability to seawater intrusion using GALDIT method: part 2- GALDIT Indicator Descriptions. In IAHS and LNEC, proceedings of the 4 the fourth interceltic colloquium on hydrology and management of water resources. University Minho.

Chachadi A.G., Lobo-Ferreira J.P. (2001). Sea water intrusion vulnerability mapping of aquifers using GALDIT method. Proceedings of the workshop on modelling in hydrogeology, 143–156.

Chachadi, A.G., & Lobo-Ferreira, J.P. (2007). Assessing aquifer vulnerability to seawater intrusion using GALDIT method: Part 2, GALDIT indicators description. *Water Celt Countries Quant Qual Clim Var*, 310, 172–180.

Chacuttrikul, P., Kiguchi, M., & Oki, T. (2018). Impacts of climate and land use changes on river discharge in a small watershed: A case study of the Lam Chi subwatershed, northeast Thailand. *Hydrological Research Letters*, 12(2), 7–13. DOI: 10.3178/hrl.12.7

Chaminé, H. I., Pereira, A. J. S. C., Teodoro, A. C., & Teixeira, J. (2021). Remote sensing and GIS applications in earth and environmental systems sciences. *SN Applied Sciences*, 3(12), 2–4. DOI: 10.1007/s42452-021-04855-3

Chanapathi, T., & Thatikonda, S. (2020). Investigating the impact of climate and land-use land cover changes on hydrological predictions over the Krishna river basin under present and future scenarios. *The Science of the Total Environment*, 721, 137736. DOI: 10.1016/j.scitotenv.2020.137736 PMID: 32169648

Change, IPCC Climate. (2019). Land: An IPCC Special Report on Climate Change. *Desertification, Land Degradation, Sustainable Land Management, Food Security, and Greenhouse Gas Fluxes in Terrestrial Ecosystems.*

Chardon, M. (1982). Martin (J.).—Le Moyen Atlas Central. Etude géomorphologique. *Revue de Geographie Alpine*, 70(4), 444–444.

Chehbouni, A., Escadafal, R., Duchemin, B., Boulet, G., Simonneaux, V., Dedieu, G., Mougenot, B., Khabba, S., Kharrou, H., Maisongrande, P., Merlin, O., Chaponnière, A., Ezzahar, J., Er-Raki, S., Hoedjes, J., Hadria, R., Abourida, A., Cheggour, A., Raibi, F., & Sobrino, J. A. (2008). An Integrated Modelling and Remote Sensing Approach for Hydrological Study in Arid and Semi-arid Regions: The SUDMED Programme. *International Journal of Remote Sensing*, 29(17–18), 5161–5181. DOI: 10.1080/01431160802036417

Chen, L.-C., Zhu, Y., Papandreou, G., Schroff, F., & Adam, H. (2018a). Encoder-decoder with atrous separable convolution for semantic image segmentation. *Proceedings of the European conference on computer vision (ECCV)*, 801–818.

Chen, L.-C., Zhu, Y., Papandreou, G., Schroff, F., & Adam, H. (2018b). Encoder-decoder with atrous separable convolution for semantic image segmentation.

Chen, , Xu, Zhu, Yuan, Suen, Guo, Xu, Li, Zhao, & Yang. (2019). Stable classification with limited sample: Transferring a 30-m resolution sample set collected in 2015 to mapping 10-m resolution global land cover in 2017. *Science Bulletin*, 64, 3. PMID: 36659725

Chen, S., Liu, J., Ma, S., Fan, Y., Jia, J., Chen, J., & Chen, F. (2024). Holocene dust storm variations across northern monsoonal Asia and arid central Asia: Contrasting impacts of climate change. *Global and Planetary Change*, 240, 104524. DOI: 10.1016/j.gloplacha.2024.104524

Chim, K., Tunnicliffe, J., Shamseldin, A. Y., & Bun, H. (2021). Assessment of land use and climate change effects on hydrology in the upper Siem Reap River and Angkor Temple Complex, Cambodia. *Environmental Development*, 39, 100615. DOI: 10.1016/j.envdev.2021.100615

Choudhury, R., Nath, B., Rahman, M. M., Medhi, S., & Dutta, J. (2024). Hydrogeochemical characteristics of groundwater contamination in Guwahati city, Assam, India: Tracing the elemental Threads. *Journal of Environmental Management*, 359(February), 120933. DOI: 10.1016/j.jenvman.2024.120933 PMID: 38696848

Chuvieco, E. (1993). Integration of linear programming and GIS for land-use modelling. *International Journal of Geographical Information Systems*, 7(1), 71–83. DOI: 10.1080/02693799308901940

Computing MNDWI using Sentinel on GEE | Life in GIS. (n.d.). Retrieved August 31, 2024, from https://www.lifeingis.com/computation-of-mndwi-in-google-earth-engine-using-sentinel-2/

Conchedda, G., Durieux, L., & Mayaux, P. (2008). An object-based method for mapping and change analysis in mangrove ecosystems. *ISPRS Journal of Photogrammetry and Remote Sensing*, 63(5), 578–589. DOI: 10.1016/j.isprsjprs.2008.04.002

Custodio, E. (2002). Aquifer overexploitation: What does it mean? *Hydrogeology Journal*, 10(2), 254–277. DOI: 10.1007/s10040-002-0188-6

Custodio, E. (2010). Coastal aquifers of Europe: An overview. *Hydrogeology Journal*, 18(1), 269–280. DOI: 10.1007/s10040-009-0496-1

Cvik, J. E. (2011). Normalized difference water index. *Remote Sensing Letters*, 2(4), 225–236.

Daki, Y., Zahour, G., Lachgar, R., & El Hadi, H. (2016). Caractérisation De La Sécheresse Climatique Du Bassin Versant D'oum Er Rbia (Maroc) Par Le Biais De L'indice De Précipitation Standardisé (SPI). *European Scientific Journal*, 12(14), 198. DOI: 10.19044/esj.2016.v12n14p198

Damnati, B. (2000). The climate of North Africa during the last interglacial-glacial cycle: A synthesis. *Journal of Archaeological Science*, 27(12), 1085–1089.

Dang, T. D., Cochrane, T. A., Arias, M. E., Van, P. D. T., & de Vries, T. T. (2016). Hydrological alterations from water infrastructure development in the Mekong floodplains. *Hydrological Processes*, 30(21), 3824–3838. DOI: 10.1002/hyp.10894

Das, S. (2019). *Extreme rainfall estimation at ungauged sites: Comparison between region-of-influence approach of regional analysis and spatial interpolation technique*. International.

de Oliveira, R. G., Júnior, L. C. G. V., da Silva, J. B., Espíndola, D. A., Lopes, R. D., Nogueira, J. S., & Rodrigues, T. R. (2021). Temporal trend changes in reference evapotranspiration contrasting different land uses in southern Amazon basin. Agricultural Water Management, 250.

Deb, P., Babel, M. S., & Denis, A. F. (2018). Multi-GCMs approach for assessing climate change impact on water resources in Thailand. *Modeling Earth Systems and Environment*, 4(2), 825–839. DOI: 10.1007/s40808-018-0428-y

Deng, Y., Jiang, W., Tang, Z., Ling, Z., & Wu, Z. (2019). Long-Term Changes of Open-Surface Water Bodies in the Yangtze River Basin Based on the Google Earth Engine Cloud Platform. *Remote Sensing (Basel)*, 11(19), 2213. DOI: 10.3390/rs11192213

Derdous, O., Bouamrane, A., & Mrad, D. (2021). Spatiotemporal analysis of meteorological drought in a Mediterranean dry land: Case of the Cheliff basin–Algeria. *Modeling Earth Systems and Environment*, 7(1), 135–143. DOI: 10.1007/s40808-020-00951-2

Derradji, F. (2015). *Intrusion marine dans les nappes côtières: cas de la nappe des graviers d'Annaba*. Extrême Nord-Est d'Algérie.

Devkota, M., & Yigezu, Y. A. (2020). Explaining yield and gross margin gaps for sustainable intensification of the wheat-based systems in a Mediterranean climate. *Agricultural Systems*, 185, 102946. DOI: 10.1016/j.agsy.2020.102946

Diani, K., Ettazarini, S., & Hahou, Y. (2023). Identification of soil erosion sites in semiarid zones: Using GIS, remote sensing, and PAP/RAC model. In *Handbook of Hydroinformatics* (pp. 169–183). Elsevier. DOI: 10.1016/B978-0-12-821961-4.00003-8

Diaz-Nieto, J., & Wilby, R. L. (2005). A comparison of statistical downscaling and climate change factor methods: Impacts on low flows in the River Thames, United Kingdom. *Climatic Change*, 69(2-3), 245–268. DOI: 10.1007/s10584-005-1157-6

Dile, Y. T., Berndtsson, R., & Setegn, S. G. (2013). Hydrological response to climate change for gilgel abay river, in the lake tana basin-upper blue Nile basin of Ethiopia. *PLoS One*, 8(10), e79296. DOI: 10.1371/journal.pone.0079296 PMID: 24250755

Diskin, M. H., & Nazimov, N. (1995). Linear Reservoir with Feedback Regulated Inlet as a Model for the Infiltration Process. *Journal of Hydrology (Amsterdam)*, 172(1–4), 313–330. DOI: 10.1016/0022-1694(95)02709-X

Dissanayaka, K. D. C. R., & Rajapakse, R. L. H. L. (2019). Long-term precipitation trends and climate extremes in the Kelani River basin, Sri Lanka, and their impact on streamflow variability under climate change. *Paddy and Water Environment*, 17(2), 281–289. DOI: 10.1007/s10333-019-00721-6

Dlamini, N. S., Kamal, M. R., Soom, M. A. B. M., Mohd, M. S. F., Abdullah, A. F. B., & Hin, L. S. (2017). Modeling potential impacts of climate change on streamflow using projections of the 5th assessment report for the Bernam River Basin, Malaysia. *Water (Basel)*, 9(3), 226. DOI: 10.3390/w9030226

Dlamini, P., Orchard, C., Jewitt, G., Lorentz, S., Titshall, L., & Chaplot, V. (2011). Controlling factors of sheet erosion under degraded grasslands in the sloping lands of KwaZulu-Natal, South Africa. *Agricultural Water Management*, 98(11), 1711–1718. DOI: 10.1016/j.agwat.2010.07.016

Dorel, M. (1993). Travail Du Sol En Bananeraies: Cas Des Andosols. Fruits, 48(2), 77–82. https://doi.org/ RTC.2409DOI: 10.62498.A

Driss, S., Brahim, B., Afaf, A., Habiba, A., Ferdaouss, L., & Abdel-Ilah, M. (2023). Mapping of Zones Vulnerable to Water Erosion Using PAP / CAR and GIS Method Upper the Aït Moulay Hmad Dam, Northeast of the Central Massif, Morocco. *Iraqi Geological Journal*, 56(1B), 181–197. DOI: 10.46717/igj.56.1B.14ms-2023-2-22

Droj, G. (2012). GIS and remote sensing in environmental management. *Journal of Environmental Protection and Ecology*, 13(1), 361–367.

Drusch, M., Del Bello, U., Carlier, S., Colin, O., Fernandez, V., Gascon, F., Hoersch, B., Isola, C., Laberinti, P., Martimort, P., Meygret, A., Spoto, F., Sy, O., Marchese, F., & Bargellini, P. (2012). Sentinel-2: Esa's optical high-resolution mission for gmes operational services. *Remote Sensing of Environment*, 120, 25–36. DOI: 10.1016/j.rse.2011.11.026

Dutta, P., Hinge, G., Marak, J. D. K., & Sarma, A. K. (2021). Future climate and its impact on streamflow: A case study of the Brahmaputra river basin. *Modeling Earth Systems and Environment*, 7(4), 2475–2490. DOI: 10.1007/s40808-020-01022-2

Eccles, R., Zhang, H., & Hamilton, D. (2019). A review of the effects of climate change on riverine flooding in subtropical and tropical regions. *Journal of Water and Climate Change*, 10(4), 687–707. DOI: 10.2166/wcc.2019.175

Eddefli, F., Tayebi, M., Hajaj, S., Khaddari, A., Ouakil, A., & El Harti, A. (2023). Hydric Erosion Mapping Enhancement in Korifla Sub-Watershed (Central Morocco). *Journal of Landscape Ecology*, 16(3), 54–75. DOI: 10.2478/jlecol-2023-0017

Edelman, Alexandra, Andrea Gelding, Elena Konovalov, Rodney McComiskie, Anne Penny, Nicholas Roberts, Shelley Templeman, Dennis Trewin, Mark Ziembicki, and Blair Trewin. 2014. 'State of the Tropics 2014 report'.

El Aoula, R., Mhammdi, N., Dezileau, L., Mahe, G., & Kolker, A. S. (2021). Fluvial sediment transport degradation after dam construction in North Africa. *Journal of African Earth Sciences*, 182, 104255. DOI: 10.1016/j.jafrearsci.2021.104255

El Assaoui, N., Sadok, A., & Merimi, I. (2021). Impacts of climate change on Moroccan's groundwater resources: State of art and development prospects. *Materials Today: Proceedings*, 45, 7690–7696. DOI: 10.1016/j.matpr.2021.03.220

El Bilali, A., Taghi, Y., Briouel, O., Taleb, A., & Brouziyne, Y. (2022). A framework based on high-resolution imagery datasets and MCS for forecasting evaporation loss from small reservoirs in groundwater-based agriculture. *Agricultural Water Management*, 262, 107434. Advance online publication. DOI: 10.1016/j.agwat.2021.107434

El- Bouhali, A., Amyay, M., & El Ouazanı Ech- Chahdi, K. (2024). Changes in water surface area of the Middle Atlas-Morocco lakes: A response to climate and human effects. *International Journal of Engineering and Geosciences*, 9(2), 221–232. DOI: 10.26833/ijeg.1391957

El Garouani, A., Aharik, K., & El Garouani, S. (2020). Water Balance Assessment Using Remote Sensing, Wet-Spass Model, CN-SCS, and GIS for Water Resources Management in Saïss Plain (Morocco). *Arabian Journal of Geosciences*, 13(15), 738. DOI: 10.1007/s12517-020-05730-y

El Ghazali, F. E., Laftouhi, N.-E., Fekri, A., Randazzo, G., & Benkirane, M. (2021a). Enhancing the Success of New Dams Implantation under Semi-Arid Climate, Based on a Multicriteria Analysis Approach: Case of Marrakech Region (Central Morocco). *Open Geosciences*, 13(1), 1494–1508. DOI: 10.1515/geo-2020-0313

El Hamidi MJ, Larabi A, Faouzi M (2021) Modeling and Mapping of coastal aquifer vulnerability to seawater intrusion using SEAWAT code and GALDIT index technique: the case of the Rmel aquifer – Larache, Morocco. E3S Web Conf 298:05002. DOI: 10.1051/e3sconf/202129805002

El Khalki, Y., & Akdim, B. (2001). Les dolines d'effondrement et les dolines-lacs des Causses du SW du Moyen Atlas (Maroc). *Karstologia : revue de karstologie et de spéléologie physique, 38*(1), 19-24. DOI: 10.3406/karst.2001.2481

El Khalki, E. M., Tramblay, Y., Hanich, L., Marchane, A., Boudhar, A., & Hakkani, B. (2021). Climate change impacts on surface water resources in the Oued El Abid basin, Morocco. *Hydrological Sciences Journal*, 66(15), 2132–2145. DOI: 10.1080/02626667.2021.1982137

El Khalki, Y., & Hafid, A. (2002). Turbidité, indicateur du fonctionnement perturbé du géosystème karstique de l'Atlas de Beni Mellal (Moyen Atlas méridional, Maroc). *Karstologia*, 40(1), 39–44. DOI: 10.3406/karst.2002.2506

El Mahmouhi, N., El Wartiti, M., & Azelmad, R. (2018). Spatial Imaging of the Water Resource and Natural Environment Coastal Gharb Basin: Downstream Part of the Hydraulic Basin of the Sebou River (Morocco). In Singh, D. N., & Galaa, A. (Eds.), *Contemporary Issues in Geoenvironmental Engineering* (pp. 169–175). Springer International Publishing. DOI: 10.1007/978-3-319-61612-4_13

El Mountassir, O., & Bahir, M. (2023). The Assessment of the Groundwater Quality in the Coastal Aquifers of the Essaouira Basin, Southwestern Morocco, Using Hydrogeochemistry and Isotopic Signatures. *Water (Basel)*, 15(9), 1769. DOI: 10.3390/w15091769

El Orfi, T. (2023). *Quantification, modélisation et usages des ressources en eau dans le Haut Bassin de l'Oum Er Rbia (Amont du barrage Ahmed El Hansali, Maroc)* [These de doctorat, Université de Lorraine]. https://theses.fr/2023LORR0265

El Qorchi, F., Yacoubi Khebiza, M., Omondi, O. A., Karmaoui, A., Pham, Q. B., & Acharki, S. (2023). Analyzing Temporal Patterns of Temperature, Precipitation, and Drought Incidents: A Comprehensive Study of Environmental Trends in the Upper Draa Basin, Morocco. *Water (Basel)*, 15(22), 3906. DOI: 10.3390/w15223906

Elair, C., Rkha Chaham, K., & Hadri, A. (2023). Assessment of drought variability in the Marrakech-Safi region (Morocco) at different time scales using GIS and remote sensing. *Water Science and Technology: Water Supply*, 23(11), 4592–4624. DOI: 10.2166/ws.2023.283

Elbadaoui, K., Mansour, S., Ikirri, M., Abdelrahman, K., Abu-Alam, T., & Abioui, M. (2023). Integrating Erosion Potential Model (EPM) and PAP/RAC Guidelines for Water Erosion Mapping and Detection of Vulnerable Areas in the Toudgha River Watershed of the Central High Atlas, Morocco. *Land (Basel)*, 12(4), 837. DOI: 10.3390/land12040837

Elfadli, K. I., & Wahab, M. M. A. (2024) Impacts of climate change on drought in Libya. In: Hydroclimatic Extremes in the Middle East and North Africa. Elsevier.

Elfoul, & Ghachi. (2018). Impact Des Actions Anthropiques Sur Le Bilan Hydrologique Dans Le Bassin Versant De La Tassaout (Amont Du Barrage Moulay Youssef) : Modele Hydrologique Orchy Ii (1986-2010). *Journal of Water and Environmental Sciences*, 2(1), 297–304.

Elghouat, A., Algouti, A., Algouti, A., Baid, S., Ezzahzi, S., Kabili, S., & Agli, S. (2024). Integrated Approaches for Flash Flood Susceptibility Mapping: Spatial Modeling and Comparative Analysis of Statistical and Machine Learning Models. A Case Study of the Rheraya Watershed, Morocco. *Journal of Water and Climate Change*, 15(8), jwc2024726. Advance online publication. DOI: 10.2166/wcc.2024.726

Elhaeesahar, M., & Masoudi, M. (2019). GIS analysis for vulnerability assessment of drought in Khuzestan province in Iran using standardized precipitation index (SPI). *Iran Agricultural Research*, 38(2), 9–16.

Elhakeem, M., & Papanicolaou, A. N. (2009). Estimation of the Runoff Curve Number via Direct Rainfall Simulator Measurements in the State of Iowa, USA. *Water Resources Management*, 23(12), 2455–2473. DOI: 10.1007/s11269-008-9390-1

Emami, F., & Koch, M. (2018). Evaluation of statistical-downscaling/bias-correction methods to predict hydrologic responses to climate change in the zarrine river basin, Iran. *Climate (Basel)*, 6(2), 30. DOI: 10.3390/cli6020030

Erostate, M., Huneau, F., Garel, E., Ghiotti, S., Vystavna, Y., Garrido, M., & Pasqualini, V. (2020). Groundwater dependent ecosystems in coastal Mediterranean regions: Characterization, challenges and management for their protection. *Water Research*, 172, 115461. DOI: 10.1016/j.watres.2019.115461 PMID: 31951946

Escobar, F. (2019). Images and masks of satellite images of water bodies for image segmentation.

Etebaai, I. (2009). *L'environnement actuel et le fonctionnement hydroclimatique de quelques systèmes lacustres dans le Moyen Atlas marocain : Cas des lacs Ifrah, Iffer et Afourgagh*. 344.

European Commission. Joint Research Centre. (2020) Global warming and drought impacts in the EU: JRC PESETA IV project : Task 7. Publications Office, LU.

Ez-zaouy, Y., Bouchaou, L., Saad, A., Hssaisoune, M., Brouziyne, Y., Dhiba, D., & Chehbouni, A. (2022). Morocco's coastal aquifers: Recent observations, evolution and perspectives towards sustainability. *Environmental Pollution*, 293, 118498. DOI: 10.1016/j.envpol.2021.118498 PMID: 34798220

Fadili, A. (2014). *Etude hydrogéologique et géophysique de l'extension de l'intrusion marine dans le Sahel de l'Oualidia (Maroc): analyse statistique, hydrochimie et prospection électrique* (Doctoral dissertation, Univérsité de Chouaïb Doukkali Faculté Des Sciences El Jadida).

Fajar, Y., Suwarsono, S., Taufik, M., & Rokhis, K. (2019). The dynamics of shoreline change analysis based on the integration of remote sensing and geographic information system (GIS) techniques in Pekalongan coastal area, Central Java, Indonesia. *Journal of Degraded and Mining Lands Management*, 6(3), 1789–1782. DOI: 10.15243/jdmlm.2019.063.1789

Ferreira, C. S. S., Seifollahi-Aghmiuni, S., Destouni, G., Ghajarnia, N., & Kalantari, Z. (2022). Soil degradation in the European Mediterranean region : Processes, status and consequences. *The Science of the Total Environment*, 805, 150106. DOI: 10.1016/j.scitotenv.2021.150106 PMID: 34537691

Forootan, E. (2023). GIS-Based Slope-Adjusted Curve Number Methods for Runoff Estimation. *Environmental Monitoring and Assessment*, 195(4), 489. DOI: 10.1007/s10661-023-11039-6 PMID: 36939939

Fowler, H. J., Blenkinsop, S., & Tebaldi, C. (2007). Linking climate change modelling to impacts studies: Recent advances in downscaling techniques for hydrological modelling. *International Journal of Climatology*, 27(12), 1547–1578. DOI: 10.1002/joc.1556

Fritz, S., See, L., McCallum, I., Schill, C., Obersteiner, M., Van der Velde, M., Boettcher, H., Havlík, P., & Achard, F. (2011). Highlighting continued uncertainty in global land cover maps for the user community. *Environmental Research Letters*, 6(4), 044005. DOI: 10.1088/1748-9326/6/4/044005

From space to soil. (n.d.). The European Space Agency (ESA). https:// www.esa.int

Fujihara, Y., Tanaka, K., Watanabe, T., Nagano, T., & Kojiri, T. (2008). Assessing the impacts of climate change on the water resources of the Seyhan River Basin in Turkey: Use of dynamically downscaled data for hydrologic simulations. *Journal of Hydrology (Amsterdam)*, 353(1-2), 33–48. DOI: 10.1016/j.jhydrol.2008.01.024

Fu, S., Liu, B., Liu, H., & Xu, L. (2011). The effect of slope on interrill erosion at short slopes. *Catena*, 84(1-2), 29–34. DOI: 10.1016/j.catena.2010.08.013

Fu, S., Wang, L., Khalil, U., Cheema, A. H., Ullah, I., Aslam, B., Tariq, A., Aslam, M., & Alarifi, S. S. (2024). Prediction of surface urban heat island based on predicted consequences of urban sprawl using deep learning: A way forward for a sustainable environment. *Physics and Chemistry of the Earth Parts A/B/C*, 135, 103682. DOI: 10.1016/j.pce.2024.103682

Gabriel, N., & Diabé, E. (2021). Évaluation de La Susceptibilité Des Sols Au Ruisseau Dans Le Bassin Versant Du Knam (Ouest Cameroun), En Appliquant La Méthode Services de Conservation Des Sols Numéro de Courbe. *Revista de Geomorfologie*, 23(1), 39–53. DOI: 10.21094/rg.2021.136

Gaitán, E., Monjo, R., Pórtoles, J., & Pino-Otín, M. R. (2020). Impact of climate change on drought in Aragon (NE Spain). *The Science of the Total Environment*, 740, 140094. DOI: 10.1016/j.scitotenv.2020.140094 PMID: 32562994

Galleguillos, M., Gimeno, F., Puelma, C., Zambrano-Bigiarini, M., Lara, A., & Rojas, M. (2021). Disentangling the effect of future land use strategies and climate change on streamflow in a Mediterranean catchment dominated by tree plantations. *Journal of Hydrology (Amsterdam)*, 595, 126047. DOI: 10.1016/j.jhydrol.2021.126047

Ganie, P. A., Posti, R., Kunal, K., Kunal, G., Bharti, V. S., Sehgal, V. K., Sarma, D., & Pandey, P. K. (2023). Modelling of the Himalayan Mountain River Basin through Hydro-Morphological and Compound Factor-Based Approaches Using Geoinformatics Tools. *Modeling Earth Systems and Environment*, 9(3), 3053–3084. DOI: 10.1007/s40808-023-01691-9

Gao, B.-C. (1995). Normalized difference water index for remote sensing of vegetation liquid water from space. *Imaging spectrometry*, 225–236.

Gao, B. (1996). NDWI—A normalized difference water index for remote sensing of vegetation liquid water from space. *Remote Sensing of Environment*, 58(3), 257–266. DOI: 10.1016/S0034-4257(96)00067-3

Garnier, J., Cicerelli, R. E., De Almeida, T., Belo, J. C. R., Curto, J., Ramos, A. P. M., Valadão, L. V., Satge, F., & Bonnet, M.-P. (2024). Water Resources Monitoring in a Remote Region: Earth Observation-Based Study of Endorheic Lakes. *Remote Sensing (Basel)*, 16(15), 2790. DOI: 10.3390/rs16152790

Garreaud, R. D., Vuille, M., Compagnucci, R., & Marengo, J. (2009). Present-day South American climate. *Palaeogeography, Palaeoclimatology, Palaeoecology*, 281(3), 180–195. DOI: 10.1016/j.palaeo.2007.10.032

Garrido-Perez, J. M., Vicente-Serrano, S. M., Barriopedro, D., García-Herrera, R., Trigo, R., & Beguería, S. (2024). Examining the outstanding Euro-Mediterranean drought of 2021–2022 and its historical context. *Journal of Hydrology (Amsterdam)*, 630, 130653. DOI: 10.1016/j.jhydrol.2024.130653

Gashaw, T., Tulu, T., Argaw, M., & Worqlul, A. W. (2018). Modeling the hydrological impacts of land use/land cover changes in the Andassa watershed, Blue Nile Basin, Ethiopia. *The Science of the Total Environment*, 619, 1394–1408. DOI: 10.1016/j.scitotenv.2017.11.191 PMID: 29734616

Gasse, F. (1991). Hydrological changes in the African tropics since the Last Glacial Maximum. *Quaternary Science Reviews*, 10(4), 379–386.

Gebeyehu, M. N. (2019). Remote Sensing and GIS application in agriculture and natural resource mangement. *International Journal of Environmental Sciences and Natural Resources*, 19(2). Advance online publication. DOI: 10.19080/IJESNR.2019.19.556009

Gharbia, R. (2023). Deep learning for automatic extraction of water bodies using satellite imagery. *Photonirvachak (Dehra Dun)*, 51(7), 1511–1521. DOI: 10.1007/s12524-023-01705-0

Gholami, V., Yousefi, Z., & Zabardast Rostami, H. (2010). Modeling of ground water salinity on the Caspian southern coasts. *Water Resources Management*, 24(7), 1415–1424. DOI: 10.1007/s11269-009-9506-2

Gillespie, T. W., Foody, G. M., Rocchini, D., Giorgi, A. P., & Saatchi, S. (2008). Measuring and modeling biodiversity from space. *Progress in Physical Geography*, 32(2), 203–221. DOI: 10.1177/0309133308093606

Giorgi, F., & Lionello, P. (2008). Climate change projections for the Mediterranean region. *Global and Planetary Change*, 63(2-3), 90–104. DOI: 10.1016/j.gloplacha.2007.09.005

Giorgoudis, M. D., Hadjimitsis, D. G., & Shiftan, Y. (2014). Sustainable Transport Planning using GIS and Remote sensing: an integrated approach. DOI: 10.1117/12.2077752

Giridhar, M. V. S. S., & Viswanadh, G. K. (2014). Runoff Estimation in an Ungauged Watershed Using RS and GIS. *Journal of Indian Association Water Works*, 3, 2009. DOI: 10.1016/j.jhydrol.2014.02.028

Giri, S., Mukhopadhyay, A., Hazra, S., Mukherjee, S., Roy, D., Ghosh, S., Ghosh, T., & Mitra, D. (2014). A study on abundance and distribution of mangrove species in Indian Sundarban using remote sensing technique. *Journal of Coastal Conservation*, 18(4), 359–367. Advance online publication. DOI: 10.1007/s11852-014-0322-3

Global Ecosystem Dynamics Investigation (GEDI). (n.d.). Earthdata, NASA. https://www.earthdata. nasa.gov/sensors/gedi

Gómez-Fernández, D., López, R. S., Zabaleta-Santisteban, J. A., Medina-Medina, A. J., Goñas, M., Silva-López, J. O., Oliva-Cruz, M., & Rojas-Briceño, N. B. (2024). Landsat images and GIS techniques as key tools for historical analysis of landscape change and fragmentation. *Ecological Informatics*, 82, 102738. Advance online publication. DOI: 10.1016/j.ecoinf.2024.102738

Goparaju, L., Ecology, V., & Foundation, N. H. (2021). Wildlife monitoring and management using Remote Sensing and GIS. ICT Tools for Knowledge Management & Control of Emerging Zoonoses & Animal Health Threats, 69-75.

Gramz, M., Batchi, M., El Bouziani, J., Azagane, M. H., El-Boukhari, A., Nafii, A., & Elharrari, Y. (2024). *Impact of climate change on surface water resources in arid zones - the case of the Ykem watershed in the central plateau of Morocco*. DOI: 10.1051/e3sconf/202448904008

Gumus, V., El Moçayd, N., Seker, M., & Seaid, M. (2024). Future projection of droughts in Morocco and potential impact on agriculture. *Journal of Environmental Management*, 367, 122019. DOI: 10.1016/j.jenvman.2024.122019 PMID: 39106801

Hadri, A., Saidi, M. E. M., & Boudhar, A. (2021). Multiscale drought monitoring and comparison using remote sensing in a Mediterranean arid region: A case study from west-central Morocco. *Arabian Journal of Geosciences*, 14(2), 118. DOI: 10.1007/s12517-021-06493-w

Hajhouji, Y. (2018). Modélisation Hydrologique Du Bassin Versant de l'oued Rheraya et Sa Contribution à La Recharge de La Nappe Du Haouz (Bassin Du Tensift, Maroc). https://ird.hal.science/ird-04351673

Hakam, O., Baali, A., Azennoud, K., Lyazidi, A., & Bourchachen, M. (2023). Assessments of Drought Effects on Plant Production Using Satellite Remote Sensing Technology, GIS and Observed Climate Data in Northwest Morocco, Case of the Lower Sebou Basin. *International Journal of Plant Production*, 17(2), 267–282. DOI: 10.1007/s42106-023-00236-5

Han, D., Kohfahl, C., Song, X., Xiao, G., & Yang, J. (2011). Geochemical and isotopic evidence for palaeo-seawater intrusion into the south coast aquifer of Laizhou Bay, China. *Applied Geochemistry*, 26(5), 863–883. DOI: 10.1016/j.apgeochem.2011.02.007

Hanich, L. (2001) Structure et fonctionnement d'un aquifère multicouche carbonaté. Exemple du bassin d'Essaouira. Guide méthodologique. Thèse d'Etat ès Sciences Naturelles, Université Cadi Ayyad, Faculté des Sciences et Techniques de Marrakech, Maroc.

Han, J., Hayashi, Y., Cao, X., & Imura, H. (2009). Application of an integrated system dynamics and cellular automata model for urban growth assessment: A case study of Shanghai, China. *Landscape and Urban Planning*, 91(3), 133–141. DOI: 10.1016/j.landurbplan.2008.12.002

Harrington, M., & Cross, M. (2015). GPS, GIS, and Google Earth. In Harrington, M., & Cross, M. (Eds.), *Google Earth Forensics* (pp. 35–47). Syngress. DOI: 10.1016/B978-0-12-800216-2.00003-9

Hashim, H. Q., & Sayl, K. N. (2022). Incorporating GIS Technique and SCS-CN Approach for Runoff Estimation in the Ungauged Watershed: A Case Study West Desert of Iraq. *Iraqi Journal of Civil Engineering*, 14(2), 1–6. DOI: 10.37650/ijce.2020.172862

Hawkins, R. H., Ward, T. J., Woodward, D. E., & Van Mullem, J. A. (2008). *Curve Number Hydrology: State of the Practice*. American Society of Civil Engineers. DOI: 10.1061/9780784410042

Heinl, M., Walde, J., Tappeiner, G., & Tappeiner, U. (2009). Classifiers vs. input variables—The drivers in image classification for land cover mapping. *International Journal of Applied Earth Observation and Geoinformation*, 11(6), 423–430. DOI: 10.1016/j.jag.2009.08.002

Hekmatnia, M., Isanezhad, A., Ardakani, A. F., Ghojghar, M. A., & Ghaleno, N. D. (2023). An attempt to develop a policy framework for the global sustainability of freshwater resources in the virtual water trade. *Sustainable Production and Consumption*, 39, 311–325. DOI: 10.1016/j.spc.2023.05.022

Held, A., Ticehurst, C., Lymburner, L., & Williams, N. (2003). High resolution mapping of tropical mangrove ecosystems using hyperspectral and radar remote sensing. Int. J. Rem. Sens., 24(13), 2739-2759.

Hermans, T., & Paepen, M. (2020). Combined inversion of land and marine electrical resistivity tomography for submarine groundwater discharge and saltwater intrusion characterization. Geophysical Research Letters, 47(3). DOI: 10.1029/2019GL085877

Herzberg, A. (1901). Die wasserversorgung einiger Nordseebader. *J. Gasbeleucht. Wasserversorg.*, 44, 815–819.

High-Resolution Operational Soil Moisture Product. (n.d.). ISRO. https://isro.gov.in

Hochberg, E. J., Andre'foue¨t, S., & Tyler, M. R. (2003). Sea surface correction of high spatial resolution Ikonos images to improve bottom mapping in near-shore environments. *IEEE Transactions on Geoscience and Remote Sensing*, 41(7), 1724–1729. DOI: 10.1109/TGRS.2003.815408

Homdee, Pongput, & Kanae. (2011). Impacts of land cover changes on hydrologic responses: A case study of Chi River Basin, Thailand. *Journal of Japan Society of Civil Engineers, Ser. B1 (Hydraulic Engineering)*, 67, I_31-I_36.

Hritz, C. (2013). A malarial-ridden swamp: Using Google Earth Pro and Corona to access the southern Balikh valley, Syria. *Journal of Archaeological Science*, 40(4), 1975–1987. DOI: 10.1016/j.jas.2012.11.017

Hssaisoune, M., Bouchaou, L., Sifeddine, A., Bouimetarhan, I., & Chehbouni, A. (2020). Moroccan Groundwater Resources and Evolution with Global Climate Changes. *Geosciences*, 10(2), 81. DOI: 10.3390/geosciences10020081

Huang, G.-B., Ding, X., & Zhou, H. (2010). Optimization method based extreme learning machine for classification. *Neurocomputing*, 74(1-3), 155–163. DOI: 10.1016/j.neucom.2010.02.019

Hu, J., Dong, H., Xu, Q., Ling, W., Qu, J., & Qiang, Z. (2018). Impacts of water quality on the corrosion of cast iron pipes for water distribution and proposed source water switch strategy. *Water Research*, 129, 428–435. DOI: 10.1016/j.watres.2017.10.065 PMID: 29179122

Hu, L., Zhang, C., Zhang, M., Shi, Y., Lu, J., & Fang, Z. (2023). Enhancing fair data services in agricultural disaster: A review. *Remote Sensing (Basel)*, 15(8), 2024. Advance online publication. DOI: 10.3390/rs15082024

Humphreys, E. R., Lafleur, P. M., Flanagan, L. B., Hedstrom, N., Syed, K. H., Glenn, A. J., & Granger, R. (2006). Summer carbon dioxide and water vapor fluxes across a range of northern peatlands. *Journal of Geophysical Research*, 111(G4), 2005JG000111. Advance online publication. DOI: 10.1029/2005JG000111

Hung, P., Van Trung, L., & Vo, P. L. (2023). Decision Support Tool for Integrated Water Resources Management Based on GIS, Remote Sensing and SWAT Model: A Case Study in the Upper Part of Dong Nai River Basin, Vietnam. *Environmental Science and Engineering*, 1, 361–388. DOI: 10.1007/978-3-031-17808-5_23

Hussain, A., Mada, R., & Kamboj, V. (2022). *Applications of remote sensing and GIS in hydrological and hydrogeological studies: integrated watershed management. In Addressing Environmental Challenges Through Spatial Planning*. IGI Global. DOI: 10.4018/978-1-7998-8331-9.ch012

Hu, Z., & Lo, C. P. (2007). Modeling urban growth in Atlanta using logistic regression. *Computers, Environment and Urban Systems*, 31(6), 667–688. DOI: 10.1016/j.compenvurbsys.2006.11.001

IOCCG. (2000). Remote Sensing of Ocean Colour in Coastal, and Other Optically-Complex Waters. Reports of the International Ocean-Colour Coordinating Group, No. 3. Dartmouth, Canada: IOCCG. https://ioccg.org/reports/rep003/

IPCC, AR5. (2013). Climate Change 2013: The Physical Science Basis. Contribution of Working Group I to the Fifth Assessment Report of the Intergovernmental Panel on Climate Change.

Islam, M., Sarker, S., Morsad, G., Khan, M. F., Chowdhury, A., & Islam, S. L. U. (2023). Spatial–temporal changes of shoreline and Vegetation: Impacts on mangrove cover along the Sundarbans area, Bangladesh. *Journal of Coastal Conservation*, 28(1), 10. Advance online publication. DOI: 10.1007/s11852-023-01016-z

Jaffal, M., Charbaoui, A., Kchikach, A., El Ghorfi, M., Khaldoun, A., Amine El Mahdi, S., Bodinier, J.-L., Yazami, O. K., Jourani, E.-S., & Manar, A. (2022). Gravity study of the Western Bahira Basin and the Gantour Phosphatic Plateau, central Morocco: Interpretation and hydrogeological implications. *Journal of African Earth Sciences*, 193, 104581. DOI: 10.1016/j.jafrearsci.2022.104581

Jahandideh-Tehrani, M., Zhang, H., Helfer, F., & Yu, Y. (2019). Review of climate change impacts on predicted river streamflow in tropical rivers. *Environmental Monitoring and Assessment*, 191(12), 1–23. DOI: 10.1007/s10661-019-7841-1 PMID: 31732799

Jahan, K., Pradhanang, S. M., & Md, A. E. B. (2021). Surface Runoff Responses to Suburban Growth: An Integration of Remote Sensing, GIS, and Curve Number. *Land (Basel)*, 10(5), 452. DOI: 10.3390/land10050452

Jain, S. K., Kumar, S., & Varghese, J. (2001). Estimation of Soil Erosion for a Himalayan Watershed Using GIS Technique. *Water Resources Management*, 15(1), 41–54. DOI: 10.1023/A:1012246029263

Janga, B., Asamani, G. P., Sun, Z., & Cristea, N. (2023). A review of practical ai for remote sensing in earth sciences. *Remote Sensing (Basel)*, 15(16), 4112. Advance online publication. DOI: 10.3390/rs15164112

Jáuregui, E. (1992). Aspects of monitoring local/regional-climate change in a tropical region. *Atmosfera*, 5.

Jellali, M. (1997). *Développement des ressources en eau au Maroc*. Www.Water.Gov.Ma/Semide/Fr/Themes/Ppne.Htm. http://www.albacharia.ma/xmlui/handle/123456789/31203

Jeon, J.-H., Lim, K. J., & Engel, B. A. (2014). Regional Calibration of SCS-CN L-THIA Model: Application for Ungauged Basins. *Water (Basel)*, 6(5), 1339–1359. DOI: 10.3390/w6051339

Jiang, D., Jones, I., Liu, X., Simis, S. G. H., Cretaux, J.-F., Albergel, C., Tyler, A., & Spyrakos, E. (2024). Impacts of droughts and human activities on water quantity and quality: Remote sensing observations of Lake Qadisiyah, Iraq. *International Journal of Applied Earth Observation and Geoinformation*, 132, 104021. DOI: 10.1016/j.jag.2024.104021

Juhász, L. (2024). Assessing publication trends in selected giscience journals. *International Journal of Geographical Information Science*, 38(8), 1443–1467. DOI: 10.1080/13658816.2024.2347306

Kamusoko, C., Aniya, M., Adi, B., & Manjoro, M. (2009). Rural sustainability under threat in Zimbabwe–simulation of future land use/cover changes in the Bindura district based on the Markov-cellular automata model. *Applied Geography (Sevenoaks, England)*, 29(3), 435–447. DOI: 10.1016/j.apgeog.2008.10.002

Kana, J. D., Ahmad Diab, D., Hervé, D. G., N'kaya, G. D. M., Meli'i, J. L., Browdon, S. N., & Lebogo, S. P. K. N. (2022). A combined GIS-based remote sensing and wireline log data analysis for water resource management in the economic capital district of Cameroon. *Sustainable Water Resources Management*, 8(5), 129. Advance online publication. DOI: 10.1007/s40899-022-00714-0

Karmaoui, A., Moumane, A., El Jaafari, S., Menouni, A., Al Karkouri, J., Yacoubi, M., & Hajji, L. (2023). Thirty Years of Change in the Land Use and Land Cover of the Ziz Oases (Pre-Sahara of Morocco) Combining Remote Sensing, GIS, and Field Observations. *Land (Basel)*, 12(12), 12. Advance online publication. DOI: 10.3390/land12122127

Kazemi Garajeh, M., Haji, F., Tohidfar, M., Sadeqi, A., Ahmadi, R., & Kariminejad, N. (2024). Spatiotemporal monitoring of climate change impacts on water resources using an integrated approach of remote sensing and Google Earth Engine. *Scientific Reports*, 14(1), 5469. DOI: 10.1038/s41598-024-56160-9 PMID: 38443699

Kerry, G. R., Montalbo, F. J. P., Das, R., Patra, S., Mahapatra, G. P., Maurya, G. K., Nayak, V., Jena, A. B., Ukhurebor, K. E., Jena, R. C., Gouda, S., Majhi, S., & Rout, J. R. (2022). *An overview of remote monitoring methods in biodiversity conservation*. Environmental Science and Pollution Research. Springer Berlin Heidelberg. DOI: 10.1007/s11356-022-23242-y

Kesgin, E., Yaldız, S. G., & Güçlü, Y. S. (2024). Spatiotemporal variability and trends of droughts in the Mediterranean coastal region of Türkiye. *International Journal of Climatology*, 44(4), 1036–1057. DOI: 10.1002/joc.8370

Khaoula, L., Abdessamad, G., Abdellah, O., Khalid, E., Rhizlane, F., & Saida, E. (2019). Application of geographical information system (GIS) for the development of climatological air temperature vulnerability maps: An example from Morocco. .DOI: 10.1002/met.1871

Khouz, A., Jaffal, M., Trindade, J., Bougadir, B., El Bchari, F., Kchikach, A., El Ghorfi, M., Ibouh, H., Jadoud, M., Kadiri, O., & Manar, A. (2023). Understanding the Deep Structure of the Essaouira Basin Using Gravity Data: Hydrogeological Inferences for a Semiarid Region in Central-Western Morocco. *Geosciences*, 13(11), 345. DOI: 10.3390/geosciences13110345

Khouz, A., Trindade, J., Santos, P. P., Oliveira, S. C., El Bchari, F., Bougadir, B., Garcia, R. A. C., Reis, E., Jadoud, M., Saouabe, T., & Rachidi, S. (2023). Flood susceptibility assessment through statistical models and HEC-RAS analysis for sustainable management in Essaouira Province, Morocco. *Geosciences*, 13(12), 382. DOI: 10.3390/geosciences13120382

Kili, M., El Mansouri, B., Chao, J., & Fora, A. A. (2006). De nouveaux éléments structuraux du complexe aquifère profond du bassin du Rharb (Maroc) : Implications hydrogéologiques. *Comptes Rendus Geoscience*, 338(16), 1194–1202. DOI: 10.1016/j.crte.2006.09.008

Koneti, S., Sunkara, S. L., & Roy, P. S. (2018). Hydrological modeling with respect to impact of land-use and land-cover change on the runoff dynamics in Godavari River Basin using the HEC-HMS model. *ISPRS International Journal of Geo-Information*, 7(6), 206. DOI: 10.3390/ijgi7060206

Koyama, C. N., Watanabe, M., Hayashi, M., Ogawa, T., & Shimada, M. (2019). Mapping the spatial-temporal variability of tropical forests by ALOS-2 L-band SAR big data analysis. *Remote Sensing of Environment*, 233, 111372. DOI: 10.1016/j.rse.2019.111372

Krois, J., & Schulte, A. (2014). GIS-Based Multi-Criteria Evaluation to Identify Potential Sites for Soil and Water Conservation Techniques in the Ronquillo Watershed, Northern Peru. *Applied Geography (Sevenoaks, England)*, 51, 131–142. DOI: 10.1016/j.apgeog.2014.04.006

Kumar, A., Kishore, B. S. P. C., Saikia, P., Deka, J., Bharali, S., Singha, L. B., Tripathi, O. P., & Khan, M. L. (2019). Tree diversity assessment and above ground forests biomass estimation using SAR remote sensing: A case study of higher altitude vegetation of North-East Himalayas, India. *Physics and Chemistry of the Earth Parts A/B/C*, 111, 53–64. DOI: 10.1016/j.pce.2019.03.007

Kumar, M., & Jha, R. (2022). Evaluating Water Resources Availability in Headwater Sub-catchments of Subarnarekha River Basin in a changing Environment using Remote Sensing and GIS Software. *Indian Journal of Environmental Protection*, 42(9), 1034–1041.

Kumar, N., Yamaç, S., & Velmurugan, A. (2015). Applications of Remote Sensing and GIS in Natural Resource Management. *Journal of the Andaman Science Association*, 20(1), 1–6.

Kundu, K., Halder, P., & Mandal, J. (2020). Forest Cover Change Analysis in Sundarban Delta Using Remote Sensing Data and GIS. .DOI: 10.1007/978-981-13-7334-3_7

Kundu, K., & Halder, P. (2019). Geomorphological Changes in Sundarbans Delta Using GIS and Remote Sensing Data. In Mandal, J., Mukhopadhyay, S., Dutta, P., & Dasgupta, K. (Eds.), *Computational Intelligence, Communications, and Business Analytics. CICBA 2018. Communications in Computer and Information Science* (Vol. 1031). Springer. DOI: 10.1007/978-981-13-8581-0_4

La sécheresse au Maghreb: Diagnostic, impacts et perspectives pour le renforcement de la résilience du secteur agricole. (2023). FAO. DOI: 10.4060/cc7126fr

Labbaci, A., Marghadi, S., Laaribya, S., & Moukrim, S. (2020). Integrating Sentinel-2 Data and PAPCAR Model to Map Water Erosion: Case of Beni Boufrah Watershed. *RJESTE*, 3(1). Advance online publication. DOI: 10.4314/rjeste.v3i1.4S

Lahcen, M., Nidal, F., & Mohamed, G. (2020). *Floristic Study and Assessment of the Environmental Factors Governing the Distribution of Riparian Plants in the Zat Sub-Basin*. Tensift Watershed. DOI: 10.35759/JAnmPlSci.v45-2.3

Laouina, A. (2006). Le litoral marocain, milieux côtier et marin. *Environnement et Territoires, Cadre Natural*, 187-216.

Lasco & Moglen. (2023). Hydrologic Conservatism as a Rationale for Selecting NRCS Initial Abstraction Ratio. Journal of Hydrologic Engineering, 28(12), 04023041. DOI: 10.1061/JHYEFF.HEENG-6016

Latrubesse, E. M., Stevaux, J. C., & Sinha, R. (2005). Tropical rivers. *Geomorphology*, 70(3-4), 187–206. DOI: 10.1016/j.geomorph.2005.02.005

Laurance, W. F. (2007). Forests and floods. *Nature*, 449(7161), 409–410. DOI: 10.1038/449409a PMID: 17898753

Lauri, H., de Moel, H., Ward, P. J., Räsänen, T. A., Keskinen, M., & Kummu, M. (2012). Future changes in Mekong River hydrology: impact of climate change and reservoir operation on discharge. Hydrology and Earth system sciences, 16(12), 4603-4619. DOI: 10.5194/hess-16-4603-2012

Layati, E., & El Ghachi, M. (2023). Quantitative analysis of the morphometric parameters of the Oued Lakhdar watershed (Morocco) : Statistical and GIS approaches. In *1st International Conference On Water And Earth Sciences (ICWES'22)* (p. 47-58). Sciendo. https://doi.org/DOI: 10.2478/9788367405270-007

Layati, E., & El Ghachi, M. (2024). Contribution of Geomatics in the Quantification of Water Erosion : Application in the Oued Lakhdar (Morocco). *Photonirvachak (Dehra Dun)*, 52(6), 1315–1327. DOI: 10.1007/s12524-024-01880-8

Le Réseau ADUM. (n.d.). Retrieved August 18, 2024, from https://adum.parisnanterre .fr/as/ed/cv.pl?mat=104963&site=adumfr

Lechner, A. M., Foody, G. M., & Boyd, D. S. (2020). Applications in Remote Sensing to Forest Ecology and Management. *One Earth*, 2(5), 405–412. DOI: 10.1016/j. oneear.2020.05.001

Lee, J., Park, M., Min, J.-H., & Na, E. H. (2023). Integrated Assessment of the Land Use Change and Climate Change Impact on Baseflow by Using Hydrologic Model. *Sustainability (Basel)*, 15(16), 12465. DOI: 10.3390/su151612465

Lehodey, P., Bertignac, M., Hampton, J., Lewis, A., & Picaut, J. (2008). El Niño Southern Oscillation and tuna in the western Pacific. *Nature*, 389(6647), 715–718.

Leifer, I., Lehr, W. J., Simecek-Beatty, D., Bradley, E., Clark, R. N., Dennison, P. E., & Roberts, D. A. (2012). State of the art satellite and airborne marine oil spill remote sensing: Application to the BP Deepwater Horizon oil spill. *Remote Sensing of Environment*, 124, 185–209. DOI: 10.1016/j.rse.2012.03.024

Lemus-Canovas M, Insua-Costa D, Trigo RM, & Miralles DG (2024) Record-shattering 2023 Spring heatwave in western Mediterranean amplified by long-term drought. NPJ Clim Atmos Sci 7:25. DOI: 10.1038/s41612-024-00569-6

Li, C., & Fang, H. (2021). Assessment of climate change impacts on the streamflow for the Mun River in the Mekong Basin, Southeast Asia: Using SWAT model. *Catena*, 201, 105199. DOI: 10.1016/j.catena.2021.105199

Li, J., Li, L., Song, Y., Chen, J., Wang, Z., Bao, Y., Zhang, W., & Meng, L. (2023). A robust large-scale surface water mapping framework with high spatiotemporal resolution based on the fusion of multi-source remote sensing data. *International Journal of Applied Earth Observation and Geoinformation*, 118, 103288. DOI: 10.1016/j.jag.2023.103288

Liu, & Curry. (2010). Accelerated warming of the Southern Ocean and its impacts on the hydrological cycle and sea ice. *Proceedings of the National Academy of Sciences*, 107, 14987-92. DOI: 10.1073/pnas.1003336107

Liu, L., Zhang, Q., Guo, Y., Li, Y., Wang, B., Chen, E., Li, Z., & Hao, S. (2024). Mapping coniferous forest distribution in a semi-arid area based on multi-classifier fusion and Google Earth Engine combining Gaofen-1 and Sentinel-1 Data : A Case Study in Northwestern Liaoning, China. *Forests*, 15(2), 288. DOI: 10.3390/f15020288

Liu, Z., Yao, Z., Huang, H., Wu, S., & Liu, G. (2014). Land Use and Climate Changes and Their Impacts on Runoff in the Yarlung Zangbo River Basin, China. *Land Degradation & Development*, 25(3), 203–215. DOI: 10.1002/ldr.1159

Li, X., Tan, L., Li, Y., Qi, J., Feng, P., Li, B., Liu, D. L., & Zhang, X. (2022). Effects of global climate change on the hydrological cycle and crop growth under heavily irrigated management–A comparison between CMIP5 and CMIP6. *Computers and Electronics in Agriculture*, 202, 107408. DOI: 10.1016/j.compag.2022.107408

Li, Z., Leong, W. J., Durand, M., Howat, I., Wadkowski, K., Yadav, B., & Moortgat, J. (2023). Super-resolution deep neural networks for water classification from free multispectral satellite imagery. *Journal of Hydrology (Amsterdam)*, 626, 130248. DOI: 10.1016/j.jhydrol.2023.130248

Long, J., Shelhamer, E., & Darrell, T. (2015). Fully convolutional networks for semantic segmentation. *Proceedings of the IEEE conference on computer vision and pattern recognition*, 3431–3440.

López-Bermúdez, F., Romero-Díaz, A., Martínez-Fernandez, J., & Martínez-Fernandez, J. (1998). Vegetation and soil erosion under a semi-arid Mediterranean climate : A case study from Murcia (Spain). *Geomorphology*, 24(1), 51–58. DOI: 10.1016/S0169-555X(97)00100-1

Loulad, S., Nguyen, T. T., Simou, M. R., Rhinane, H., & Buerkert, A. (2023). Monitoring rural-urban transformation in the coastal region of Rabat-Sale-Kenitra, Morocco. *PLoS One*, 18(8), e0290829. DOI: 10.1371/journal.pone.0290829 PMID: 37651348

Lwin, K. K., & Murayama, Y. (2011). Computers, Environment and Urban Systems Modelling of urban green space walkability : Eco-friendly walk score calculator. *Computers, Environment and Urban Systems*, 35(5), 408–420. DOI: 10.1016/j.compenvurbsys.2011.05.002

Lyons, M.B., Phinn, S.R., & Roelfsema, C.M. (2012). Long term land cover and seagrass mapping using Landsat and object-based image analysis from 1972 to 2010 in the coastal environment of South East Queensland, Australia. *ISPRS J. Photogramm. Remote Sens., 71*, 34-46.

Mabrouki, Benchrifa, Ennouhi, Azoulay, Bencheikh, Rachiq, El-Moustaqim, Al-Jadabi, Azrour, Fatima-zahra, & El Hajjaji. (2023). Geographic Information System for the Study of Water Resources in Chaâba El Hamra, Mohammedia (Morocco). *Lecture Notes in Networks and Systems* 635 LNNS:469-74. .DOI: 10.1007/978-3-031-26254-8_67

Ma, C., Li, Y., Li, X., & Gao, L. (2020). Evaluation of groundwater sustainable development considering seawater intrusion in Beihai City, China. *Environmental Science and Pollution Research International*, 27(5), 4927–4943. DOI: 10.1007/s11356-019-07311-3 PMID: 31840220

Mack, M. J. (1995). HER—Hhydrologic Evaluation of Runoff; The Soil Conservation Service Curve Number Technique as an Interactive Computer Model. *Computers & Geosciences*, 21(8), 929–935. DOI: 10.1016/0098-3004(95)00029-8

Magny, M. (2013). Holocene climate changes in the central Mediterranean as recorded by lake-level fluctuations at Lake Accesa (Tuscany, Italy). *Quaternary Science Reviews*, 66, 66–79.

Mahajan, S., & Martinez, J. (2021). Water, water, but not everywhere: Analysis of shrinking water bodies using open access satellite data. *International Journal of Sustainable Development and World Ecology*, 28(4), 326–338. DOI: 10.1080/13504509.2020.1851803

Mahendra, R. S., Mohanty, P. C., Bisoyi, H., Kumar, T. S., & Nayak, S. (2011, April). (20110. Assessment and management of coastal multi-hazard vulnerability along the Cuddalore- Villupuram, east coast of India using geospatial techniques. *Ocean and Coastal Management*, 54(4), 302–311. DOI: 10.1016/j.ocecoaman.2010.12.008

Mahmoud, M. T., Mohammed, S. A., Hamouda, M. A., & Mohamed, M. M. (2020). Impact of Topography and Rainfall Intensity on the Accuracy of IMERG Precipitation Estimates in an Arid Region. *Remote Sensing (Basel)*, 13(1), 13. DOI: 10.3390/rs13010013

Maimouni, S., Bannari, A., El-Harti, A., & El-Ghmari, A. (2012). "Indices spectraux et" spectral angle mapper" pour la cartographie des risques d'érosion hydrique à partir des données ali eo-1. Teledetection, 10(4), 213–23. https://hal.science/hal-01948890

Maiti, S., & Bhattacharya, A. K. (2009). Shoreline change analysis and its application to prediction: A remote sensing and statistics based approach. *Marine Geology*, 257(1-4), 11–23. DOI: 10.1016/j.margeo.2008.10.006

Malhi, Y., & Wright, J. (2004). Spatial patterns and recent trends in the climate of tropical rainforest regions. *Philosophical Transactions of the Royal Society of London. Series B, Biological Sciences*, 359(1443), 311–329. DOI: 10.1098/rstb.2003.1433 PMID: 15212087

Malik, S., Pal, S. C., Chowdhuri, I., Chakrabortty, R., Roy, P., & Das, B. (2020). Prediction of Highly Flood Prone Areas by GIS Based Heuristic and Statistical Model in a Monsoon Dominated Region of Bengal Basin. *Remote Sensing Applications: Society and Environment*, 19, 100343. DOI: 10.1016/j.rsase.2020.100343

Marcé, R., & Armengol, J. (2009). Water Quality in Reservoirs Under a Changing Climate. In Sabater, S., & Barceló, D. (Eds.), *Water Scarcity in the Mediterranean. Springer Berlin Heidelberg, Berlin* (pp. 73–94). DOI: 10.1007/698_2009_38

Marouane, M., Dahchour, M., Dousset, S., & El Hajjaji, S. (2014) Monitoring the Risk of Nitrate Pollution in Mnasra Groundwater Under Field Condition–Morocco. Moroccan Journal of Chemistry 2. https://doi.org/DOI: 10.48317/IMIST.PRSM/MORJCHEM-V2I4.2137

Martinez, A. I., & Labib, S. M. (2023). Demystifying Normalized Difference Vegetation Index (NDVI) for Greenness Exposure Assessments and Policy Interventions in Urban Greening. *Environmental Research*, 220, 115155. DOI: 10.1016/j.envres.2022.115155 PMID: 36584843

Martinsen, O. (1994). Mass movements. In Maltman, A. (Ed.), *The Geological Deformation of Sediments* (pp. 127–165). Springer Netherlands. DOI: 10.1007/978-94-011-0731-0_5

Martins, V. S., Kaleita, A., Barbosa, C. C. F., Fassoni-Andrade, A. C., Felipe de Lucia, L., & Novo, E. M. L. M. (2019). Remote sensing of large reservoir in the drought years: Implications on surface water change and turbidity variability of Sobradinho reservoir (Northeast Brazil). *Remote Sensing Applications: Society and Environment*, 13, 275–288. DOI: 10.1016/j.rsase.2018.11.006

Matomela, N., Tianxin, L., Morahanye, L., Bishoge, O. K., & Ikhumhen, H. O. (2019). Rainfall-runoff estimation of Bojiang lake watershed using SCS-CN model coupled with GIS for watershed management. *J Appl Adv Res*, 4(1), 16–24. DOI: 10.21839/jaar.2019.v4i1.263

Mauser, W., & Bach, H. (2009). PROMET–Large scale distributed hydrological modelling to study the impact of climate change on the water flows of mountain watersheds. *Journal of Hydrology (Amsterdam)*, 376(3-4), 362–377. DOI: 10.1016/j.jhydrol.2009.07.046

Mazigh, N., Taleb, A., Bilali, A. E., & Ballah, A. (2022). The Effect of Erosion Control Practices on the Vulnerability of Soil Degradation in Oued EL Malleh Catchment using the USLE Model Integrated into GIS, Morocco. *Trends in Sciences*, 19(2), 2059. DOI: 10.48048/tis.2022.2059

Mbilinyi, B. P., Tumbo, S. D., Fm Mahoo, H., & Mkiramwinyi, F. O. (2007). GIS-Based Decision Support System for Identifying Potential Sites for Rainwater Harvesting. *Physics and Chemistry of the Earth Parts A/B/C*, 32(15–18), 1074–1081. DOI: 10.1016/j.pce.2007.07.014

McFeeters, S. (1996). The use of the normalized difference water index (ndwi) in the delineation of open water features. *International Journal of Remote Sensing*, 17(7), 1425–1432. DOI: 10.1080/01431169608948714

Mdiker, N. A. B. I. L., El Achheb, A., Mandour, A., Younsi, A., El Maliki, S., & Outeyeb, B. (2009). Contribution à l'étude de la salinisation de la nappe côtière de sahel El Haouzia région d'El Jadida au Maroc. *Afrique Science: Revue Internationale des Sciences et Technologie, 5*(2).

Meißl, G., Klebinder, K., Zieher, T., Lechner, V., Kohl, B., & Markart, G. (2023). Influence of Antecedent Soil Moisture Content and Land Use on the Surface Runoff Response to Heavy Rainfall Simulation Experiments Investigated in Alpine Catchments. *Heliyon*, 9(8), e18597. Advance online publication. DOI: 10.1016/j.heliyon.2023.e18597 PMID: 37560701

Meteorological Service of Catalonia. (2024) L'any 2023, el segon més càlid, i dels més secs, URL https://static-m.meteo.cat/wordpressweb/wp-content/uploads/2024/01/02141659/ABCA_2023_ABM_des.pdf

Milano, M. (2012). *The Mediterranean Region under Climate Change*. IRD Éditions.

Ming, Alison, Isobel Rowell, Sam Lewin, Robert Rouse, Thomas Aubry, and Emma Boland. 2021. 'Key messages from the IPCC AR6 climate science report'.

Mirzavand, M., Ghasemieh, H., Sadatinejad, S. J., & Bagheri, R. (2020). An overview on source, mechanism and investigation approaches in groundwater salinization studies. *International Journal of Environmental Science and Technology*, 17(4), 2463–2476. DOI: 10.1007/s13762-020-02647-7

Mishra, S. K., Kansal, A. K., & Aggarwal, N. (2012). Assessment of Design Runoff Curve Number for a Watershed. *Water Practice & Technology*, 7(4), wpt2012066. Advance online publication. DOI: 10.2166/wpt.2012.066

Mitsova, D., Shuster, W., & Wang, X. (2011). A cellular automata model of land cover change to integrate urban growth with open space conservation. *Landscape and Urban Planning*, 99(2), 141–153. DOI: 10.1016/j.landurbplan.2010.10.001

Mohammed, J., Abdelaziz, M., Al Mazini, I., Magrane, B., El Mansouri, B., & Mohamed, E. B. (2022). Hydrogeophysical and hydrochemical assessment of groundwater quality of the Rharb shallow aquifer (NW Morocco). *Arabian Journal of Geosciences*, 15(7), 660. DOI: 10.1007/s12517-022-09956-w

Molle, F., & Tanouti, O. (2017). Squaring the circle: Agricultural intensification vs. water conservation in Morocco. *Agricultural Water Management*, 192, 170–179. DOI: 10.1016/j.agwat.2017.07.009

Monition, L. (1953). *Sur l'intérêt de sondages dans la plaine d'Akermoud*. Rap. Inéd. MTPC/DH/DRE.

Moravec, V., Markonis, Y., Rakovec, O., Svoboda, M., Trnka, M., Kumar, R., & Hanel, M. (2021). Europe under multi-year droughts: How severe was the 2014–2018 drought period? *Environmental Research Letters*, 16(3), 034062. DOI: 10.1088/1748-9326/abe828

Moreno-de-las-Heras, M., Bochet, E., Vicente-Serrano, S. M., Espigares, T., Molina, M. J., Monleón, V., Nicolau, J. M., Tormo, J., & García-Fayos, P. (2023). Drought conditions, aridity and forest structure control the responses of Iberian holm oak woodlands to extreme droughts: A large-scale remote-sensing exploration in eastern Spain. *The Science of the Total Environment*, 901, 165887. DOI: 10.1016/j.scitotenv.2023.165887 PMID: 37517715

Moujabber, M. E., Samra, B. B., Darwish, T., & Atallah, T. (2006). Comparison of different indicators for groundwater contamination by seawater intrusion on the Lebanese coast. *Water Resources Management*, 20(2), 161–180. DOI: 10.1007/s11269-006-7376-4

Moumane, A., Al Karkouri, J., Benmansour, A., El Ghazali, F. E., Fico, J., Karmaoui, A., & Batchi, M. (2022). Monitoring long-term land use, land cover change, and desertification in the Ternata oasis, Middle Draa Valley, Morocco. *Remote Sensing Applications: Society and Environment*, 26, 100745. DOI: 10.1016/j.rsase.2022.100745

Moumane, A., El Ghazali, F. E., Al Karkouri, J., Delorme, J., Batchi, M., Chafiki, D., & Karmaoui, A. (2021). Monitoring spatiotemporal variation of groundwater level and salinity under land use change using integrated field measurements, GIS, geostatistical, and remote-sensing approach: Case study of the Feija aquifer, Middle Draa watershed, Moroccan Sahara. *Environmental Monitoring and Assessment*, 193(12), 769. DOI: 10.1007/s10661-021-09581-2 PMID: 34735624

Musy, A., & Higy, C. (2004). Hydrologie: Une Science de La Nature. Vol. 21. PPUR presses polytechniques. DOI: 10.4000/geocarrefour.1794

Nadi, P. A., & Murad, A. K. (2017). Reviewing the use of Geographic Information System (GIS) to measure sustainable urban transport performance. Journal of Geoscience, Enineering. *Environmental Technology*, 2(2), 171–178. PMID: 28934911

Nadjla, B., Abdellatif, D., & Assia, S. (2021). Mapping of the groundwater vulnerability to saline intrusion using the modified GALDIT model (Case: The Ain Temouchent coastal aquifer,(North-Western Algeria)). *Environmental Earth Sciences*, 80(8), 319. DOI: 10.1007/s12665-021-09614-6

Näschen, K., Diekkrüger, B., Evers, M., Höllermann, B., Steinbach, S., & Thonfeld, F. (2019). The impact of land use/land cover change (LULCC) on water resources in a tropical catchment in Tanzania under different climate change scenarios. *Sustainability (Basel)*, 11(24), 7083. DOI: 10.3390/su11247083

Nelson, L. (2021). Predictive Digital Soil Mapping. White Paper prepared for LWS 548.

Nguyen, T. T., Ngo, H. H., Guo, W., Nguyen, H. Q., Luu, C., Dang, K. B., & Zhang, X. (2020). New approach of water quantity vulnerability assessment using satellite images and GIS-based model: An application to a case study in Vietnam. Science of the Total Environment, 737, 139784.

Nhi, P. T. T., Khoi, D. N., Nguyen, T. T. T., Van Ty, T., & Fang, S. (2022). Hydrological impacts of future climate and land use/cover changes in the Lower Mekong Basin: A case study of the Srepok River Basin, Vietnam. *Environmental Monitoring and Assessment*, 194(S2), 768. DOI: 10.1007/s10661-022-10175-9 PMID: 36255530

Okou, F. A. Y., Tente, B., Bachmann, Y., & Sinsin, B. (2016). Regional erosion risk mapping for decision support: A case study from West Africa. *Land Use Policy*, 56, 27–37. DOI: 10.1016/j.landusepol.2016.04.036

Omrania, S., El Khodrani, N., Zouahri, A., Douaik, A., Iaaich, H., Lahmar, M., & El Hajjaji, S. (2019). Evaluation of the Groundwater Quality of the M'nasra, Kenitra (Morocco). *Materials Today: Proceedings*, 13, 1092–1101. DOI: 10.1016/j.matpr.2019.04.076

Ongoma, V., Driouech, F., Brouziyne, Y., Chfadi, T., Epule Epule, T., Tanarhte, M., & Chehbouni, A. (2024). Morocco's climate change impacts, adaptation and mitigation—A stocktake. *Regional Environmental Change*, 24(1), 14. Advance online publication. DOI: 10.1007/s10113-023-02176-2

Oo, H. T., Zin, W. W., & Thin Kyi, C. C. (2020). Analysis of streamflow response to changing climate conditions using SWAT model. *Civil Engineering Journal*, 6(2), 194–209. DOI: 10.28991/cej-2020-03091464

Orfi, T. E. (n.d.). *Quantification, modélisation et usages des ressources en eau dans le Haut Bassin de l'Oum Er Rbia (Amont du barrage Ahmed El Hansali, Maroc).*

Osman, K. T. (2014). Soil Erosion by Water. In K. T. Osman, *Soil Degradation, Conservation and Remediation* (p. 69-101). Springer Netherlands. DOI: 10.1007/978-94-007-7590-9_3

Ouallali, A., Moukhchane, M., Aassoumi, H., Berrad, F., & Dakir, I. (2016). The Mapping of the Soils' Degradation State by Adaptation the PAP/RAC Guidelines in the Watershed of Wadi Arbaa Ayacha, Western Rif, Morocco. *Journal of Geoscience and Environment Protection*, 04(07), 77–88. DOI: 10.4236/gep.2016.47009

Ouassanouan, Y., Fakir, Y., Simonneaux, V., Kharrou, M. H., Bouimouass, H., Najar, I., Benrhanem, M., Sguir, F., & Chehbouni, A. (2022). Multi-Decadal Analysis of Water Resources and Agricultural Change in a Mediterranean Semiarid Irrigated Piedmont under Water Scarcity and Human Interaction. *The Science of the Total Environment*, 834, 155328. DOI: 10.1016/j.scitotenv.2022.155328 PMID: 35452720

Ouharba, E. L. H., Mabrouki, J., & Triqui, Z. E. L. A. (2024). Assessment and Future Climate Dynamics in the Bouregreg Basin, Morocco – Impacts and Adaptation Alternatives. *Ecological Engineering and Environmental Technology*, 25(3), 51–63. DOI: 10.12912/27197050/177823

Ozdemir, S., Akbulut, Z., Karsli, F., & Kavzoglu, T. (2024). Extraction of water bodies from high-resolution aerial and satellite images using visual foundation models. *Sustainability (Basel)*, 16(7), 2995. DOI: 10.3390/su16072995

Pant, H. & Verma, J. & Surya, S. (2022). Application of Remote Sensing and GIS in Biodiversity Management. Green Technology: A Roadmap for Sustainable Development-2020, 1-18.

Pant, H., Verma, J., & Richhariya, S. (2022). Wildlife Conservation and Management using GIS and Remote Sensing. Geospatial Technology and its Applications in Resource Management, 279-286.

Parizi, E., Hosseini, S. M., Ataie-Ashtiani, B., & Simmons, C. T. (2019). Vulnerability mapping of coastal aquifers to seawater intrusion: Review, development and application. *Journal of Hydrology (Amsterdam)*, 570, 555–557. DOI: 10.1016/j.jhydrol.2018.12.021

Parra, L. (2022a). Applied sciences Remote Sensing and GIS in Environmental Monitoring.

Parra, L. (2022b). Remote Sensing and GIS in Environmental Monitoring. *Applied Sciences (Basel, Switzerland)*, 12(16), 8045. Advance online publication. DOI: 10.3390/app12168045

Perez-Torres, W. I., Uman-Flores, D. A., Quispe-Quispe, A. B., Palomino-Quispe, F., Bezerra, E., Leher, Q., Paixão, T., & Alvarez, A. B. (2024). Exploratory analysis using deep learning for water-body segmentation of perus high-mountain remote sensing images. *Sensors (Basel)*, 24(16), 5177. Advance online publication. DOI: 10.3390/s24165177 PMID: 39204871

Petchprayoon, P., Blanken, P. D., Ekkawatpanit, C., & Hussein, K. (2010). Hydrological impacts of land use/land cover change in a large river basin in central–northern Thailand. *International Journal of Climatology*, 30(13), 1917–1930. DOI: 10.1002/joc.2131

Phinyoyang, A., & Ongsomwang, S. (2021). Optimizing Land Use and Land Cover Allocation for Flood Mitigation Using Land Use Change and Hydrological Models with Goal Programming, Chaiyaphum, Thailand. *Land (Basel)*, 10(12), 1317. DOI: 10.3390/land10121317

Pierdicca, R., & Paolanti, M. (2022). Geoai: A review of artificial intelligence approaches for the interpretation of complex geomatics data. *Geoscientific Instrumentation, Methods and Data Systems*, 11(1), 195–218. DOI: 10.5194/gi-11-195-2022

Pijl, A., Reuter, L. E. H., Quarella, E., Vogel, T. A., & Tarolli, P. (2020). GIS-based soil erosion modelling under various steep-slope vineyard practices. *Catena*, 193, 104604. DOI: 10.1016/j.catena.2020.104604

Pinder, G. F. (1973). A Galerkin-finite element simulation of groundwater contamination on Long Island, New York. *Water Resources Research*, 9(6), 1657–1669. DOI: 10.1029/WR009i006p01657

Potapov, P., Li, X., Hernandez-Serna, A., Tyukavina, A., Hansen, M. C., Kommareddy, A., Pickens, A., Turubanova, S., Tang, H., Silva, C. E., Armston, J., Dubayah, R., Blair, J. B., & Hofton, M. (2021). Mapping global forest canopy height through integration of GEDI and Landsat data. *Remote Sensing of Environment*, 253(October), 112165. DOI: 10.1016/j.rse.2020.112165

Powers, D. (2008). Evaluation: From precision, recall and f-factor to roc, informedness, markedness & correlation. *Mach. Learn. Technol., 2.*

Prasad, N., Plants, A., Semwal, M., Plants, A., & Roy, P. S. (2015). Recent Advances in Lichenology. *Remote Sensing and GIS for Biodiversity Conservation.*, (January). Advance online publication. DOI: 10.1007/978-81-322-2181-4

Prestele, R., Arneth, A., Bondeau, A., de Noblet-Ducoudré, N., Pugh, T. A. M., Sitch, S., Stehfest, E., & Verburg, P. H. (2017). Current challenges of implementing anthropogenic land-use and land-cover change in models contributing to climate change assessments. *Earth System Dynamics: ESD*, 8(2), 369–386. DOI: 10.5194/esd-8-369-2017

Pulido-Leboeuf, P. (2004). Seawater intrusion and associated processes in a small coastal complex aquifer (Castell de Ferro, Spain). *Applied Geochemistry*, 19(10), 1517–1527. DOI: 10.1016/j.apgeochem.2004.02.004

Qalmoun, A., Bouzrarf, K., & Belqat, B. (2022). Assessment of the ecological status of the Oum Er-rabie River basin (Central Morocco) through physicochemical, bacteriological parameters and biotic indices. *Biologia*, 77(9), 2533–2547. DOI: 10.1007/s11756-022-01128-1

Qanza, H., Maslouhi, A., Abboudi, S., Mustapha, H., & Hmimou, A. (2019). Unsaturated Soil Hydraulic Properties Identification using Numerical Inversion and In-Situ Experiments from Mnasra Area, Morocco. *KSCE Journal of Civil Engineering*, 23(11), 4949–4959. DOI: 10.1007/s12205-019-0420-8

Qi, P., Xu, Y. J., & Wang, G. (2020). Quantifying the individual contributions of climate change, dam construction, and land use/land cover change to hydrological drought in a marshy river. *Sustainability (Basel)*, 12(9), 3777. DOI: 10.3390/su12093777

Quirós, E., & Fragoso-Campón, L. (2024). Spatial analysis of remote sensing and meteorological indices in a drought event in southwestern Spain. *Theoretical and Applied Climatology*, 155(5), 3757–3770. DOI: 10.1007/s00704-024-04846-5

Rafik, A., Bahir, M., Beljadid, A., Chehbouni, A., Dhiba, D., & Ouhamdouch, S. (2022). The combination of the quality index, isotopic, and GIS techniques to assess water resources in a semi-arid context (Essaouira watershed in Morocco). *Groundwater for Sustainable Development*, 17, 100768. DOI: 10.1016/j.gsd.2022.100768

Raible, C. C. (2010). Mediterranean climate variability over the last centuries: A review. *Climate of the Past*, 6(1), 99–114.

Raj, A., Jhariya, M. K., Yadav, D. K., Banerjee, A., & Meena, R. S. (2019). Soil for Sustainable Environment and Ecosystems Management. In Jhariya, M. K., Banerjee, A., Meena, R. S., & Yadav, D. K. (Eds.), *Sustainable Agriculture, Forest and Environmental Management* (pp. 189–221). Springer Singapore. DOI: 10.1007/978-981-13-6830-1_6

Ramat, G., Santi, E., Paloscia, S., Fontanelli, G., Pettinato, S., Santurri, L., Souissi, N., Da Ponte, E., Abdel Wahab, M. M., Khalil, A. A., Essa, Y. H., Ouessar, M., Dhaou, H., Sghaier, A., Hachani, A., Kassouk, Z., & Lili Chabaane, Z. (2023). Remote sensing techniques for water management and climate change monitoring in drought areas: Case studies in Egypt and Tunisia. *European Journal of Remote Sensing*, 56(1), 2157335. DOI: 10.1080/22797254.2022.2157335

Rani, M., Rehman, S., Sajjad, H., Chaudhary, B. S., Sharma, J., Bhardwaj, S., & Kumar, P. (2018). Assessing coastal landscape vulnerability using geospatial techniques along VizianagarameSrikakulam coast of Andhra Pradesh, India. *Natural Hazards*, 94(2), 711–725. DOI: 10.1007/s11069-018-3414-9

Rhoujjati, N., Hanich, L., Bouchaou, L., Patris, N., N'da, A. B., & Chehbouni, A. (2021). Isotopic Tracers to Assess the Snowmelt Contribution to the Groundwater Recharge: A Case from the Moroccan High and Middle Atlas Mountains. *Arabian Journal of Geosciences*, 14(23), 2611. DOI: 10.1007/s12517-021-08737-1

Ritchie, J. C., Schiebe, F. R., Cooper, C. M., & Harrington, J. A.Jr. (1994). Chlorophyll measurements in the presence of suspended sediment using broad band spectral sensors aboard satellites. *Journal of Freshwater Ecology*, 9(2), 197–206. DOI: 10.1080/02705060.1994.9664887

Robinson, N. P., Allred, B. W., Jones, M. O., Moreno, A., Kimball, J., Naugle, D., Erickson, T., & Richardson, A. (2017). A Dynamic Landsat Derived Normalized Difference Vegetation Index (NDVI) Product for the Conterminous United States. *Remote Sensing (Basel)*, 9(8), 863. DOI: 10.3390/rs9080863

Rodier, J., Legube, B., & Merlet, N. (2009). *L'Analyse de l'eau 9e édition. Entièrement mise à jour*. Dunod.

Rojas, O. (2021). Next generation agricultural stress index system (ASIS) for agricultural drought monitoring. *Remote Sensing (Basel)*, 13(5), 959. DOI: 10.3390/rs13050959

Ronneberger, O., Fischer, P., & Brox, T. (2015). U-net: Convolutional networks for biomedical image segmentation. *Lecture Notes in Computer Science*, 9351, 234–241. DOI: 10.1007/978-3-319-24574-4_28

Rossi, L., Naumann, G., Gabellani, S., & Cammalleri, C. (2023). A combined index to characterize agricultural drought in Italy at municipality scale. *Journal of Hydrology. Regional Studies*, 47, 101404. DOI: 10.1016/j.ejrh.2023.101404

Saâdi, Z., & Maslouhi, A. (2003). Modeling nitrogen dynamics in unsaturated soils for evaluating nitrate contamination of the Mnasra groundwater. *Advances in Environmental Research*, 7(4), 803–823. DOI: 10.1016/S1093-0191(02)00055-2

Saaty. (1984). The Analytic Hierarchy Process: Decision Making in Complex Environments. In Quantitative assessment in arms control. Springer. DOI: 10.3390/conservation3010003

Sab u, D. A., & Gh Şerban, P. (2023). Combining Radar Quantitative Precipitation Estimates (QPEs) with Distributed Hydrological Model for Controlling Transit of Flash-Flood Upstream of Crowded Human Habitats in Romania. *Natural Hazards*, 116(1), 1209–1238. DOI: 10.1007/s11069-022-05718-9

Salma, K., Ahmed, A., Abdellah, A., & Akram, E. G. (2023). Quantification of Water Erosion Using Empirical Models RUSLE and EPM in the Rheraya Basin in the High Atlas of Marrakech. *Disaster Advances*, 16(5), 19–28. DOI: 10.25303/1605da019028

Salvan, H.M. (1984). Les formations évaporitiques du Trias marocain. Problèmes stratigraphiques, paléogéographiques et paléoclimatologiques. Quelques réflexions. *Revue de Géologie Dynamique et de Géographie Physique*, 25, 87–203.

Samal, D. R., & Gedam, S. (2021). Assessing the impacts of land use and land cover change on water resources in the Upper Bhima river basin, India. *Environmental Challenges*, 5, 100251. DOI: 10.1016/j.envc.2021.100251

Sanad, H., Mouhir, L., Zouahri, A., Moussadek, R., El Azhari, H., Yachou, H., Ghanimi, A., Oueld Lhaj, M., & Dakak, H. (2024a). Assessment of Groundwater Quality Using the Pollution Index of Groundwater (PIG), Nitrate Pollution Index (NPI), Water Quality Index (WQI), Multivariate Statistical Analysis (MSA), and GIS Approaches: A Case Study of the Mnasra Region, Gharb Plain, Morocco. *Water (Basel)*, 16(9), 1263. DOI: 10.3390/w16091263

Sanad, H., Moussadek, R., Mouhir, L., Oueld Lhaj, M., Dakak, H., El Azhari, H., Yachou, H., Ghanimi, A., & Zouahri, A. (2024b). Assessment of Soil Spatial Variability in Agricultural Ecosystems Using Multivariate Analysis, Soil Quality Index (SQI), and Geostatistical Approach: A Case Study of the Mnasra Region, Gharb Plain, Morocco. *Agronomy (Basel)*, 14(6), 1112. DOI: 10.3390/agronomy14061112

Sardar, P., & Samadder, S. R. (2021). Understanding the dynamics of landscape of greater Sundarban area using multi-layer perceptron Markov chain and landscape statistics approach. Ecological Indicators. *Ecological Indicators*, 106914. Advance online publication. DOI: 10.1016/j.ecolind.2020.106914

Sawassi, A., Khadra, R., & Crookston, B. (2024). Water Banking as a Strategy for the Management and Conservation of a Critical Resource: A Case Study from Tunisia's Medjerda River Basin (MRB). *Sustainability (Basel)*, 16(9), 3875. DOI: 10.3390/su16093875

Sayad, A., Chakiri, S., Martin, C., Bejjaji, Z., & Echarfaoui, H. (2011). Effet des conditions climatiques sur le niveau du lac Sidi Ali (Moyen Atlas, Maroc). *Physio-Géo*, 5(Volume 5), 251–268. DOI: 10.4000/physio-geo.2145

Sayl, K. N., Mohammed, A. S., & Ahmed, A. D. (2020). GIS-Based Approach for Rainwater Harvesting Site Selection. B IOP Conference Series: Materials Science and Engineering. DOI: 10.1088/1757-899X/737/1/012246

Scull, P., Franklin, J., Chadwick, O. A., & McArthur, D. (2003). Predictive soil mapping: A review. *Progress in Physical Geography*, 27(2), 171–197. DOI: 10.1191/0309133303pp366ra

Sebbar, A., Ilmen, R., Mohammed, H., Hassan, F., & Badri, W. (2019, December 12). *Caractérisation de la sécheresse climatique au Maroc.*

Seddique, A. A., Masuda, H., Anma, R., Bhattacharya, P., Yokoo, Y., & Shimizu, Y. (2019). Hydrogeochemical and isotopic signatures for the identification of seawater intrusion in the paleobeach aquifer of Cox's Bazar city and its surrounding area, south-east Bangladesh. *Groundwater for Sustainable Development*, 9, 100215. DOI: 10.1016/j.gsd.2019.100215

See, L. M., & Fritz, S. (2006). A method to compare and improve land cover datasets: Application to the GLC-2000 and MODIS land cover products. *IEEE Transactions on Geoscience and Remote Sensing*, 44(7), 1740–1746. DOI: 10.1109/TGRS.2006.874750

Sharma, S. K., Misra, S. K., & Singh, J. B. (2020). The role of GIS-enabled mobile applications in disaster management: A case analysis of cyclone Gaja in India. *International Journal of Information Management*, 51, 102030. DOI: 10.1016/j.ijinfomgt.2019.10.015

Sherif, M. M., & Hamza, K. I. (2001). Mitigation of seawater intrusion by pumping brackish water. *Transport in Porous Media*, 43(1), 29–44. DOI: 10.1023/A:1010601208708

Sherif, M., Liaqat, M. U., Baig, F., & Al-Rashed, M. (2023). Water resources availability, sustainability and challenges in the GCC countries: An overview. *Heliyon*, 9(10), e20543. DOI: 10.1016/j.heliyon.2023.e20543 PMID: 37817990

Shivakumar, B. R., & Rajashekararadhya, S. V. (2018). Investigation on Land Cover Mapping Capability of Maximum Likelihood Classifier: A Case Study on North Canara, India. *Procedia Computer Science*, 143, 579–586. DOI: 10.1016/j.procs.2018.10.434

Shrestha, S., Bhatta, B., Talchabhadel, R., & Salvatore, G. P. V. (2022). Integrated assessment of the landuse change and climate change impacts on the sediment yield in the Songkhram River Basin, Thailand. *Catena*, 209, 105859. DOI: 10.1016/j.catena.2021.105859

Simonneaux, V., Cheggour, A., Deschamps, C., Mouillot, F., Cerdan, O., & Le Bissonnais, Y. (2015). Land use and climate change effects on soil erosion in a semi-arid mountainous watershed (High Atlas, Morocco). *Journal of Arid Environments*, 122, 64–75. DOI: 10.1016/j.jaridenv.2015.06.002

Singh, A., Meena, G. K., Kumar, S., & Gaurav, K. (2019). Evaluation of the penetration depth of L- And S-band (NISAR mission) microwave SAR signals into ground. URSI Asia-Pacific Radio Science Conference, AP-RASC. DOI: 10.23919/URSIAP-RASC.2019.8738217

Singh, K. V., Setia, R., Sahoo, S., Prasad, A., & Pateriya, B. (2015). Evaluation of NDWI and MNDWI for assessment of waterlogging by integrating digital elevation model and groundwater level. *Geocarto International*, 30(6), 650–661. DOI: 10.1080/10106049.2014.965757

Singh, S. K., & Marcy, N. (2017). Comparison of simple and complex hydrological models for predicting catchment discharge under climate change. *AIMS Geosciences*, 3(3), 467–497. DOI: 10.3934/geosci.2017.3.467

Singkran, N., Tosang, J., Waijaroen, D., Intharawichian, N., Vannarart, O., Anantawong, P., Kunta, K., Wisetsopa, P., Tipvong, T., Janjirawuttikul, N., Masthawee, F., Amornpatanawat, S., & Kirtsaeng, S. (2015). Influences of land use and climate changes on hydrologic system in the northeastern river basin of Thailand. *Journal of Water and Climate Change*, 6(2), 325–340. DOI: 10.2166/wcc.2014.127

Soergel, B., Kriegler, E., Weindl, I., Rauner, S., Dirnaichner, A., Ruhe, C., Hofmann, M., Bauer, N., Bertram, C., Bodirsky, B. L., Leimbach, M., Leininger, J., Levesque, A., Luderer, G., Pehl, M., Wingens, C., Baumstark, L., Beier, F., Dietrich, J. P., & Popp, A. (2021). A sustainable development pathway for climate action within the UN 2030 Agenda. *Nature Climate Change*, 11(8), 656–664. DOI: 10.1038/s41558-021-01098-3

Sokolova, M., & Lapalme, G. (2009). A systematic analysis of performance measures for classification tasks. *Information Processing & Management*, 45(4), 427–437. DOI: 10.1016/j.ipm.2009.03.002

Solanlki, H. U., Dwivedi, R. M., & Nayak, S. R. (2001). Application of Ocean Colour Monitor chlorophyll and AVHRR SST for fishery forecast: Preliminary validation results off Gujarat coast, northwest coast of India. *Indian Journal of Geo-Marine Sciences*, 30(3), 132–138.

Soria-lara, J. A., Arce-ruiz, R. M., Arranz-lópez, A., & Ariza-Álvarez, A. (2020). Environmental Impact Assessment for transport projects : A review of technical and process-related issues. *Standard Transport Appraisal Methods. Elsevier Inc.*, 6(1), 255–285. Advance online publication. DOI: 10.1016/bs.atpp.2020.07.002

Soulaimani, A., & Ouanaimi, H. (2011). Anti-Atlas et Haut Atlas, Circuit Occidental. Nouveaux Guides Géologiques et Miniers Du Maroc. *Notes et Mémoires Du Service Géologique Du Maroc*, 3, 9–72.

Stonestrom, D. A., Scanlon, B. R., & Zhang, L. (2009). Introduction to special section on impacts of land use change on water resources. *Water Resources Research*, 45(7), 45. DOI: 10.1029/2009WR007937

Sun, D., Gao, G., Huang, L., Liu, Y., & Liu, D. (2024). Extraction of water bodies from high-resolution remote sensing imagery based on a deep semantic segmentation network. *Scientific Reports*, 14(1), 14604. DOI: 10.1038/s41598-024-65430-5 PMID: 38918493

Surekha, J. N., & Gupta, N. (2023). Ndr-unet: Segmentation of water bodies in remote sensing using nested dense. DOI: 10.21203/rs.3.rs-3159538/v1

Sur, K., Verma, V. K., & Pateriya, B. (2021). Surface water estimation at regional scale using hybrid techniques in GEE environment-A case study on Punjab State of India. *Remote Sensing Applications: Society and Environment*, 24, 100625. DOI: 10.1016/j.rsase.2021.100625

Syvitski, Cohen, Kettner, & Brakenridge. (2014). How important and different are tropical rivers? An overview. *Geomorphology*, 227, 5-17.

Taha, Z. A., & Hadi, A. (2019). Performance evaluation metrics for semantic image segmentation: A comprehensive survey. *Information (Basel)*, 10(10), 319.

Tahouri, J., Sadiki, A., Karrat, L., Johnson, V. C., Chan, N. W., Fei, Z., & Kung, H. T. (2022). Using a modified PAP/RAC model and GIS-for mapping water erosion and causal risk factors : Case study of the Asfalou watershed, Morocco. *International Soil and Water Conservation Research*, 10(2), 254–272. DOI: 10.1016/j.iswcr.2021.07.003

Talabi, A. O., Abdu-Raheem, Y. A., Afolagboye, L. O., Oguntuase, M. A., & Akinola, O. O. (2020). Hydrogeochemistry of shallow groundwater in Ado-Ekiti Area, Southwestern Nigeria. *Groundwater for Sustainable Development*, 11, 100386. DOI: 10.1016/j.gsd.2020.100386

Talbi, M. (n.d.). *Impact Des Stress Hydriques Sur Les Potentiels De Dayet Aoua (Moyen Atlas - Maroc)*. Retrieved August 18, 2024, from https://www.academia.edu/20886804/IMPACT_DES_STRESS_HYDRIQUES_SUR_LES_POTENTIELS_DE_DAYET_AOUA_MOYEN_ATLAS_MAROC_

Tamiminia, H., Salehi, B., Mahdianpari, M., & Goulden, T. (2024). State-wide forest canopy height and aboveground biomass map for New York with 10 m resolution, integrating GEDI, Sentinel-1, and Sentinel-2 data. Ecological Informatics, 79(April), 102404. .DOI: 10.1016/j.ecoinf.2023.102404

Tamiminia, H., Salehi, B., Mahdianpari, M., Quackenbush, L., Adeli, S., & Brisco, B. (2020). Google Earth Engine for geo-big data applications: A meta-analysis and systematic review. *ISPRS Journal of Photogrammetry and Remote Sensing*, 164, 152–170. DOI: 10.1016/j.isprsjprs.2020.04.001

Tanarhte, M., De Vries, A. J., Zittis, G., & Chfadi, T. (2024). Severe droughts in North Africa: A review of drivers, impacts and management. *Earth-Science Reviews*, 250, 104701. DOI: 10.1016/j.earscirev.2024.104701

Tang, X., Zang, Z., Lin, H., Wang, X., & Wen, Z. (2023). Using a Vegetation Index to Monitor the Death Process of Chinese Fir Based on Hyperspectral Data. *Forests*, 14(12), 2444. DOI: 10.3390/f14122444

Tarekegn, N., Abate, B., Muluneh, A., & Dile, Y. (2021). Modeling the impact of climate change on the hydrology of Andasa watershed. *Modeling Earth Systems and Environment*, 1–17.

Teixeira, A. M. D. A., Batista, L. V., Silva, R. M. D., Freitas, L. M. T., & Santos, C. A. G. (2024). Dynamic monitoring of surface area and water volume of reservoirs using satellite imagery, computer vision and deep learning. *Remote Sensing Applications: Society and Environment*, 35, 101205. Advance online publication. DOI: 10.1016/j.rsase.2024.101205

Teodoro, A.C. (2018) Optical Satellite Remote Sensing of the Coastal Zone Environment-An Overview. Intech open.

Teutschbein, C., & Seibert, J. (2010). Regional climate models for hydrological impact studies at the catchment scale: A review of recent modeling strategies. *Geography Compass*, 4(7), 834–860. DOI: 10.1111/j.1749-8198.2010.00357.x

Thi, N. Q., Govind, A., Le, M.-H., Linh, N. T., Tran, T. M. A., Hai, N. K., & Ha, T. V. (2023). Spatiotemporal characterization of droughts and vegetation response in Northwest Africa from 1981 to 2020. *The Egyptian Journal of Remote Sensing and Space Sciences*, 26(3), 393–401. DOI: 10.1016/j.ejrs.2023.05.006

Thomas, T., Rangel-Buitrago, N., Phillips, M. R., Anfuso, G., & Williams, A. T. (2015). Mesoscale morphological change, beach rotation and storm climate influences along a macrotidal embayed beach. *Journal of Marine Science and Engineering*, 3(3), 1006–1026. DOI: 10.3390/jmse3031006

Thornes, J. B., & Wainwright, J. (2004). *Environmental Issues in the Mediterranean*. Routledge. DOI: 10.4324/9780203495490

Tian, L., Wu, X., Tao, Y., Li, M., Qian, C., Liao, L., & Fu, W. (2023). Review of Remote Sensing-Based Methods for Forest Aboveground Biomass Estimation: Progress, Challenges, and Prospects. *Forests*, 14(6), 1–31. DOI: 10.3390/f14061086

Tola, T. L., Zhang, K., Chukalla, A. D., Liu, L., Ezaz, G. T., Elsadek, E. A., & Abdallah, M. (2024). Spatiotemporal lakes surface area changes over 35 years and potential causes in the Central Rift Valley, Ethiopia. *Journal of Hydrology. Regional Studies*, 54, 101863. Advance online publication. DOI: 10.1016/j.ejrh.2024.101863

Tramblay Y, Llasat MC, Randin C, & Coppola E (2020) Climate change impacts on water resources in the Mediterranean. Reg Environ Change 20:83, s10113-020-01665-y. DOI: 10.1007/s10113-020-01665-y

Tramblay, Y. (2012). Impact of climate change on extreme rainfall in Morocco: New uncertainties and new vulnerabilities. *Journal of Hydrology (Amsterdam)*, 479, 1–19.

Tramblay, Y., Koutroulis, A., Samaniego, L., Vicente-Serrano, S. M., Volaire, F., Boone, A., Le Page, M., Llasat, M. C., Albergel, C., Burak, S., Cailleret, M., Kalin, K. C., Davi, H., Dupuy, J.-L., Greve, P., Grillakis, M., Hanich, L., Jarlan, L., Martin-StPaul, N., & Polcher, J. (2020). Challenges for drought assessment in the Mediterranean region under future climate scenarios. *Earth-Science Reviews*, 210, 103348. DOI: 10.1016/j.earscirev.2020.103348

Tripathy, B. R., Seenipandi, K., Sajjad, H., Joshi, P. K., Chaudhary, B. S., & Kumar, P. (2018). Monitoring of seasonal variability and movement of suspended sediment concentrations along the Thiruvananthapuram coast, southern India, using the Landsat OLI sensor. *Ocean Science*, 14(5), 1085–1092. DOI: 10.5194/os-14-1085-2018

Ukpere, D. R., & Ottah, C. (2017). Remote Sensing and Gis Application in Environmental Management. *Journal of Environmental Management*, 1(69), 5–24.

UN World Water Development Report 2020 – Water and Climate Change | UNESCO. (n.d.). Retrieved August 18, 2024, from https://www.unesco.org/en/wwap/wwdr/2020

van Vliet, M. T. H., Franssen, W. H. P., Yearsley, J. R., Ludwig, F., Haddeland, I., Lettenmaier, D. P., & Kabat, P. (2013). Global river discharge and water temperature under climate change. *Global Environmental Change*, 23(2), 450–464. DOI: 10.1016/j.gloenvcha.2012.11.002

Vicente-Serrano, S. M., Peña-Gallardo, M., Hannaford, J., Murphy, C., Lorenzo-Lacruz, J., Dominguez-Castro, F., López-Moreno, J. I., Beguería, S., Noguera, I., Harrigan, S., & Vidal, J.-P. (2019). Climate, irrigation, and land cover change explain streamflow trends in countries bordering the northeast Atlantic. *Geophysical Research Letters*, 46(19), 10821–10833. DOI: 10.1029/2019GL084084

Villani, L., Castelli, G., Yimer, E. A., Chawanda, C. J., Nkwasa, A., Van Schaeybroeck, B., Penna, D., van Griensven, A., & Bresci, E. (2024). Impacts of climate change and vegetation response on future aridity in a Mediterranean catchment. *Agricultural Water Management*, 299, 108878. DOI: 10.1016/j.agwat.2024.108878

Wagner, P. D., & Waske, B. (2016). Importance of spatially distributed hydrologic variables for land use change modeling. *Environmental Modelling & Software*, 83, 245–254. DOI: 10.1016/j.envsoft.2016.06.005

Walega, A., Amatya, D. M., Caldwell, P., Marion, D., & Panda, S. (2020). Assessment of Storm Direct Runoff and Peak Flow Rates Using Improved SCS-CN Models for Selected Forested Watersheds in the Southeastern United States. *Journal of Hydrology. Regional Studies*, 27, 100645. DOI: 10.1016/j.ejrh.2019.100645

Wang, J., Lin, Q., Meng, S., Huang, H., & Liu, Y. (2024). Individual Tree-Level Monitoring of Pest Infestation Combining Airborne Thermal Imagery and Light Detection and Ranging.

Wang, M., Shi, W., & Son, S. (2015). Evaluation of MODIS SWIR and NIR-SWIR atmospheric correction algorithms using AERONET measurements. *Remote Sensing of Environment*, 159, 232–248. DOI: 10.1016/j.rse.2014.12.002

Wang, Q., Song, K., Xiao, X., Jacinthe, P.-A., Wen, Z., Zhao, F., Tao, H., Li, S., Shang, Y., Wang, Y., & Liu, G. (2022). Mapping water clarity in North American lakes and reservoirs using Landsat images on the GEE platform with the RGRB model. *ISPRS Journal of Photogrammetry and Remote Sensing*, 194, 39–57. DOI: 10.1016/j.isprsjprs.2022.09.014

Wang, W., Shi, K., Wang, X., Zhang, Y., Qin, B., Zhang, Y., & Woolway, R. I. (2024). The impact of extreme heat on lake warming in China. *Nature Communications*, 15(1), 70. Advance online publication. DOI: 10.1038/s41467-023-44404-7 PMID: 38167519

Waqas, M., Bonnet, S., Wannasing, U. H., Hlaing, P. T., Lin, H. A., & Hashim, S. (2023). Assessment of Advanced Artificial Intelligence Techniques for Flood Forecasting. *2023 International Multi-disciplinary Conference in Emerging Research Trends (IMCERT)*, 1-6. DOI: 10.1109/IMCERT57083.2023.10075119

Waqas, M., Humphries, U. W., Wangwongchai, A., Dechpichai, P., & Ahmad, S. (2023). Potential of Artificial Intelligence-Based Techniques for Rainfall Forecasting in Thailand: A Comprehensive Review. *Water (Basel)*, 15(16), 2979. DOI: 10.3390/w15162979

Waqas, M., Shoaib, M., Saifullah, M., Naseem, A., Hashim, S., Ehsan, F., Ali, I., & Khan, A. (2021). Assessment of advanced artificial intelligence techniques for streamflow forecasting in Jhelum River Basin. *Pakistan Journal of Agricultural Research*, 34(3), 580. DOI: 10.17582/journal.pjar/2021/34.3.580.598

Wei, Y., Liu, Z., Zhang, Y., Cui, T., Guo, Z., Cai, C., & Li, Z. (2022). Analysis of gully erosion susceptibility and spatial modelling using a GIS-based approach. *Geoderma*, 420, 115869. DOI: 10.1016/j.geoderma.2022.115869

Werf, G. R., Randerson, J. T., Giglio, L., Collatz, G. J., Mu, M., Kasibhatla, P. S., Morton, D. C., DeFries, R. S., Jin, Y., & van Leeuwen, T. T. (2010). Global Fire emissions and the contribution of deforestation, savanna, forest, agricultural, and peat fires. *Atmospheric Chemistry and Physics*, 10.

Werner, A. D., & Simmons, C. T. (2009). Impact of sea-level rise on sea water intrusion in coastal aquifers. *Groundwater*, 47(2), 197–204. DOI: 10.1111/j.1745-6584.2008.00535.x PMID: 19191886

Winkler, K., Fuchs, R., Rounsevell, M., & Herold, M. (2021). Global land use changes are four times greater than previously estimated. *Nature Communications*, 12(1), 2501. DOI: 10.1038/s41467-021-22702-2 PMID: 33976120

Wu, F., Yang, X., Cui, Z., Ren, L., Jiang, S., Liu, Y., & Yuan, S. (2024). The impact of human activities on blue-green water resources and quantification of water resource scarcity in the Yangtze River Basin. *The Science of the Total Environment*, 909, 168550. DOI: 10.1016/j.scitotenv.2023.168550 PMID: 37979857

Xhafa, P. D. S. (2015). Geographic Information Systems (GIS). *Urban Planning*, 4138(April), 85–92.

Xing, J., Sun, S., Huang, Q., Chen, Z., & Zhou, Z. (2024). Application of Geoinformatics in Forest Planning and Management.

Xue, H., & Sigg, L. (1999). Comparison of the complexation of Cu and Cd by humic or fulvic acids and by ligands observed in lake waters. *Aquatic Geochemistry*, 5(4), 313–335. DOI: 10.1023/A:1009679819002

Xu, H. (2006). Modification of normalised difference water index (NDWI) to enhance open water features in remotely sensed imagery. *International Journal of Remote Sensing*, 27(14), 3025–3033. DOI: 10.1080/01431160600589179

Yahya, B., Ahmed, K. A., & Saeed, A. M. (2023). Engineering and Technology Quarterly Reviews.

Yang, S., Zhao, B., Yang, D., Wang, T., Yang, Y., Ma, T., & Santisirisomboon, J. (2023). Future changes in water resources, floods and droughts under the joint impact of climate and land-use changes in the Chao Phraya basin, Thailand. *Journal of Hydrology (Amsterdam)*, 620, 129454. DOI: 10.1016/j.jhydrol.2023.129454

Yan, T., Bai, J., Arsenio, T., Liu, J., & Shen, Z. (2019). Future climate change impacts on streamflow and nitrogen exports based on CMIP5 projection in the Miyun Reservoir Basin, China. *Ecohydrology & Hydrobiology*, 19(2), 266–278. DOI: 10.1016/j.ecohyd.2018.09.001

Yeh, A. G.-O. (1991). The development and applications of geographic information systems for urban and regional planning in the developing countries. *International Journal of Geographical Information Systems*, 5(1), 5–27. DOI: 10.1080/02693799108927828

Yeh, A. G.-O., & Chow, M. H. (1996). An integrated GIS and location-allocation approach to public facilities planning. *Computers, Environment and Urban Systems*, 20(4-5), 339–350. DOI: 10.1016/S0198-9715(97)00010-0

Zalidis, G., Stamatiadis, S., Takavakoglou, V., Eskridge, K., & Misopolinos, N. (2002). Impacts of agricultural practices on soil and water quality in the Mediterranean region and proposed assessment methodology. *Agriculture, Ecosystems & Environment*, 88(2), 137–146. DOI: 10.1016/S0167-8809(01)00249-3

Zghibi, A., Merzougui, A., Mansaray, A. S., Mirchi, A., Zouhri, L., Chekirbane, A., Msaddek, M. H., Souissi, D., Mabrouk-El-Asmi, A., & Boufekane, A. (2022). Vulnerability of a Tunisian coastal aquifer to seawater intrusion: Insights from the GALDIT Model. *Water (Basel)*, 14(7), 1177. DOI: 10.3390/w14071177

Zhang, H., Wang, B., Liu, D. L., Zhang, M., Feng, P., Cheng, L., Yu, Q., & Eamus, D. (2019). Impacts of future climate change on water resource availability of eastern Australia: A case study of the Manning River basin. *Journal of Hydrology (Amsterdam)*, 573, 49–59. DOI: 10.1016/j.jhydrol.2019.03.067

Zhang, J., Steele, M., Lindsay, R., & Schweiger, A. (2017). What drove the dramatic retreat of Arctic sea ice during summer (2007). *Geophysical Research Letters*, 34(10), L18502. Advance online publication. DOI: 10.1029/2007GL031527

Zhang, K., Douglas, B. C., & Leatherman, S. P. (2016). Global warming and coastal erosion. *Climatic Change*, 134(1-2), 249–264. DOI: 10.1007/s10584-017-2079-9

Zhang, X., Ge, Y., Chen, J., Ling, F., Wang, Q., Du, D., & Xiang, R. (2023). Highquality super-resolution mapping using spatial deep learning. *iScience*, 26(6), 106875. DOI: 10.1016/j.isci.2023.106875 PMID: 37288344

Zhao, B., Wu, J., Han, X., Tian, F., Liu, M., Chen, M., & Lin, J. (2023). An improved surface water extraction method by integrating multi-type priori information from remote sensing. *International Journal of Applied Earth Observation and Geoinformation*, 124, 103529. DOI: 10.1016/j.jag.2023.103529

About the Contributors

Mouhcine Batchi is a prominent academic figure specializing in the intricate realms of Geographic Information Systems (GIS), Remote Sensing, and human-environment interactions. Holding the esteemed position of Vice-Dean in Charge of Academic Affairs at the Faculty of Humanities and Social Sciences, University of Ibn Tofail, Dr. Batchi plays a pivotal role in shaping academic endeavors and fostering intellectual growth within the institution. Education: Dr. Batchi completed his doctoral studies in Geography at Ibn Tofail University in Kenitra, Morocco, in 2014. His doctoral dissertation titled "Agro-spatial changes and risks of environmental degradation in the coastal zone of Gharb (North-Western Morocco)" showcased his profound insights into the dynamic interplay between agricultural practices and environmental sustainability. Prior to his doctoral studies, he earned a Master's degree in Environmental Geography from Ibn Tofail University, where his thesis focused on the spatial estimation of water erosion utilizing GIS and remote sensing techniques. Appointments: In addition to his current role as Vice-Dean in Charge of Academic Affairs, Dr. Batchi has held several key positions within the academic sphere. Since 2016, he has served as a Professor in the Department of Geography at the University of Ibn Tofail, where his expertise and dedication have contributed significantly to the academic community. Furthermore, he assumed the role of Head of the Department of Geography in January 2021, further solidifying his commitment to academic leadership and excellence. Dr. Batchi's contributions extend beyond his immediate department, as he has been an active member of the Ibn Tofail University Board, serving on both the pedagogical and scientific commissions. Teaching Experience: Dr. Batchi's passion for education is evident through his extensive teaching experience at both undergraduate and master's degree levels. At the undergraduate level, he has taught a diverse array of courses ranging from Cartographic Semiology to Mediterranean Geography, equipping students with fundamental knowledge and analytical skills crucial for geographical studies. Additionally, he has supervised numerous undergraduate students, guiding them through their academic journey and fostering their research capabilities. At the master's degree level, Dr. Batchi has imparted specialized knowledge in Geomatics, Geographic Information Systems (GIS), and Remote Sensing, providing students with advanced tools and methodologies essential for cutting-edge research and practical applications. Research Projects: Dr. Batchi's scholarly pursuits extend beyond the confines of the classroom, as evidenced by his involvement in various research projects. Noteworthy among these is his participation in the realization of the Kenitra SDAU project (Urban Development Plan) from 2012 to 2014, demonstrating his practical engagement in addressing urban development challenges. Additionally, he has been an integral member of the SHS N°2 research project, focusing on spatial and environmental changes in the coastal zone of Gharb (North West Morocco) in the context of climate change. His research endeavors underscore his commitment to addressing pressing environmental issues and advancing knowledge in the field of geography. Dr. Mouhcine Batchi's contributions to academia, research, and institutional leadership exemplify his unwavering dedication to excellence and innovation in geographical studies. His multifaceted expertise and visionary leadership continue to enrich the academic community at the University of Ibn Tofail and beyond.

Adil Moumane is an expert in geospatial analysis, remote sensing, and environmental management. He holds a significant academic and research background in the fields of Geographic Information Systems (GIS) and Remote Sensing, with a focus on applications in water resources management, land degradation, and environmental monitoring. Moumane has contributed to numerous projects and publications that integrate advanced spatial analysis techniques with environmental sustainability challenges, particularly in arid and semi-arid regions.

* * *

Jamal Al Karkouri is a prominent geographer and academic leader, currently serving as the dean (doyen) of the Faculty of Human and Social Sciences at Ibn Tofail University in Kenitra, Morocco. His significant contributions to the field of geography encompass various topics, including spatial dynamics, land use change, and environmental issues affecting Morocco's diverse geographic regions.

Fatima Ezzahra El Ghazali is a hydrologist with expertise in using Geographic Information Systems (GIS) and remote sensing for water resource management, particularly in arid and semi-arid regions. Her work often focuses on identifying suitable sites for dam construction and improving water management strategies in regions where surface water is scarce.

Nourddine Enneya is affiliated with the Laboratory of Informatics Systems and Optimization at Ibn Tofail University in Kenitra, Morocco. He specializes in computer science, informatics, and optimization methods. His research focuses on developing and applying optimization techniques to solve complex problems in informatics systems.

Phyo Thandar Hlaing is a Ph.D. student who is studying her second year happily and passionately. Phyo Thandar Hliang received her Bachelor of Civil Engineering from Technological University (Meiktila), Myanmar, and her Master of Water Resources Engineering from Yangon Technological University, Myanmar. Her research interests focus on climate change and water resources. She is passionate about work related to hydrological modeling, GIS, and Remote Sensing.

Usa Wannasingha Humphries is an Associate Professor at the Department of Mathematics, King Mongkut's University of Technology Thonburi (KMUTT), Bangkok, Thailand. She completed her bachelor's degree in mathematics from Prince of Songkla University, Thailand, in 1991. She obtained her master's degree in applied mathematics from KMUTT, Thailand, in 1994 and her PhD in applied mathematics from the University of Exeter, England, in 2000. Her research interests are in the fields of oceanic modeling, climate, and applied mathematics.

Abdellah Khouz is a researcher focused on hydrogeology and geophysics, particularly in studying groundwater resources in Morocco. His work involves using gravity data to investigate the deep structure of the Essaouira Basin, with hydrogeological implications for arid and semi-arid regions. He is known for contributing to the understanding of faults and groundwater systems in the region, making his research significant for water resource management and protection in Morocco.

|**Jorge Trindade** holds a PhD in Physical Geography and is an assistant professor at Universidade Aberta and a researcher in the Coastal and Fluvial Systems Research Group at the Centre for Geographical Studies, University of Lisbon.

Muhammad Waqas is a Ph.D. student who is determined and passionate in his second year of study. Muhammad Waqas received his BSc degree in Agricultural Engineering from Bahauddin Zakariya University, Pakistan, in 2013 and an M.Sc. (Hons) in Agricultural Engineering (Hydrology & Water Resources) from the same university in 2018. His research interests focus on Artificial Intelligence and Machine Learning applications in Hydrology and Water Resources. His works clearly show that he is passionate about conducting research and devoted to positively impacting people's lives.

Index

M

N

O

P

R

S

www.ingramcontent.com/pod-product-compliance
Lightning Source LLC
LaVergne TN
LVHW081533300926
849161LV00006B/1408
9798369396513